Terra Incognita

Terra Incognita

A NAVIGATION AID FOR ENERGY LEADERS

Christopher E.H. Ross and Lane E. Sloan

Copyright © 2007 by
PennWell Corporation
1421 South Sheridan Road
Tulsa, Oklahoma 74112-6600 USA

800.752.9764
+1.918.831.9421
sales@pennwell.com
www.pennwellbooks.com
www.pennwell.com

Director: Mary McGee
Managing Editor: Marla Patterson
Production/Operations Manager: Traci Huntsman
Production Editor: Tony Quinn
Book Designer: Susan E. Ormston Thompson
Cover Designer: Karla Pfeifer

Library of Congress Cataloging-in-Publication Data

Ross, Christopher E. H.
  Terra incognita : a navigation aid for energy leaders / Christopher E.H. Ross and Lane E. Sloan.
     p. cm.
  ISBN 978-1-59370-109-3 (hardcover)
  1. Energy industries--Management. 2. Energy policy. I. Sloan, Lane E. II. Title.
  HD9502.A2R67 2006
  333.79--dc22

                                                                    2007000489

Printed in the United States of America

4 5   10 09 08

Contents

Acknowledgments

This book ultimately owes its existence to countless people in the energy industry—colleagues, clients, and others connected in various ways to this most fascinating of industries—who have shared their thoughts and ideas with the authors over the past 35–40 years.

The original idea for the book was driven by the Global Energy Management Institute (UH-GEMI), a center of excellence at the University of Houston's Bauer College of Business, as a text in support of an MBA course on energy strategy and leadership. This course would not have been possible without the support of Associate Dean for Graduate Programs David Shields and Director Dan Currie of the Center for Executive Development. We also appreciate the important contribution from UH-GEMI Program Manager Jason Davila in designing key graphics for chapter 1.

Many of the conceptual frameworks described in the first part of the book come from friends and colleagues at CRA International, Inc. We are particularly grateful for the wholehearted support we received from CEO Jim Burrows and chief strategy officer Arnie Lowenstein.

Most of all, we want to recognize the tremendous cooperation we received from senior executives in the industry. The second half of the book summarizes more than 20 interviews that shed light on their strategies, leadership, and execution approaches and confirms their fundamental integrity and decency. Chevron was particularly generous in providing access, not only to CEO Dave O'Reilly, but also to Chief Technology Officer Don Paul and to Pat Yarrington, head of policy, government and public affairs. Burlington Resources was also generous in their commitment, granting valuable interviews with CEO Bobby Shackouls, CFO Steve Shapiro and Ellen DeSanctis, head of strategy and investor relations. Valero also allowed us to "double dip" with stimulating interviews with retiring CEO and founder Bill Greehey, and head of refining Rich Marcogliese.

Running an energy company is a tough job, and we were honored that so many CEOs were prepared to spend time with us: Clarence Cazalot of Marathon, John Wilder of TXU, Aubrey McClendon of Chesapeake, Duane Radtke of Dominion Exploration and Production, Frank Chapman of BG Group, Helge Lund of Statoil, Steve Lowden of Suntera, Gwyn Morgan of EnCana, and Rick George of Suncor were all open, frank, and responsive to our questions. We were particularly honored to be able to interview Abdallah Jum'ah, CEO of Saudi Aramco, who provided a fascinating national oil company perspective.

In addition, we benefited from the insightful comments of the chief strategists of BP (Nick Butler), Total (Ian Howat), and ConocoPhillips (John Lowe). In each case, we gained an understanding of how these companies became super-majors and the essentials of their companies' management processes. A few of Nick Butler's and Ian Howat's comments strongly reflected the views of their CEOs, Lord Browne of BP and Thierry Desmarest of Total, and we asked and gratefully received permission to attribute the comments accordingly. Finally, we received a helpful written response to our questions from Jeroen Van der Veer, CEO of Shell.

We greatly appreciate the willingness of these senior industry executives to spend time on our project. Our methodology was to transcribe the interviews, separate the comments into sets of ideas, and compile the ideas into themes that are described in chapters 7, 8, and 9. We do not expect that all the interviewees will fully subscribe to the themes that we have developed, which are the responsibility of the authors. Nor do our respective institutions, CRA International, Inc. and the University of Houston, necessarily agree with the views and conclusions presented in the book, which again are the authors' responsibility. We have tested the ideas with numerous colleagues and contacts and they seem to be useful. We certainly are very grateful for all the feedback we have received. We would also like to recognize the important contributions of our editor, Chuck McCabe.

Finally, we must pay tribute to the patience of our families who supported our endeavors wholeheartedly throughout the process of writing this book. We dedicate the result to Helen, David, Stephen, and Alexander Ross and to Diane, Jonathan, and Natalie Sloan.

Prologue

Energy is the fundamental building block for modern industrialized economies, the very lifeblood of developed economies. When it is available, we take it entirely for granted, like water flowing tamely from the tap. When availability fails, our lives are disrupted and restricted, the veneer of comfort and accommodation quickly erodes, and economic activity grinds to a halt. Public discourse and consumer attention seldom turn to energy until an aberration occurs, such as a blackout, supply shortage or price run-up. While typically far from the forefront of public policy debate in developed countries, energy can quickly escalate to the top of the agenda because of its vital importance to economic security. By contrast, energy is always high among public policy concerns for developing countries, such as China and India, because it fuels the growth of their economies and elevation of their living standards.

Not only does energy play a central role in economic security, it is equally fundamental to national security through the prosecution or prevention of war in all its facets, from the critical role in troop mobility and in equipment and materiel logistics, to safeguarding the strong economy required to fund military expenditures. Access to energy resources were critical strategic issues for both the First and Second World Wars and since then, large producing countries have periodically deployed "the energy weapon" by withholding supplies. Additionally, the hydrocarbon sources of energy transformed into petrochemicals also play a vital role in the development of many consumer products—plastics, styrene, fertilizers—the list is endless. Whether through a heart valve, styrene cup, or plentiful food, this petrochemical chain is central to the quality of our lives. It is not surprising, then, that energy becomes uniquely intertwined with geopolitics.

The late Dr. Richard Smalley, Nobel laureate in nanotechnology for the discovery and characterization of the Buckminsterfullerene molecule, or the "buckyball," prominently professed in his latter years that energy was the most significant challenge of the 21ˢᵗ century. Central to his discussion was the projection of a world population increase from over six billion people today to 10 billion by 2050, coupled with the need to resolve the huge disparity in standards of living between developed and underdeveloped countries. Dr. Smalley's thesis was that solutions to the major challenges facing the world, such as water, food, and terrorism, depended on energy, but not vice versa. We may expect that energy will retain its vital importance in human affairs for the foreseeable future, and, indeed, to become even more central to the well-being and advancement of people everywhere.

Energy is, in fact, a vast, complex business—the biggest in the world. The energy industry satisfies vital societal needs of mobility and power through an extensive value chain, stretching from the many primary sources, such as coal, nuclear, oil, and natural gas, through transformations to secondary, "consumable" forms, such as electricity and motor gasoline that travel through a web of distribution networks and outlets to eventually reach the consuming public. And perhaps uniquely, the resource side of the energy industry today largely rests on a depleting source, fossil fuels. As an example, oil and gas producers face decline rates of 5% to 30% per year in their production rates as their assets in the ground deplete. It is a major task to replace reserves that have been produced, let alone add sufficient reserves to allow production growth for expanding economies.

Multitudes of companies and countries make up the energy industry, and the value-chain relationships are not always transparent. For example, the retail gasoline stations that bear the names ExxonMobil, Shell, BP, Chevron, and so forth, are generally operated by independent businessmen called dealers. They are not employees of the companies whose brand they operate under. There are many small, independent producing companies that drill the majority of the wells for oil and gas in the United States, more than 30,000 of them. At the same time, some of the largest private companies in the world are integrated energy companies. But the ownership of the most important oil and gas reserves across the globe is controlled by national oil companies (NOCs). The economic interests of producers to maximize the value of their resources for the benefit of their people do not necessarily coincide with the economic interests of privately owned companies for rapid pay-out of investments, or of consumers who want ample supplies of cheap energy at all times. Coal companies are rarely connected to integrated oil and gas companies, but both compete to provide fuel to electric power companies. Nuclear plants are generally owned by power companies. Many U.S. electric utility companies are highly regulated, whereas in some states and other countries they are largely deregulated. This intricate web of regulatory and competitive interactions culminates in an energy marketplace.

Clearly the energy industry is neither a simple business nor is it well understood. Moreover, the earth is a dynamic living system, and the utilization and transformation of energy sources impacts this ecosystem, which brings additional public policy issues to the forefront. Certainly, global warming is high on the agenda of world leaders. The modern energy executive must focus not only on providing financial results, but also protecting the environment and demonstrating social responsibility.

Given the significance of energy in the global society, the importance of strategic energy leadership cannot be overstated. However, this is a focus that has largely been ignored in the literature. There are innumerable books on leadership and strategy, yet they are without specific concentration and application to the energy industry, excepting the occasional biography of an influential leader such as John D. Rockefeller. A number of books analyze the energy industry, but generally from a historical, geopolitical perspective, often raising issues of public policy and the interactions between consuming and producing nations. These analyses are largely outside views looking in, not an inside perspective looking out, which is the vantage point of this book.

Moreover, the overriding purpose of this book is to provide a strategic dimension through analytical frameworks that integrate leadership and strategy directly into an energy context. We believe this integration brings much greater insights with deeper impact than the sum of studying the individual dimensions of energy history and business strategy in isolation. We bring this integration alive through the statements of individual energy leaders discussing how they run their companies. These statements also provide a more realistic face to the industry than that portrayed in the media. Finally, our premise is that energy strategic leadership is a topic of importance to people in the business, and is also essential for all who wish to understand the policy options of modern society.

Energy strategic leadership framework

The overarching analytical framework of this book comprises three fundamental strategic activities: *Strategic Assessment*, *Strategy Formulation*, and *Strategy Implementation*. We have sandwiched this analytical framework between a first chapter that draws out lessons from the past, and a final chapter that provides the authors' summary of the leadership challenges of the future.

In the early stages of man eons ago, the first source of controllable energy was fire for warmth and cooking. The historical evolution since then was of little consequence until the advent of the industrial revolution in the 1700s, which depended on the availability of commercial energy sources. Thus, this book begins our journey there, only three centuries ago. In this regard, our analysis of events from that time forward will focus on the dominant energy source prevalent during a given period, such as wood was at the beginning of the industrial revolution, until it quickly lost its market position to coal, which in turn gave way to oil.

The activity of *Strategic Assessment* involves the dynamics and drivers of the energy industry. Ultimately, these forces lead to an understanding of how value is created in the energy industry, which becomes the foundation for *Strategy Formulation*. Strategies turn into actions through *Strategy Implementation*. Leaders orchestrate, guide, and monitor this three stage process with the ultimate objective of creating value: value to customers, value to shareholders, and value to other stakeholders.

Building on our learning from the past, the second chapter continues the *Strategic Assessment* journey by examining the role of national oil companies, which are the current dominant providers of the reigning major energy source today, crude oil. In chapter 3 the analysis moves to the future in searching for insights that will impact the formulation of strategies. Key parameters affecting energy demand and supply alternatives are explored within a backdrop of policy externalities.

The focus then shifts in chapter 4 to understanding the perspectives and needs of energy company stakeholders. Geopolitics and energy are joined at the hip; societal needs continue to grow and evolve, setting an expectations framework that companies and countries must consider in envisioning forward plans.

With an understanding of the major drivers impacting the energy industry, the next step involves *Strategy Formulation*. The groundwork is set in chapter 5 by reflecting on how a persuasive investor value proposition can be developed by an energy company for its shareholders. Then, chapter 6 provides an overall strategic framework the leader must think through, including clarifying the corporate vision, developing strategies to further strengthen existing business while leveraging core competencies into new business opportunities.

From this strategic thinking, we describe how current leaders think about their challenges and accomplishments in the next three chapters. In chapter 7 the leader sets the course by making specific choices on strategic direction that will create the most value. With strategy set, the task of executing the strategy begins, as outlined in chapter 8. The leader is instrumental in all aspects of the strategic framework but must be particularly engaged in the execution phase. Chapter 9 describes the critical aspects of leadership that emerged from our interviews: many of them timeless, but some influenced by the prevailing business environment.

Finally, chapter 10 ventures into the authors' own thinking about the next-generation energy leader and attempts to tie together all the elements of the total strategic process from *Strategic Assessment, Strategy Formulation* and *Strategy Implementation*. This final chapter, then, wraps up the two major objectives of this book: 1) to help the readers become the next generation energy leader, or 2) to provide insights for those outside the energy industry on how it functions.

Our conclusions are, we believe, quite provocative for energy industry leaders and also to those from outside the industry. We believe that the global energy complex is at the early stages of a phase change. The rise to materiality of China and India creates enormous potential demand that cannot be met by conventional oil from traditional places. At the same time, national oil companies have matured, and some are going global, increasing competitive intensity for new resources. Investors are demanding growth from their energy company investments, and international oil companies have been failing to deliver it.

We accept the guidance of Dwight D. Eisenhower: "If a problem cannot be solved, enlarge it." The solution to the larger problem statement is an interconnected network of diverse energy resources, transformed into transportation fuels and electricity that satisfy societal needs for mobility and power. This larger system will present opportunities for new business models and new strategic directions, will pose challenges for strategy execution, and will create a need for great leadership. We hope this book will stimulate discussion on how to nurture and develop great leaders for the energy industry at this critical time

1 Learning from the Past

Introduction

The story of energy is the story of mankind's development. It is the story of population growth and the technologies that enabled it; the story of the economic growth that followed and the private capital firms that financed and organized it; and the story of the governments that promoted and sponsored the process, seeking to shape and control it for the benefit of their citizens. It is a complex story that develops through non-linear relationships leading to cycles and "punctuated evolution" as sudden events and breakthroughs disrupt historical trends. Endlessly interesting, full of drama and larger-than-life personalities, swept along in the ebb and flow of history, poorly understood and often taken for granted, the story of energy continues to fascinate and surprise those of us in the industry who try to make sense of it.

Our historical voyage in this chapter begins with the early industrial revolution, when harnessing of commercial fuels supplants human and animal effort. We will then enter the world of King Coal, and journey on to Big Oil's 20th century domain. While in the land of oil, we will follow the Rise of OPEC in the 1970s and track the inevitable market response that slowly and steadily changed the market fundamentals through the 1970s and 1980s. We will explore the poorly understood province of oil price formation before describing the down cycle in the oil patch that harrowed the industry in the late 1980s and 1990s. We will sidetrack into technology's ongoing role before landing in the exciting and unexplored territory of the new up-cycle and phase change in global energy supplies.

We will trace the co-evolution of energy and technology, in which advancements in one area open opportunities in the other for beneficial adaptation and development. We will examine the schizophrenic relationships between governments and energy companies, alternating from strong mutual support to deeply adversarial posturing, according to changing national priorities. We observe that cartels only work some of the time; each success bears within it the seeds of the cartel's destruction as high prices trigger adaptive demand and supply responses. The latest plot twist in our story involves increasing societal demands that are forcing energy companies to change their behaviors and practices.

Throughout this journey, we draw lessons that will serve as navigational aids for our energy future. Some of the situations these lessons describe will undoubtedly be repeated with twists and turns as the future unfolds. The wise explorer refers occasionally to existing maps, no matter how imperfect, when entering unfamiliar territory. As we see it today, the journey ahead seems fraught with many perils that will have to be overcome in the search for new opportunities.

The Early Industrial Revolution

As our modern technologies have evolved since the early days of the industrial revolution, energy sources have played a fundamental role every step of the way. New technologies create paradigm shifts in the way society functions. Observing these past transitions equips us to identify and understand potential future discontinuities. With these insights, energy leaders can choose the strategies and develop plans that have the greatest likelihood of success.

Birth of industrialization

Energy has contributed to living standards since humans began using fire for warmth and cooking. Wood was the main global fuel for centuries, and was also used to construct homes, ships, furniture, and so forth. However, expanding human activity began to take an early toll on the environment and fuel supply. Around the 1570s, deforestation in Britain created a wood scarcity crisis and provoked a national security threat for Her Majesty's wooden-vessel navy during the rule of Elizabeth I. The rulers reacted by passing laws limiting the taking of wood. Britain was fortunate to have a plentiful supply of coal, which had been growing in use since the 12th century. Coal demand surged, avoiding an energy crisis, and by the end of Elizabeth's rein in 1603, coal was the main source of fuel for Britain.[1] This important process would later be labeled interfuel substitution.

The industrial revolution emerged after the English Civil War in the 17th century, bringing the end of feudalism. European colonial expansion strengthened trade and provided capital for industrial development. Technical innovation developed in parallel with financial innovation, such as the use of private capital to fund public stock companies. The resulting prosperity provided a surplus to fund further technical and commercial innovation.

New agricultural technologies lowered sustenance food costs and freed up people to pursue mercantile opportunities, such as textiles for clothing. The migration of workers into cities coincided with opportunities for employment in factories. Energy played a fundamental role, providing heat for the smelting of iron and brass as well as for domestic uses. As early as 1746, an integrated brass mill at Warmley provided housing for workers on site.[2]

Social impacts: Malthus to Maslow

The transition from a household-based economy to an industrial economy brought on new social challenges. There is considerable debate about the progress in living standards during this early industrial period; Thomas Malthus' *Essay on Population* (1798) suggested the darker side: "The increasing wealth of the nation has little or no tendency to better the conditions of the labouring poor. They have not, I believe, a greater command of the necessaries and conveniences of life, and a much greater proportion of them than at the period of the Revolution [1688] is employed in manufactures and crowded together in close and unwholesome rooms."[3]

Others argue that the shift was largely from distasteful work in the mean and servile tasks of the pre-industrial farms to equally distasteful work in the factories or workshops, but that in every day life, "the people ate better, were clothed better, and generally led healthier and more productive lives."[4] What is key to learn here is that a society, made up of individuals with various stakeholder roles, responds to change based on the current driving needs of its members. Abraham Maslow's well known hierarchy of needs, depicted in figure 1.1, offers a model to illustrate this process. British society during the early industrial age was largely focused on the second level of Maslow's hierarchy—security and safety—having largely satisfied survival needs. In other words, the driver was largely economic security, not the content and conditions of the work, in this very early stage of industrialization.

TIME FRAME	MASLOW'S NEED LEVEL	ENERGY IMPACT
21st Century ??	Self Actualization	Sustainable World Personal Freedom
2nd Half of 20th Century	Ego	Economic Prosperity Tailored Goods & Services Communication Flexibility
1st Half of 20th Century	Socialization	Improved Working Conditions and Standard of Living Flexible Mobility
19th Century	Safety & Security	Reduced Physical Toil Economic livelihood Greater Living & Mobility
Prior to Industrial Revolution	Survival	Warmth & Cooking Agrarian

Figure 1.1: As society climbs Maslow's hierarchy of needs, the energy impact evolves

The population in England and Wales was about five million in 1700 and grew between 17% and 18% in the first half of the century. By 1800, the population had grown to over nine million, which was a more than 50% increase in the second half century. Although the life of those at the bottom of the economic pyramid was dismal, the rising population testifies to higher aggregate living standards with fewer people dying from starvation and disease. The growth in cities was remarkable. Around 1700, London had a population of slightly over a half million inhabitants; some 70 years later, it contained one-sixth of the whole British population. Obviously, the transition to a modern industrial manufacturing system coupled with technological innovations in agriculture had a profound impact on population growth and where people lived.[5]

Machines + energy = industry

Some historians introduce the industrial revolution with a discussion of the invention of new machines and new machine tools. But machines have been around for a long time. The lathe, for example, is traceable back a great many centuries, so it is not just a question of new machines. The profound change involved the use of energy: the new machines were powered by an energy source fueling an engine, not by water or wind, animals, or manual labor. The landmark industrial invention was the first modern steam engine built by Thomas Newcomen at the beginning of the 1700s. With substantial improvements in design by James Watts and John Wilkinson in the second half of the 18th century, steam engines powered rotary drive-shafts for machinery in numerous applications.

In 1804, the first steam-engine tramway locomotive hauled a load of 10 tons of iron, 70 men, and nine wagons about nine miles. This led in 1835, through the ingenuity of George Stephenson, to the first railroad to carry both goods and passengers, on a line between Stockton and Darlington in Britain.[7] Railroads played a key role in the growth of inland trade by reducing the time and cost of transportation. Similarly, the steamboat also developed during this period beginning with Robert Fulton's *Clermont* carrying passengers on the Hudson River in 1807, promoting trade growth as well as personal travel. A few decades later, steamships were making transatlantic crossings, further developing global trade. So from the very early days of the industrial revolution, energy was applied to transportation and the two became interlinked. Steam engines could be fueled by either coal or wood, but coal soon proved to be the preferred fuel.

As the industrial revolution progressed, steam power found its way into the mass production of goods. While it did not create the factory system, it significantly changed the scale, the process, and the location of factories. Whereas early factories, mainly cotton mills, were powered by waterwheels, by 1830 Manchester was the epicenter of coal-driven, steam-driven cotton mills. The city had more than seven cotton mills with more than 1000 workers each and another 75 with hundreds of workers each.[8]

Iron production also began to depend on coal, which enabled much larger blast furnaces and made Britain the most efficient and cheapest iron producer. The ample supply of competitively advantaged iron helped Britain build its industry at home and the empire abroad. Coal demand surged, quadrupling from 1842 and 1856. Industrialization was energy intensive. The monumental technological breakthrough of the industrial revolution was the marriage of machines and energy to reduce man's toil and improve productivity. The steam engine was the harbinger of the paradigm shift from an agrarian to an industrial society.

Other scientific discoveries that would lead to new ways to harness energy also emerged early in the industrial revolution. Benjamin Franklin proved that lightning is electricity with his famous kite experiment on June 15, 1752.[9] This began electricity-based innovation. William Cooke and Charles Wheatstone installed the first electric telegraph line along the Great Western Railway in 1839 to track the position of trains. It was also offered to the public as a communication service, making it the first use of electricity in a commercial enterprise. Electricity revolutionized communications and forever altered the way business is transacted on planet Earth.

The remarkable inventor and entrepreneur Thomas Edison essentially created the electric power industry in the late 19th century. His innovations led to a practical incandescent light bulb in 1879 that was quickly accepted as superior to kerosene, gas, and candles for illuminating houses, offices, and streets. In 1882, Edison opened in New York City the first commercial electric power generating plant in the United States. The plant burned about 10 pounds of coal to generate one kilowatt hour of electricity, a far more efficient rate than its predecessors, and at an initial cost of about 24 cents per kWh.[10] Coal was the primary fuel source for the burgeoning power industry, and even today fuels about 50% of U.S. electricity generation.

King Coal: The First Strategic Energy Source

Coal deposits provided a concentration of mineable fuel with energy potential far exceeding that of wood scattered through forests. One half ton of coal produced as much energy as two tons of wood and was cheaper. As such, coal was much more amenable to mass production, and its concentration provided for stronger distribution networks, even though coal was rather bulky. The low density of biomass sources continues to limit their economic utility even today, despite their renewability.

In fostering industrialization, Britain unveiled the power of coal while the rest of the world was relying on wood as its primary energy source. Industrialization quickly spread throughout Europe, and, as in Britain, the population became more urban with half of the people living in cities around the beginning of the 20th century. However, the free spirit and inventiveness of America also drove 19th century industrial growth. Together, Britain and the United States had over 50% of the world's manufacturing capacity in 1870, with Britain holding a small edge on the U.S. By 1913, however, America was home to about 36% of total global manufacturing capacity, almost twice as much as any other country.[12] Hence, to understand the dynamics of this time period our focus shifts to activities in the United States.

The demand for coal accelerated further with innovations in the process of making steel. William Kelly, an American, patented the idea for a system in which air or steam blows the carbon out of pig iron; however, he went bankrupt in 1857. Henry Bessemer, an Englishman who had been working on the same idea, purchased Kelly's patent and eventually developed a process for making steel cheaply. His works at Sheffield produced ordnance for the military and steel rails. The open-hearth method developed by Siemens-Martin around the same time was also important because it was less sensitive to the iron ore characteristics.

The U.S. Civil War increased demand for coal to fuel the needs of the federal army, navy, and military suppliers. This drove up the price for eastern anthracite coal about 45% above the 1860 level. The Confederate Congress organized the Niter and Mining Bureau, which opened new coal minefields in North Carolina and Alabama. We can see from this that energy had become an integral part of the prosecution of war,[13] and also that wartime tends to cause fuel shortages.

Coal barons, early leaders

As demand accelerated and coal became a big business, it spawned the first generation of energy leaders. Their vision and strategy shaped and structured an industry that began in unruly fragmentation. Transportation was a critical piece of the value chain; coal was often shipped to market by rail or barge. Driven by the expansion of coal trade, various railroads integrated backward into coal mining, either by purchasing coal tracts directly or leasing them through subsidiary firms.[14] Thus began the drive to consolidate energy production and its delivery to consumers.

One of the first coal consolidators was Franklin B. Gowen. He began his career apprenticing with a coal trader, then switched to law and became legal counsel to the Reading Railroad. A man of charismatic personality, Gowen was running the Reading by the age of 33. Coal was a major cost for Gowen's railroad, and he believed that fierce competition and the forces of unionization made the coal industry unstable. Coal price volatility was not good for Gowen's business.

Initially, he fought unions indirectly by raising freight rates to coal operators who appeared to be acquiescing to union demands. However, Gowen felt he could control the unions more effectively by owning the mines. The Philadelphia and Reading Rail Road Coal and Iron Company proceeded to spend $40 million buying up most of the coal mines in its portion of the anthracite region, some 125,000 acres. Anthracite had emerged as a particularly attractive fuel coal for iron making.[15]

Gowen also organized the first major energy cartel by forging a price-fixing agreement among the other railroads and some independent mine owners in the other two portions of the anthracite district.[16] As Barbara Freese points out in her book, *Coal: A Human History*: "It is no accident that the classic board game Monopoly includes the Reading and other coal railroads."[17] Railroads could gain some control over pricing through forward or backward integration. Gowen pursued this strategy avidly, but stumbled when he accumulated too much debt. With Reading unable to pay its bills, J.P. Morgan, a prominent New York banker, was brought in to reorganize the railroad. Gowen was forced out in 1886 and purportedly committed suicide three years later. Morgan went on to become one of the most successful power brokers in this era of consolidators who saw competition as wasteful.[18]

Competition was also fierce in bituminous coal. Henry Clay Frick emerged as an energy baron by focusing on producing coke, an important ingredient in making steel, from bituminous coal. Coke is made by heating the coal until its gases and tars are removed. Frick and Co. bought out other bituminous coal mines and became the largest producer of coke in the world with over 12 thousand coke ovens and 40 thousand acres of coal.[19] By deploying technology and pursuing a strategy selectively focused on the steel market, Frick gained the lofty title of "Coal King."[20]

Frick entered a partnership with Andrew Carnegie that was eventually reorganized into Carnegie Steel Company in 1892, with Frick heading it up. This was the largest steel company in the world with a value of $25 million and represented a forward integration of the value chain that differed in character from the railroads' logistical integration. Henry Frick, much like Gowen, used a heavy hand with the mill workers, culminating in the bloody Homestead Works strike. Carnegie and Frick disagreed over these tactics, and Frick left Carnegie Steel in 1899.[21] In 1901, Frick joined with J.P. Morgan to buy Carnegie Steel Company for $500 million and established U.S. Steel Corporation, valued at an incredible $1.4 billion.[22]

Coal's troubled triumph

With the increasing demand for iron and steel, coal production grew dramatically. From around 1870 to 1880, Pittsburgh's coal production increased by 300% to an annual output of 13 million tons.[23] And around the mid-1880s, as depicted in figure 1.2, coal became the primary fuel in the United States, surpassing wood with over 50% of the total energy supplied. By 1910, coal truly had become king, providing around 75% of America's supplied energy.

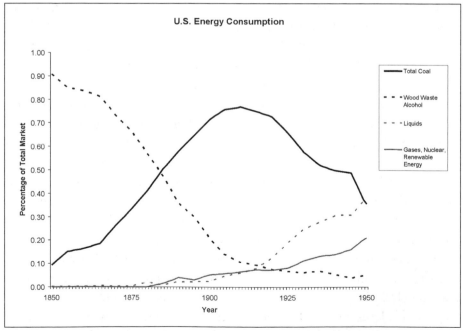

Figure 1.2: Coal became the strategic energy resource in the 19th century. *Source: Energy Information Administration*

The evolution of coal as the dominant energy source provides a very important lesson and vital insight for understanding today's situation with crude oil. While wood was the primary energy source for most of the 19th century, the strategic fuel from the early 1800s on was clearly coal, because it was setting the pace and driving economic and technological change. It was the fuel at the margin. To envision the future and set strategic options, energy leaders must understand the emerging strategic energy source that is setting the pace at the margin.

Coal's rise to dominance required new technology, which has always played a critical role in the economics and operating practices of the energy industry. Beginning shortly after the Civil War, coal mining companies struggled to gain a competitive edge through technology improvements in the intensely competitive U.S. industry. Between 1870 and 1900, coal mining technology went from manual labor with pick and shovel at shallow depths to black powder blasting and electric cutting machines at much greater depths.[24]

The new technologies industrialized coal mining and reduced the miners from skilled workers to machine tenders subject to a sharp division of labor, impersonal supervisory hierarchies, and poor working conditions. All this exacerbated the relationship between management and workers, strengthening the union movement.

These changes brought unintended consequences that might be considered a strange perversion of the capitalistic free enterprise system. Struggling to gain competitive advantage through technological innovations, the coal companies instead created a vicious spiral in the industry's fortunes.[25] Economic cycles ravaged coal prices in 1873, when demand fell by 50%, and later in the collapse of 1893 that resulted in a 25% drop in demand.[26] Wages accounted for 60 to 70% of realized value in the coal industry during the 1870 to 1900 period, and trimming labor costs by adopting new technology became the clear target for management in a commodity industry where profitability was driven by the cost structure.[27] In many cases, operators that failed to pursue the new mining techniques became unprofitable and fell by the economic wayside.

At the same time, the formation of strong unions counterbalanced to some degree the cost savings operators were pursuing. To offset this force, coal companies pursued further technological advances. But the expanding capacity resulting from newer technologies came at a time when interfuel competition from oil and natural gas was on the rise, limiting profitability improvement.[28] To offset this, the coal barons pursued greater industry concentration, as typified by Gowen and Frick. However, this increased economic power triggered corrective responses from government and organized groups such as labor unions, as well as market realignments.

In 1902, under the direction of John Mitchell, the United Mine Workers, by then the nation's biggest union, called a major strike involving nearly 150,000 anthracite coal miners. The bituminous coal producers were not able to offset the loss of five million tons per month of anthracite production. This strike lasted 160 days and provoked what may well have been the United States first energy crisis. President Theodore Roosevelt intervened to broker a deal between Mitchell and J.P. Morgan, still in control of the railroads, to settle the strike. Energy had become such an integral part of the economy that it was intertwined with public policy and executive actions.[29]

Crude Oil Industry's Formative Stage

Coal's many advantages over wood made it the strategic and eventually dominant energy source in the 19th century. However, as the 20th century unfolded, two fundamental forces began to shift the role of strategic energy source from coal to crude oil. These forces are environmental concerns and new technologies.

Pollution hampers coal

Air pollution from burning coal was a significant problem for London even in the 1600s. By the 1700s, the black soot filtered the sun, compromised health, and permeated a noxious stench. The poet William Blake noted the effect on the city's buildings toward the end of the 18th century in a poem of social commentary titled "London:" "And the chimney sweeper's cry / Every blackening church appalls." Blake's contemporary William Wordsworth, in a poem written in 1803 extolling the momentary beauty of a London morning, includes mention of "the smokeless air," apparently a rare occurrence. Truly, environmental issues around coal have a long history. While the distaste for coal was immense, the need for the warmth it provided lies at the survival level of Maslow's hierarchy of needs. So people chose to burn coal.

In the latter part of the 19th century, coal displaced wood in U.S. homes, as it had in Britain several centuries earlier. Although coal towns seemed willing to sacrifice their air for the city's livelihood, a growing antipollution movement was stirring in the kitchens. Women's clubs, like the Ladies Health Protective Association of Pittsburg, took on smoke abatement along with water and sewage issues. Although still unable to vote, women lobbied for municipal laws to ban release of dense smoke. Issues around health effects were a prominent part of the debate. As early as 1861, Chicago and Cincinnati enacted laws to improve air quality, an initiative which a century later spanned the continent with the passage of the Clean Air Act in 1963.[31] American society had moved up Maslow's hierarchy from basic security needs into the realm of social needs. The lesson for energy leaders is this: the public's needs change, and when energy companies are not satisfying these needs, the public will turn to the government for policy action. In a more prosperous society, coal's negatives opened the way for other fuels.

Early days of oil

Before moving forward, let's take a brief excursion back in time to look at the early drivers of the oil industry. Much rich history and geopolitical analysis exists in outstanding books like Daniel Yergin's *The Prize: The Epic Quest for Oil, Money, and Power*. But our purpose in looking back is to identify some key learning for energy company leaders in the early 21st century. Colonel Drake's discovery of Pennsylvania crude oil in 1859 with a 69-foot well began oil's rise to the starring role in the energy industry. Suddenly, men could search for oil as they had prospected for gold. The quest for liquid "black gold" would eventually dwarf the pursuit of the mineral variety.

The magic of oil was that it could be easily refined into several commercial products, or cuts of the barrel. Oil was a much more adaptable and flexible commodity than coal and offered greater utility given the technologies of the period. It was the kerosene "cut of the barrel" that created the initial market for petroleum as the demand for artificial illumination grew.

Apart from candles, artificial lighting in the 19th century was provided by whale oil and "town gas" from coal. "The coal would be baked at gasworks on the edge of town to drive off the gases, which were then piped beneath the streets into the street lights and people's homes."[32] Town gas was affordable for middle- and upper-class families, but there was a need for a reliable, relatively cheap lamp fuel for the masses. Prices for sperm whale oil, the standard for high quality illumination, were rising with the increased demand and reduced numbers of whales. During this same period, coal and shale rock processes were developed to produce "coal-oils" which became competitive substitutes. But growing crude supply and competitive pricing made kerosene from oil attractive.

By 1902, however, 18 million light bulbs lit up America. Electricity captured the urban lighting market, and kerosene demand leveled out and increasingly came from rural America. Similarly, Europe became electrified. But while the oil industry's initial kerosene market was under pressure from a preferred substitute, the electric light bulb, automobile use was creating a new market for the oil industry.

New engines drive oil's growth

Around 1905, the gasoline-powered car emerged as the victor over electric and steam vehicles. The growth in automobile ownership was staggering; registrations in the United States grew from some 8,000 in 1900 to 902,000 by 1912. Gasoline sales first exceeded kerosene sales around 1910. Additionally, the use of fuel oil in boilers of factories, trains, and ships was expanding rapidly. This set the stage for the significant role oil would play in World War I.[33]

World War I provides us the first signs that planet earth was becoming a global village. The ability to move armies and navies over vast territories with some speed underscored the role that fuels had begun to play in world politics. "Strategists for all the major powers increasingly perceived oil as a key military asset, due to the adoption of oil-powered naval ships, horseless army vehicles such as trucks and tanks, and even military airplanes. Use of crude oil during the war increased so rapidly that a severe shortage developed in 1917–1918."[34] It was clear from World War I that oil had become the strategic energy source. While coal met the largest share of overall energy demand until around 1950, it was no longer the energy pacesetter after WW I. Although there is still much to learn from the coal industry post-1920, we must now look to petroleum for the key 20th century energy industry lessons. One man shaped the early industry structure of the oil industry, John D. Rockefeller.

John D. Rockefeller, oil titan

It did not take long for a powerful, monopolizing leader like Gowen and Frick to emerge in the new oil industry and begin to tame the chaos and disorder of a commodity business experiencing wild boom and bust cycles. John D. Rockefeller started in the refining side of the business and bought out his partner, Maurice Clark, in 1865. After experiencing the over-capacity in refining, Rockefeller conceived a plan to fully consolidate oil refining into one controllable behemoth. By 1879, Rockefeller's Standard Oil Company controlled 90% of America's refining capacity. More than that, Standard Oil also controlled the pipelines and gathering systems and dominated transportation. As with coal, transportation and the railroads were critical.

For a number of years, Standard Oil stayed out of the exploration and production (E&P) business, sticking to transportation, refining and marketing, because upstream E&P was considered too risky. "The lives of the drillers turned quickly from excitement to misery, as the prices fell. The year after the first discovery, the price of oil was $20 a barrel; at the end of the next year it was 10 cents a barrel, and sometimes a barrel of oil was literally cheaper than a barrel of water."[35] However, Rockefeller concluded that having a fully integrated operation would provide further insulation against the volatility of the market and ensure greater overall control. By 1891, Standard Oil produced a quarter of America's total output of crude oil.[36] The strategy of integrating to enhance control, creating efficiencies throughout the value chain, and buffering the volatility in the different business segments through diversification would prove to be the business model for the coming century.

Oil producers were incensed by the control that Rockefeller wielded on the midstream and downstream parts of the oil value chain. To counteract the public outcry, and in many respects deceive the interrogators while carrying out the strategy of market dominance, the Standard Oil Trust Agreement was put in place in 1892. The trust allowed the establishment of a central entity to coordinate and optimize the workings of the various operating entities. Creating economies of scale was part of a strategy to be the low-cost producer. This fundamental strategy still pervades the energy industry.[37]

Growth of oil competition

Economic opportunities engender competition. Russian oil producers are critical players in today's energy marketplace and were formidable also in the early days of the oil industry. The Swedish Nobel family created Nobel Brothers Petroleum Producing Company and built a vast network of wells, transportation, and refineries in Russia. By the mid 1880's, Russian oil production, particularly from the prolific Baku fields, was about one-third that of the United States. The Nobels were producing half of all Russian kerosene. In 1886, the Rothchilds entered the scene with the Black Sea Petroleum Company (Benito) and soon became the second-largest Russian competitor with Europe as a principal target market. By 1891, Russia's share of the export trade in illuminating oil was 29%.[38]

Transportation is a fundamental economic parameter in creating value for an energy source. The Rothchilds needed transportation to open up world markets and eventually came to an arrangement with Marcus Samuel, a British trader, to sell Benito's kerosene east of the Suez Canal. Samuel commissioned the design and construction of technologically advanced ships to carry the kerosene, the first being named the *Murex*, for a type of seashell. Ninety percent of the tanker passages through the Suez Canal were made by Samuel's group, opening up access to Asia and creating significant competition for Standard Oil. Samuel formed Shell Transport and Trading in 1897.[39]

Royal Dutch Petroleum began oil development in the Dutch East Indies and had established an integrated operation from oil production through a pipeline into a refinery by 1892. The business grew rapidly and attracted the interest of both Marcus

Samuel and Standard Oil. Henri Deterding came to head Royal Dutch and, like Rockefeller, was repulsed by the wild fluctuations in oil prices. He saw the need for unity among new oil companies to protect Royal Dutch from Standard Oil. In 1907, Deterding and Samuel finally reached an agreement to form the Royal Dutch Shell Group, wherein Royal Dutch held 60% of the stock in the operating companies and Shell the remaining 40%.[40] This dual shareholding continued until 2005 when the stock was merged under Royal Dutch/Shell.

During this same period, Standard Oil was busy trying to bring virtually all oil production under one system in negotiations with the Nobels and Rothchilds. Interestingly, the Nobels and Rothchilds were able to provide a common front for Russian producers, but Standard Oil, with some 85% to 90% control of America oil, was unable to bring in the independent American refiners and producers, and the proposed agreement collapsed.[41]

Competition continued to emerge at the beginning of the new century. The British had their eyes on Persia, now known as Iran. William Knox D'Arcy, a former gold mining risk-taker, entered negotiations with Shah Muzaffar al-Din and was successful in gaining the first concession in the Middle East in May 1901. Following a seven-year search, the venture struck oil in May 1908 in Masjet Soleiman, Iran. The concession was incorporated as the Anglo-Persian Oil Company, which went public in April 1909. The British government under Prime Minister Churchill made the unusual move of buying 51% of the company in 1914. The driving concern was securing sufficient oil for the Royal Navy, an urgent issue given the political difficulties at the time. The company later became British Petroleum in 1954, a development that is more fully chronicled in the next chapter.[42]

Back in the United States, domestic competitors emerged with the discovery of oil in California and Texas. In 1910, California represented some 22% of total world production, more than any foreign nation. The Texas oil boom began with the discovery of Spindletop in 1901 by Captain Anthony Lucas and James Guffey. Several entities grew out of this gusher field. Gulf Oil emerged from Guffey's interest combined with oil from Oklahoma. William Mellon was Gulf's visionary who understood the need to develop an integration strategy around the stranglehold that Standard Oil had on the midstream and downstream businesses. A former Standard Oil employee, Joseph Cullinan, was the entrepreneur who started Texaco, building off a Spindletop lease position of former Texas governor James Hogg's group.

Government's expanding role

We have seen from the early dynamics of the coal industry that the federal government would intervene in the energy industry to ensure economic and national security. But beyond managing the macro-economy and orchestrating geo-political strategy, government also took on the role of protecting consumers and employees while promoting fairness in the market place. While competitive pressures had been reducing Standard Oil's market dominance, the public clearly disliked this powerful entity. By the end of the 19th century, after Rockefeller had ceded his leadership to John Archbold, the assault on Standard Oil gained force.[42]

Standard Oil's image sustained severe damage from articles in *McClure's* magazine starting in 1902 written by journalist Ida Minerva Tarbell. "Month after month, she spun the story of machination and manipulation, of rebates and brutal competition, of the single-minded Standard and its constant war on the injured independents."[44] Eventually, her 24 articles were consolidated into a 1904 book titled *The History of the Standard Oil Company.*

The public demanded protection from the perceived power of Standard Oil. The role fell to the federal government and President Theodore Roosevelt, the trust-buster. In 1906, the Roosevelt administration sued Standard Oil under the Sherman Antitrust Act of 1890, charging it with conspiring to restrain trade. The federal court found in favor of the government in 1909 and ordered the trust dissolved, but Standard Oil appealed. In May 1911 the final verdict came down: Standard Oil would be broken up over the next six months.

The separate entities emerging from the breakup of Standard Oil were an impressive group of companies, as figure 1.3 indicates. Equally impressive is the successful recombining of these companies as they have responded to the industry imperative of market stability through integration and controlled competition.

The Standard Oil breakup offers several lessons. As chapter 4 will elaborate, in a democratic society the public is the ultimate judge of the value being created by the energy industry and the companies it comprises. The public will influence governmental actions based on its assessments of the industry's behavior and performance. Ultimately, secretive and manipulative actions will emerge into view, and if the actions are considered harmful to society, the penalties will be severe. Powerful forces beget countervailing forces to create a balance. Public relations matter. From the earliest days, energy companies have dealt poorly with the public and the press, and have tended to treat politicians simply as power brokers.

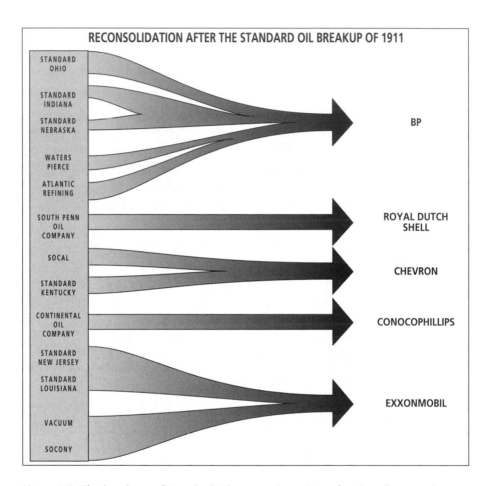

Figure 1.3: The break-up of Standard Oil spawned a variety of major oil companies

Big Oil Emerges

The drive for market stability and control did not vanish with the breakup of Standard Oil, but it did subside during World War I when oil prices were increasing in real terms. Shortages led to an Inter-Allied Petroleum Conference in 1918 to coordinate supplies and shipments among the Allies. Exxon and Royal Dutch/Shell made the system work; the US provided about 80% of the Allies wartime oil supplies. Allied efforts to interdict Germany's oil supply were largely successful and contributed to Germany's surrender in November of 1918. The victorious nations drove for strategic control of oil in dividing the spoils.

Ensuring energy security strengthened Big Oil

Worldwide oil shortages between 1918 and 1920 put energy security on national agendas. The United States pressed for the principle of an "open door," meaning equal access to oil regions for U.S. capital and business. The Americans protested the way the British and French carved up Mesopotamia, ignoring the open door principle. In response, the U.S. Congress passed the Mineral Leasing Act of 1920, denying access to drilling rights on public lands to foreign companies whose governments prevented similar access to American companies.

The British government became more accommodating. In July 1928 an agreement for the Iraqi opportunities was hammered out that included a "self-denying" clause whereby all participants would work jointly in the region containing most of the former Ottoman Empire. The scope was defined by the red line on the map in figure 1.4.[45] BP, Royal Dutch/Shell, Compagnie Francaise Petrole, and Exxon/Mobil, received 23.75%, with 5% going to Gulbenkian of the Turkish Petroleum Company. The "self-denying" clause, which came to be called the "Red Line Agreement," covered the Middle-East apart from Kuwait, Iran, Israel and Trans-Jordan. While conflicts arose, the agreement set the framework for Middle-East oil development.

Oil price collapse drives industry cooperation

There was much positioning among countries and companies after the war. By the early 1920's, international price wars were breaking out. Russia's revolutionary turmoil led to aggressive competition for its oil; all the bickering, particularly between Royal Dutch/Shell and Mobil, led to savage price cutting in India that spread worldwide. Oil discoveries were mounting in the U.S., from Los Angeles to Oklahoma.

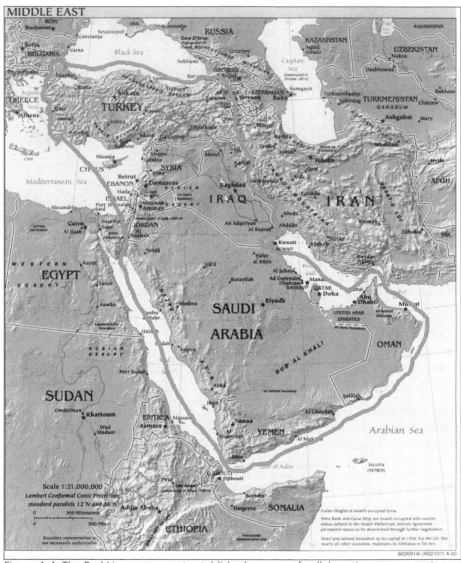

Figure 1.4: The Red Line agreement established an area of collaboration among major oil companies. *Source: Mount Holyoke College International Relations Program*

Mexico's production was increasing, and by 1921 it was the second largest producer in the world with an annual output of 193 million barrels. As figure 1.5 shows, real prices fell from almost $25 per barrel in 2004 dollars in 1920 to near $5 per barrel in the latter part of that decade. The companies were growing weary.

Early in 1928, Exxon, Royal Dutch/Shell, and BP began to hold secret meetings to rectify their destructive international competition. Behind the scenes, the British government had been

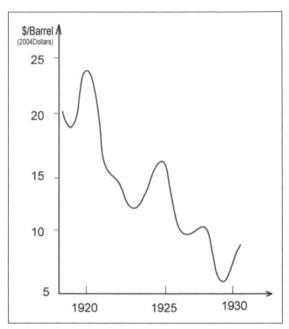

Figure 1.5: Collapsing oil prices led to cartel behaviors. *Courtesy of WTRG Economics: www.wtrg.com*

encouraging industry collaboration for its own self-interest. In August, industry leaders met at Achnacarry Castle in the highlands of Scotland to bring truce to the oil wars. In early September, the three originators had approved a statement of principles that became known as the "As Is" agreement for market shares outside the U.S. Deterding of Royal Dutch/Shell and Walter Teagle of Exxon were proving to be the next-generation Rockefellers of the industry.

The agreement began "with the basic rule: 'The acceptance by the units of their present volume of business and their proportion of any future increase in production.'... To protect American oil, the oil from anywhere else in the world was fixed at the price in the Gulf of Mexico, whence most United States oil was shipped overseas, plus the standard freight charges for shipping the oil from the Gulf to its market." Fifteen other American companies approved the 'cartel arrangement,' including the other four companies (Mobil, Texaco, Gulf, and Chevron) that with Exxon, Royal Dutch/Shell, and BP became known as the Seven Sisters.

Government's cooperative role in price stability

The "As Is" agreement, designed to support oil prices, proved ineffective when the super-giant field in East Texas, later known as the Black Giant, was discovered in 1930. "The East Texas boom, coming at a time of both depression and overproduction, produced a new glut of oil, when it was least of all welcome to the big companies, and it looked like a return to the wild old days of individualism before the Big Hand of Rockefeller reached out. . . . Prices fell down to 10 cents a barrel, and gas stations competed with free chickens to lure customers."[47]

Overproduction was driving prices below cost and had the appearance of industrial suicide. East Texas producers were a conglomeration of independents, some very famous like H.L. Hunt, but all outside the "as is" agreement. It became imperative to create a regulatory system that would control production and stabilize prices.

The Texas Railroad Commission was given the task and devised a scheme based on preventing the kind of physical waste of oil that had occurred at Spindletop. Gaining the cooperation of the independents proved difficult, until finally in August 1931, Texas governor Ross Sterling announced that East Texas was in a "state of insurrection," brought in National Guardsmen and Texas Rangers, and shut down oil production. Unsurprisingly, prices rose. The Texas legislature then enacted a market pro-ration system late in 1932. Unfortunately, the Texas Railroad Commission set the quotas too high, and illegal overproduction known as "hot oil" continued to depress prices.

The Federal government, believing that the flood of oil would bring disastrous results to the industry and the country, intervened to stop the movement of "hot oil" through federal-state cooperation, including the passage of the Connally Hot Oil Act and later the Interstate Oil Compact. The Bureau of Mines was to prepare demand estimates for the coming period and suggest voluntary quotas for the states, which they largely followed. The Texas Railroad Commission became much more effective. Additionally, with production being "controlled" in the United States, a tariff was enacted to prevent a flood of oil from foreign producers. The resulting system proved effective for stabilizing oil prices, which averaged around $1 per barrel from 1934 to 1940. Beyond the concern for market stabilization, the government was further extending its role to that of a good steward of the country's national resources and the TRC precedent is often cited by OPEC in support of its market interventions.

Middle East oil expands the group

The Red Line and "as is" agreements solidified the status of Exxon, Royal Dutch Shell, and BP positions as super-majors. New Middle East discoveries would elevate Chevron, Texaco, Gulf, and Mobil to form the elite Seven Sisters. Those remnants of the former Standard Oil confined more to U.S. domestic operations, such as AMOCO and Conoco, found their roles more limited for lack of a Middle East presence.

Both Gulf Oil and BP were very interested in obtaining a concession in Kuwait. The struggle eventually led to the formation of a joint company in December

1933 called Kuwait Oil Company, which would be a 50-50 joint venture. About a year later, Sheikh Ahmad of Kuwait signed a 75-year concession with the jointly owned company.

Earlier, in May 1933, Chevron, which had been successful in Bahrain and was not bound to the "Red Line" agreement, signed with Saudi Arabia for a concession in al-Hasa. Concerned about marketing and distribution of its Mid-East production, Chevron partnered with Texaco for its extensive marketing network in Africa and Asia. The California-Texas company (CalTex) was formed by pooling their assets east of the Suez, including the Saudi Arabian concessions.

The Kuwait Oil Company struck oil in the Burgan field in southeastern Kuwait in early 1938. This major find further strengthened oil company interest in the Mid-East. Shortly thereafter, the oil world changed forever when Chevron drilled into the amazingly prolific Arab Zone in its Saudi Arabian concession and formed Aramco. Such news brought not only U.S. and British interests to the desert kingdom, but also the Germans, Japanese, and Italians. The Japanese, offering a significant sum of money, won a concession in the Neutral Zone between Saudi Arabia and Kuwait.

Everette DeGolyer, the preeminent geologist of his day, made a speech on Saudi Arabian oil reserves to a Texas group in 1940 that correctly premised the future: "No such galaxy of fields of the first magnitude over such a wide area has been developed previously in the history of the oil industry . . . I will be rash enough to prophesy that the area we have been considering will be the most important oil producing region in the world within the next score of years." DeGolyer proved an oil prophet; today Saudi Arabia stands at the pinnacle of world energy supply with about 25% of the world's proved oil reserves. With such a strong position, the kingdom has become the swing producer, maintaining spare capacity to buffer supply fluctuations and moderate price movements.

Chevron and Texaco were concerned about the enormous scale of the capital and markets needed to develop their Arabian concession after World War II. There were both business and political risks. The U.S. government encouraged Aramco to bring in additional partners. Exxon and Mobil were the logical choices. Not surprisingly, there were lots of issues in accomplishing this, including dissent within Chevron and the matter of the Red Line Agreement, which still bound both Exxon and Mobil. After much negotiation, the Red Line Agreement was cancelled and Exxon and Mobil became partners in Aramco in March 1947.

Crude Oil's Dominance and Geopolitics

As the industrialized world turned increasingly to crude oil to meet strategic energy needs, access to oil and control of supplies became issues of great geopolitical significance. No nation, no matter how obscure, escaped the interest of western powers if it was believed to have substantial crude oil prospects.

World War II

Oil was even more central to World War II than it had been to the First World War. Access to oil was both a contributing cause of WW II and a strategic objective of great urgency. Ultimately, the availability of oil would prove instrumental in determining the victors and the vanquished.

Hitler's game plan

Hitler knew that Germany's lack of oil had been a critical factor in World War I. But Germany was rich in coal resources and had developed gasification and liquefaction processes that produced a high-grade liquid fuel from coal. By the late 1930s, coal provided 90% of Germany's energy requirements. Hitler envisioned synthetic fuel technology as a means to reduce dependence on foreign oil and launch his military campaign. These synthesizing technologies are enjoying a renaissance today for transportation fuels as conventional oil resources deplete.

Nevertheless, Hitler saw oil as "the vital commodity of the industrial age and for economic power. He read about it, he talked about it, he knew the history of the world's oil fields. If the oil of the Caucasus—along with the 'black earth,' the farmlands of the Ukraine—could be brought into the German empire, then Hitler's New Order would have within its borders the resources to make it invulnerable."[49] Hitler did not succeed in Russia, however, and his only hope then lay with the North African campaign led by General Rommel to capture the oil fields of the Middle East. But short fuel supplies hampered that campaign, which was eventually defeated. The only thing left for Germany was the production of synthetic fuels, but bombs destroyed those plants. In time, lack of fuel prevented Germany even from protecting the homeland.

Japanese oil vulnerability

The Japanese knew their lack of energy resources created vulnerabilities. Their imperialistic ambitions of the 1930s would require oil, but at the time they imported 80% of their supplies from the United States. Thus, they went after the East Indies fields. As the war progressed and Japan allied itself with Hitler and Mussolini, the U.S. embargoed oil to Japan.

The Allies concentrated on limiting Japan's oil supplies by destroying oil tankers. This severely hampered the Japanese offensive, based as it was on naval power. Eventually, lack of fuel would drive the Japanese into desperate acts, including kamikaze pilots, which proved fruitless in the end. Fueled by more than sufficient oil supplies, the American retaliation was relentless. The stubborn Japanese even turned to pine roots as a fuel source.

National security linked to energy security

World War II forged inseparable links between energy security, national security, and a nation's ability to wage war. During the war, both the British and American governments took control of oil in their countries, diverting supplies to the military while rationing civilian uses. Britain merged the downstream activities of its oil companies into a single national monopoly. Among measures taken by the U.S. was a declaration that "the defense of Saudi Arabia is vital to the defense of the United States," accompanied by direct aid to the oil-rich kingdom.

The key lesson here is that governments will go to great lengths to manage energy supplies during a time of crisis. With national security dependent on energy supplies, energy becomes, in effect, a state resource, and issues of free-market competition and economic benefits take a back seat. National security is fundamental to survival on Maslow's hierarchy of needs and will take precedence over economic well-being when a serious threat appears.

Oil Reigns in Second Half of the 20ᵗʰ Century

In the years after World War II, oil became the dominant energy source, as shown in figure 1.6. Coal-fueled railroads lost market share to freight trucks that ran on gasoline or diesel fuel. The railroads themselves began switching to diesel locomotives, driving coal out of the transportation sector entirely. Further, the continuing saga of labor problems coupled with the need for improved safety drove up coal production costs.

It was not coal's losses, however, that made oil dominant. Gasoline demand increased by over 40% from 1945 to 1950. "America was in love once again with the automobile, and now consumers had the means to carry on the romance. In 1945, 26 million cars were in service; by 1950, 40 million."[50] At least historically, we can see a correlation between the dominant fuel source and the major mode of transportation.

Over time, the strategic energy source becomes the dominant energy supplier. This happened first in the case of coal and then with crude oil. Coal's share of U.S. energy consumption had peaked by 1910, and after a plateau period began its precipitous drop following World War I. We can also observe in figure 1.6 that liquids' share of energy consumption peaked in the late 1970s. This is strong evidence that another shift in the strategic energy source may well be in the making if the coal paradigm holds true for crude oil.

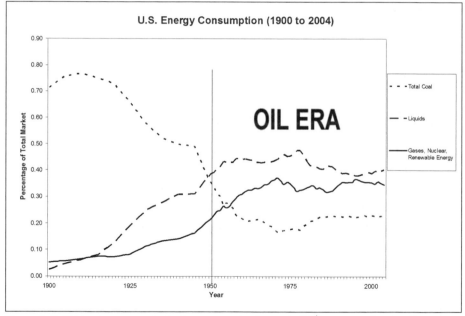

Figure 1.6: Oil took over as strategic resource in the early 20th century. *Source: Energy Information Administration*

The regime of the seven sisters, established through various agreements and the assistance of government agencies such as the Texas Railroad Commission, led to a period of very stable oil prices from the mid-1930s to the early 1970s, as figure 1.7 shows. The oil price was high enough to encourage energy investments but low enough to enable immense economic growth in developed countries, particularly during the period after World War II.

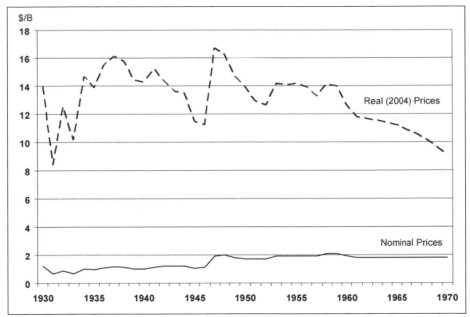

Figure 1.7 Oil prices were stable from 1930 to 1970. *Source: Energy Information Administration*

This twist to the free market system, involving loosely knit cooperation interspersed with government control and the punitive overhang of anti-trust laws, proved to be very effective for developed countries like the United States.

Resource nationalism and the rise of OPEC

Several major shifts in the structure and operation of energy markets since the 1950s began with the formation of the Organization of the Petroleum Exporting Countries, or OPEC. This action began the transfer of control over oil production and pricing from Big Oil, closely associated with consuming countries, to oil producing nations. The rulers of producing countries began to assert their authority as stewards for their nation's resources as oil moved center stage in wealth generation and geopolitics.

While OPEC was officially formed on September 17, 1960, various precedents had set the stage for its founding. Precious resources were nationalized by Russia in 1917, Argentina in 1922, and Mexico in 1938, and other countries had threatened such action. Controlling oil resources meant not only economic benefits, but also political clout for those countries with abundant reserves.

Although the Middle East dominates the current membership of OPEC, it was the prickly nationalism of Latin America that advanced government control, with

Venezuela leading the way. In 1943, Venezuela established the "50-50" principal under which the government's share of the oil wealth would increase to equal that of the oil companies. In 1945, Venezuela set another precedent by deciding to take some of its royalties in oil rather than money, with the intent of capturing income from downstream transportation, refining and marketing. Venezuela became a place where anyone could buy oil through direct negotiation.

By 1950, Saudi Arabia had established its own form of a "50-50" deal. In Iran, however, the situation turned out quite differently. Iranian nationalism was thwarted by a U.S.-engineered coup that put Shah Reza Pahlavi in control and established an oil consortium that included all of the Big Oil Seven Sisters plus CFP. Actual ownership of the oilfields remained with the Iranian National Oil Company, but the consortium had rights to buy the oil under favorable terms and restrict production.[52]

Against a backdrop of rising Pan-Arab nationalism promoted by the Egyptian leader Gamal Abdel Nasser, the relationships between the producing countries and the IOCs ran smoothly during the 1950s, with production growing and the companies "posting" acceptable prices. However, in 1959, BP cut its posted price. This quickly led Saudi Arabia, Venezuela, Kuwait, Iraq and Iran into a cooperative agreement to defend the price structure.[53] Then in August 1960, Exxon cut its posted price without warning. By mid-September, the creation of OPEC was completed.

OPEC did not have much impact during the 1960s other than deterring the oil companies from unilateral actions. The surplus oil on world markets and political rivalries kept OPEC weak. And although the third Arab-Israeli war, the Six Day War, in June 1967 triggered an Arab oil embargo against the U.S., UK, and Germany, it was ineffective and was lifted in September.

The tide turned in OPEC's favor in the 1970s. Muammar al-Qaddafi seized power in Libya in 1969 and fashioned himself after Egypt's Nasser. As rising energy demand reduced the crude oil surplus, Qaddafi pressured the 21 oil companies operating in Libya for better terms. By September 1970, he had won a 30 cent increase in posted price coupled with a 20% increase in royalties and taxes. This was a critical tipping point, as Daniel Yergin explains: "The Libyan agreements decisively changed the balance of power between the governments of producing countries and the oil companies. . . . It not only abruptly reversed the decline in the real price of oil, but also reopened the exporters' campaign for sovereignty and control over their oil resources."[54]

In October 1973, the fourth Arab-Israeli war began. Oil companies were again negotiating prices with the Arab producers, who finally made the decision on their own. The oil world convulsed as the Gulf States raised posted oil prices by 70% to over $5 per barrel. The continued escalation in the war brought the ultimate oil weapon into play—an embargo against the United States. With tight supplies, the embargo triggered a panic, and the hoarding that resulted drove spot prices over $17 per barrel.[55]

Eventually, OPEC decreed a new price of $11.65 based on Saudi light crude as a marker. Further nationalization of oil assets followed quickly. Control over oil production and pricing had fully shifted from the Western IOCs to producing countries, primarily in the Middle East. The OPEC cartel had come of age.[56] But perhaps more striking, the NOCs like Saudi Aramco were on their way to becoming the new "Big Oil."

Consuming countries respond and markets react

Perhaps not surprisingly, many Americans held the oil companies responsible for the embargo, shortages, and price increases. This is a real lesson for energy executives. U.S. senator Henry "Scoop" Jackson focused on obscene profits in Senate committee hearings. The federal government responded by erecting a bureaucratic nightmare of price controls, entitlements, and allocations, setting the stage for what later became the Windfall Profits tax of 1980. In the late 1970s, President Carter introduced an energy program he proclaimed the "moral equivalent of war;" however, it did not prove very effective. Based on these events, we can confidently predict that political turbulence will erupt whenever energy security comes under pressure, as is happening in the current up-cycle, but may be flawed by the desire to punish the oil company messengers rather than address root causes.

The overthrow of the shah of Iran in 1979 triggered a second energy shock. Iranian oil production plummeted and oil prices shot up. Energy was again at the center of geopolitical maneuvering, highlighted by the Soviet Union's invasion of Afghanistan. This led President Carter in January 1980 to assert what became known as the Carter Doctrine: "Let our position be absolutely clear. An attempt by any outside force to gain control of the Persian Gulf region will be regarded as an assault on the vital interests of the United States of America, and such an assault will be repelled by any means necessary, including military force."[57] The doctrine explicitly acknowledged the inseparability of energy supply security from national and economic security.

However, new oil supplies from non-OPEC countries, including domestic resources, soon strengthened the U.S. energy security position. The Alaskan pipeline was delivering almost two million barrels per day of North Slope oil in the 1980s. Mexico began to exploit major oil reserves found in the early 1970s both onshore and offshore, enabling oil exports once again. Most importantly, the North Sea provided a prolific opportunity for the IOCs, who had been pushed out of the Middle East.

As we have learned before, petroleum investment responds to higher energy prices. In this case, IOC investment developed new oil reserves in non-OPEC countries. The overall energy situation benefited also from significant interfuel competition from nuclear plants and a rejuvenated coal industry.

In the frenzy of the late 1970s and early 1980s, forecasters were predicting oil prices of $100 per barrel or more within a decade. These predictions would prove totally mistaken as still other factors came into play beyond the new oil supplies. While many government policy initiatives were ineffective, conservation measures in developed countries did help to reduce demand. Additionally, nuclear power emerged for a time as an alternative energy source. Over a 15-year period, the French installed 56 nuclear reactors that not only satisfied their own electricity needs, but also provided exports to other European countries.[58] Many other countries followed, and from very little capacity in the 1960s, world nuclear output had grown to around 300 gigawatts by the mid-1980s.[59]

The blow that finally buried the lofty price forecasts was the global economic downturn. Sustained higher energy prices brought on severe worldwide inflation. To combat inflation, Federal Reserve chairman Paul Volker tightened the U.S. money supply. The reduction triggered the worst recession since the Great Depression of the 1930s, with a trough in 1980 and a deeper trough in 1982, taming energy demand. By 1983 the call on OPEC oil was reduced by a whopping 13 million barrels a day. The energy market suddenly had a massive supply overhang. OPEC was losing control.

Here another lesson unfolds: large price increases resulting from supply disruptions stimulate corrective actions. Two major energy shocks in a 10-year period created a huge systemic response. A second lesson is to recognize that long lead times are needed between a triggering event (the 1973 embargo), the response (fuel efficiency standards in 1975), and the result (lower fuel demand and higher oil supply in the 1980s). This systemic lag stems from the slow turnover in consuming sector capital stock and the long lead-time for supply investments.

For OPEC, loss of market control coupled with economies dependent on oil revenues made the cartel less and less effective. Despite setting member production quotas, OPEC discipline was weak and violations were common. The price support role fell to Saudi Arabia, whose oil revenues had dropped from $119 billion in 1981 to $26 billion in 1985. Concurrently, Saudi influence within the cartel was waning. In June 1985, King Fahd warned that Saudi Arabia would secure its own interests if the cheating and discounting continued. Likewise, the Saudi's were propping up the price everywhere, particularly for the British North Sea production.

By the end of the year, the Saudis had switched from a price support strategy to a market share strategy. With a significant supply overhang, this shift caused prices to crash. West Texas Intermediate dropped from its peak of $31.75 per barrel to $10 at the end of November 1985. OPEC was in disarray. The competitive market place was back driving prices. Before continuing this story into the down cycle, it is important to understand how price formation works in a commodity market such as oil.

Oil Price Formation and Energy Market Developments

Unfortunately, the dynamics of price formation in a competitive commodity market are poorly understood by the public, and often willfully misrepresented by grandstanding politicians and special interests. Any time prices rise rapidly, accusations of manipulation and price gouging quickly follow, based on inferred market control and collusion. A rapid price rise also creates anxiety about adequacy of supplies, throwing into question the availability of oil reserves to meet future demands.

Much of the public seems to hold a simplistic model of oil price formation derived from utility pricing, under which companies incur costs, slap on a profit margin, and set the price that consumers are obliged to pay. That model in no way represents the reality of competitive commodity markets, particularly oil, where the economics of supply and demand drive prices. When oil supply is short, prices rise to encourage an increase in supplies and discourage consumers' use. Conversely, when supply is long, prices fall to discourage further investment, slow down drilling, and reduce production; at the same time, consumers are encouraged to use more by buying bigger cars and houses. It's as simple as that.

Demand drives supply

Discussion of oil prices should focus on understanding demand and what can be done to bring supply into balance with demand. In the 1930s, as we have learned, there were actually hearings and legislation aimed at rectifying low prices. Since 1970, even though the industry was in dire straits from the mid-1980s, hearings have occurred only when prices were high. In hearing after hearing and study after study, allegations of price gouging have been disproved. In a rational world, it would be clear that the allegations are a smokescreen to evade a political responsibility to look deeply at root causes, such as denying industry access to public lands to increase supply, and at measures designed to improve fuel use efficiency or to develop substitute fuels.

Only when the situation appears extremely serious, such as in 1979 and again in 2006, do policymakers focus attention on systemic issues, including demand.

People do not purchase energy for the pleasure of the experience. It is a necessity, and consumers deeply resent high, volatile prices and threats to supply. They need energy to drive their cars to work every day, to keep food chilled in their refrigerators, and to heat and cool their homes. Alternatives are not readily available. In economic terms, this translates into an inelastic short term demand curve, which means that demand does not change significantly as prices rise or fall. When price is plotted against volume, the curve is very steep in the short term indicating that it takes a large price move to induce a small demand change. Demand is more elastic in the longer term, as consumers will replace inefficient vehicles and appliances with more efficient ones over time. But because that takes time, the impact on prices in the short term is small.

Meanwhile, Adam Smith's "invisible hand" of the market is working its own magic. Significantly higher prices stimulate new investment, which generates new energy supply and eventually brings down prices. When prices drop drastically, society enjoys the short-term benefits, but the industry constricts investment because of reduced cash flows and marginal profitability, resulting eventually in less supply in the future. The long cycle time in the energy industry for investment to turn into consumable energy creates lags, and the end result is the potential for cyclical boom and bust periods. The period of 1973 through 1985 was a boom with cyclically high prices; the period 1986 through 2002 was a period of cyclically low prices.

The general public does not understand these price-setting dynamics in part because the reality of any particular situation is often difficult to explain, with many factors woven together. It usually is not just the rising price of crude oil. Refinery capacity utilization, boutique environmental requirements, product supply disruptions, weather conditions, inventories, and myriad other circumstances can create a price run up on crude oil products, either geographically focused or widespread.

Supply conditions drive price

Oil supply, on the other hand, is much more elastic as prices rise and fall. The supply curve can be much flatter except at the extremes. As long as suppliers have some spare capacity, producers are generally willing to bring on stream and produce the additional available capacity as oil production costs are largely fixed and there is a margin of profit. However, there are points of discontinuity. For example, crude oil prices had to drop to the $10 range in the mid-1980s to cause stripper wells producing less than 10 barrels a day to shut down because they were uneconomic. On the other side of the supply curve, there is a point when supply runs out, and then the curve becomes very inelastic indicating that even sharp price increases cannot induce much increased production. When supply is very tight or scarce, any increase in demand results in a sharp increase in prices. This is the current situation with increasing global oil demand particularly from China and India. Since there is not enough supply for everyone, someone wanting the fuel must eventually get pushed out of the market by the higher price. This is all rather basic economics, but it is nevertheless not well understood. It does not take a colluding cartel to cause such price spikes. What it takes is insufficient short-term supply to meet demand. Figure 1.8 portrays these different conditions, from a substantial supply glut, as in 1985, to stable commodity pricing as in 1960, to short supply as in 1980.

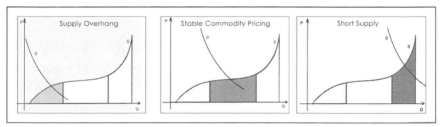

Figure 1.8: Commodity prices move sharply when supplies are very short or very long

We have also learned that while society does not like periods of high price and short supply, the industry does not like low price periods which lead to austerity and eventually to employee lay-offs, for some bankruptcy, and for most little incentive to invest in growth. Historically, wild price fluctuations have not been appreciated either by society and government or by the industry.

Market regimes controlling supply

Certainly the public and government officials have an inherent distrust of the oil industry and believe that it colludes to set prices. As we have learned, the interaction of supply and demand set the price, not the oil industry. But, reductions in supply can lead to price increases. Therefore, the fundamental issue is whether there is an orchestrated restriction on oil supply and whether or not it is harmful to society.

Over time, various regimes have been established, beginning with Rockefeller and including governmental support, to stabilize the oil market and prices. This is accomplished by restricting supply to keep it below levels that would otherwise be available at the prevailing price. Figure 1.9 depicts the periods covered so far in our analysis. On the graph, the proxy for crude price is average U.S. wellhead price.

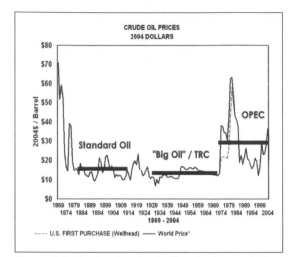

Figure 1.9: The level and volatility of prices varied with the prevailing market regime

After the Rockefeller/ Standard Oil regime, which was considered monopolistic, the designation Big Oil refers simply to the broad array of agreements among oil companies to stabilize prices by holding their market share or controlling production, starting with the "As Is" agreement in the late 1920s and followed shortly by the "Red Line Agreement." These agreements were not effective, however, without the Texas Railroad Commission's (TRC) pro-rationing and later a US import fee on crude and products. This "Big Oil/TRC" period lasted until the rapid demand growth in the 1970s, along with the peaking of U.S. production and the Arab oil embargo and production cutbacks, established the conditions for the OPEC cartel era.

Figure 1.9 indicates that the world oil price actually declined from 1920 to 1970 in real terms from a peak of over $20 per barrel to the mid-teens during the "Big Oil"/TRC regime. The price remained remarkably stable in real terms even with the significant economic growth of the 1950s and 1960s. Spare capacity existed throughout this period, which helped stabilize prices; the tradeoff was a price somewhat above the marginal cost of supply. However, this margin above the marginal cost of supply encouraged investment, which created further spare capacity.

The ability to control a market depends rather directly on market share. When only a few players have a large market share, and thus the ability to influence price, we have what economists refer to as an oligopoly. When these players cooperate and act in concert, their pooled market shares gives them great influence over price. The "As Is" agreement is one example of oligopoly.

Cohesiveness refers to the degree to which an oligopoly acts as a "unified entity." A cohesive oligopoly can exercise effective market control and earns the designation "cartel." Non-cohesive oligopolies, such as U.S. airlines, have little market influence beyond the individual companies' market shares. Cohesiveness in effect magnifies an oligopoly's impact on prices. Figure 1.10 shows the interaction of oligopoly share and cohesiveness as they work to create market concentration. The iso-curve shows equivalent relationship between the two variables .

The four attempts at oligopolic oil market control have varied in their effectiveness and the responses they caused. The early Standard Oil regime was by far the most concentrated but brought on severe anti-trust action. The graph suggests that the period of 1935 through the 1970s saw market control roughly equivalent to OPEC in 1973, although the level of market share and cohesiveness varied. The U.S. government responded in various ways to the reign of Big Oil, and government also collaborated in controlling the market. There were a few interventions and many threats, but acceptance; moderate prices and excess supply clearly deflected aggressive anti-trust actions.

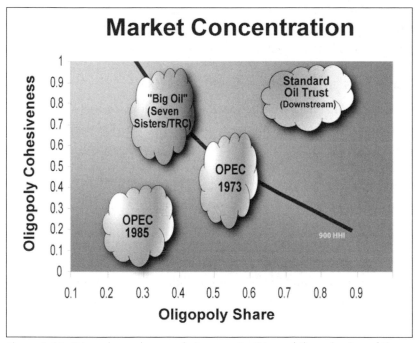

Figure 1.10: Prices depend on market concentration and the cohesion of the leading institutions

When foreign countries took control through OPEC, U.S. military, diplomatic and economic influence with the member nations replaced anti-trust action as a means to moderate oligopolistic practices, particularly with the largest OPEC producer, Saudi Arabia. Even for the mighty OPEC, however, effectiveness depended on cohesion and market share, and the cartel was much less influential in 1985 than it had been during the 1970s. OPEC cohesiveness has always been impacted by the tension within each member state between revenue generation and resource conservation. Low prices create anxieties about revenues, weaken cohesiveness, and tempt members to over-produce. When prices are high, cohesiveness returns and resource conservation receives greater priority.

We can perhaps learn from this that the cyclical volatility of prices is greater when the members of a cartel managing supplies are nations and face broader social obligations than corporations. This higher volatility is readily observable from figure 1.9.

The oligopolistic oil regimes have coexisted with various forms of government regulation and intervention, which we have examined in some detail. Natural gas and power markets, however, particularly in the United States, have experienced direct governmental price setting controls. They were not governed by market forces of supply and demand; natural gas producers sold at prescribed prices, and electric utilities lived with regulated returns. These regulated energy markets have been opening up, and that is the focus of our interest.

Natural gas and power deregulation

Natural gas interstate pipelines were controlled in 1938. The Federal Power Commission (FPC), which administered the law, also acted to stabilize prices at the wellhead, a practice that was upheld by the Supreme Court in 1954. During the Carter Administration, the Natural Gas Policy Act of 1978 provided for a phased decontrol of prices at the wellhead, concluding in 1993. During the phased decontrol period, natural gas prices declined by 24%, delivering an estimated cumulative savings to retail gas consumers of $100 billion.[61] The Federal Energy Regulatory Commission (FERC) Order 636 unbundled pipeline transportation and merchant services.

The result of deregulation was the emergence of an open market for natural gas and a huge expansion of drilling in the Gulf of Mexico, leading to low prices, substantial benefits to industrial and household consumers, and accelerating demand growth. An unintended consequence was the still-birth of the U.S. LNG industry, because negotiated price formulas were higher than market prices. It would be 20 years before LNG reemerged as a significant supply source for the U.S. market, driven by the surge in combined-cycle gas-fired power plants in the 1990s.

The U.S. electric power industry was free from federal regulation until the 1930s. The Public Utility Holding Company Act of 1935 provided for the regulation of utility holding companies, while the Federal Power Act established FPC regulation of utilities involved in interstate wholesale transmission and sale of power.[62] These laws resulted in regional monopolies supplying power through their own plants and transmission lines. In 1978, Congress passed the Public Utility Regulatory Policy Act (PURPA) to encourage alternative fuels for generating electricity. PURPA required utility companies to buy power from independent power producers at a price approved by regulators. FERC was given the authority to grant independent power producers equal access to the utilities transmission grid through the Energy Policy Act of 1992. FERC Order 888 in 1996 kept utilities from limiting the entry of lower-priced generation into their grid, while FERC Order 2000 called for the voluntary creation of regional transmission organization (RTOs) to replace state control and operation of the transmission grid.

These changes set the stage for state deregulation of the power industry. Between 1996 and 2000, 24 states and the District of Columbia passed preparatory legislation.[63] Regions that decontrolled power generation attracted large investments in highly efficient, environmentally benign, combined-cycle, gas-fired generation units. The benefits of deregulation were initially apparent, until rising demand for natural gas put unsupportable stress on natural gas supplies.

Ercot (Texas) and PJM (Pennsylvania, New Jersey, Maryland) regions crafted careful and considered deregulation rules, which have proven resilient to changing market conditions. California was also one of the early adopters of deregulation in 1996. Unfortunately, California's complex supply logistics coupled with flaws in market design enabled companies like Enron to take unfair advantage through market manipulations.

Strangely, investor-owned utilities such as Southern California Edison and Pacific Gas and Electric were obligated to sell power to other private companies, becoming buyers of wholesale electric power. At the same time, they could not pass on to customers the increases in prices from their open market purchases. They also could not enter into long-term contracts with power generators, requiring them to buy on the short-term market. This all set an enticing environment for illegal activity, which some companies could not resist. Traders are programmed to look for and exploit market anomalies, and by doing that they contribute to an efficient market. However, they are not supposed to create market anomalies. Enron and other companies have been fined heavily, and several paid the ultimate corporate price of bankruptcy for their trading exploitations.

The first lesson for energy leaders is that strategies based on poorly designed regulations can appear lucrative for a while, but are rarely sustainable. The second lesson is that some players in newly formed markets learn the hard way about what is considered acceptable societal behavior.

The more fundamental lesson involves freedom in the marketplace. Oil companies have generally avoided the electric power business, citing the deep regulatory environment. This is mildly ironic, because the oil companies face significant regulations in all parts of their business, including at times price controls. The returns of regulated utilities are often more attractive than free market returns for the oil part of the value chain, and they are made without the pressure of competition. John Wilder, chairman and CEO of the power utility TXU, gives an interesting perspective on price controls: "We have studied price controls and can't find one successful case study. Every price control has led to shortages. Resources aren't getting allocated properly. Markets can allocate resources more efficiently than governments. You either believe in market-based Adam Smith solutions, or you believe in Lenin solutions."

The Oil Patch Down-Cycle

We left the historical discussion of crude oil with prices collapsing to $10 per barrel at the end of 1985 in the face of a supply glut. Because commodity industries, including energy, are driven by costs, a pricing down-cycle brings major cost cutting efforts, often involving restructurings and consolidations to increase economies of scale. The industry under these depressed conditions responds with strategies consistent with the mature stage of a product life cycle.

Market disequilibrium brings consolidation and downsizing

The first wave of 1980s restructurings was driven by perceived disparity between the market value of a company, as determined by its share price, and the inherent worth of its underlying petroleum assets, or its reserves. Oilman and corporate raider T. Boone Pickens believed that reserves in the ground were worth more than the value of companies on the stock market. Some of this was likely the result of companies becoming bloated and less efficient during the up-cycle from 1973 to 1985. Another factor was the reduced prospects for growth as the up-cycle momentum slowed. A third fundamental factor was the economic downturn.

A wave of acquisitions and mergers ensued, the largest of which involved Chevron devouring Gulf Oil Corporation for an all cash offer of $13.2 billion. Texaco acquired Getty Oil, provoking Pennzoil, which had also sought to buy Getty, into filing a huge lawsuit against its larger rival. Royal Dutch/Shell bought out the minority shareholders in Shell Oil Company. BP eventually acquired all of Sohio (Standard of Ohio), which had a significant cash flow from the Alaskan North Slope. There were share buy-back programs from companies like Exxon. Arco, Phillips, and Unocal leveraged their balance sheets to forestall raiders. The bottom line was a competitive shakeout. And so another lesson comes clear for energy leaders: major systemic changes bring about industry restructuring and competitive realignment.

During the downturn, many oil companies displayed early 1980's price forecasts of $100 per barrel against actual realized prices, tacking on slogans like "lest we forget." Such slogans became an industry mantra and continue to govern decision-making in the new millennium. Companies recalibrated strategies to the fundamentals of commodity businesses, striving for cost leadership. This required major downsizing. The exuberance of the late 1970s and early 1980s was replaced by the somber realization that organizational fat must be sliced back severely. Massive layoffs reduced workforce levels in the U.S. petroleum industry from a peak of 800,000 in the early 80s to around 300,000 in 2000. The traditional idea of big oil companies providing lifetime careers gave way to uncertainty and career interruption.

The U.S. drilling rig count for oil and gas projects, a traditional index to the health of the oil industry, fell from a peak of 4500 in 1982 to around 1000 for the second half of the 1980s. Return on equity after the 1986 crash declined to a moribund 7% for the integrated oil majors, substantially below the S&P 500 average of around 12%. For surviving U.S. independent oil companies, however, it was a bloodbath with average returns dropping below a negative 25%. New skyscrapers in Houston, the Energy Capital of the World, stood vacant.

But for the average American, energy slid off the agenda except for environmental concerns. Gasoline lines and fears about future oil supplies were a thing of the past when lower prices appeared at the pump. The public outrage over high energy prices did not, however, yield to gratitude when prices dropped, nor to sympathy as the industry went through its gut-wrenching downsizing.

This experience holds a lesson from Maslow's hierarchy of needs. Energy mainly satisfies lower level needs of everyday living and only indirectly affects the higher level ego needs of the American "me" society. Energy does not motivate consumers when it is readily available and affordable; it is simply a hygiene factor. But when energy is scarce and impacts people's wallets, it becomes a major de-motivator, bringing out anger and resentment. Rational explanations will not satisfy the emotional needs the consuming public feels at the time of their discontent.

Real-time information mitigates, shocks, and aids markets

Iraq's invasion of Kuwait in August 1990 delivered another energy shock, and nearly doubled crude prices to $40 per barrel. In this case, however, no long gasoline lines or panic ensued in the United States, which some argue was due to a market unencumbered with the allocations or controls that existed in previous shocks.[64] Energy demand was also weakening with signs of an economic recession. The U.S. policy of using the Strategic Petroleum Reserve, if necessary, dampened fears of a supply shortage. National and economic security was too important to let the invasion of Kuwait stand unanswered, and the public seemed to understand this and was willing to tolerate higher gasoline prices for a while.

The massive attack that occurred on January 17, 1991, would bring a temporary oil price spike from $30 to $40 per barrel, but the price quickly dropped to $20 per barrel when the outcome was obvious. A new lesson here involves the information revolution and the real-time window it provides on current events. The world has become flat,[65] and the instantaneous availability of information allows businesses to assess risks quickly and more accurately.

Real-time information also affects the dynamics of financial markets, driving investor perspectives from annual or longer timeframes to quarterly and shorter. Financial analysts began a drum-beat in the 1990s for improved return on investment as the lower energy price environment drove down the industry's profitability. Return on equity for major energy producers generally lagged below the S&P industrials from 1985 through 2000.[66] Analysts' expectations demanded quarterly performance improvements from the long-cycle energy business and judged the companies on returns, not growth. Oil companies reduced investments, beginning the process of lowering the reserves' replenishment rate.

Energy companies built their strategies to improve operational efficiency and economies of scale, restructure the business portfolio, and invest selectively in projects that provided early high returns. Workforce downsizing continued through organizational de-layering, improved productivity from information technology, and further consolidations.

New round of mergers and acquisitions

The competitive situation had changed dramatically from the days of the "Seven Sisters" as OPEC and non-OPEC NOCs became big players, particularly in the upstream and in emerging energy markets such as China and India. This opened the door for the super-majors to reconsolidate parts of the former Standard Oil Empire (figure 1.11). This repositioning strengthened the American and European IOCs in a time of changing geopolitical dynamics.

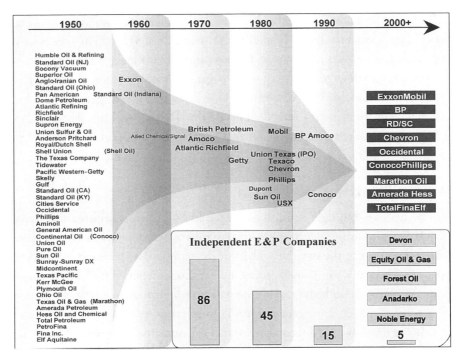

Figure 1.11: The petroleum sector has seen massive consolidation in the past half century. *Courtesy of J. S. Herold, Inc.*

BP continued its acquisition strategy, buying Amoco in 1998 and promising to deliver $2 billion in pretax cost savings while commenting that "international competition in the industry is already fierce and will grow more acute as new players emerge."[67] BP Amoco then targeted ARCO and concluded the acquisition in 2000 after agreeing to sell ARCO's Alaskan businesses to Phillips Petroleum. The French company Total merged with Belgian Petrofina, then the combined company announced in July 1999 its merger with the second French former NOC, Elf Acquitane. TotalFina's chairman Thierry Desmarest said, "I believe it is necessary to join forces to assure continued growth and to take our place as an oil major of the first rank at a time when the industry is restructuring on a global scale."[68]

The biggest merger was Exxon and Mobil, also announced in 1999, when crude prices had approached $10 per barrel. Citing cost savings and longer-term strategic benefits, Exxon's chairman Lee Raymond said, "The merger will allow ExxonMobil to compete more effectively with the recently combined multinational companies and the large state-owned oil companies that are rapidly expanding outside their home areas."[69] To complete the merger with Mobil, the FTC required Exxon to divest various businesses. These divestitures became significant opportunities for other players.

The consolidation continued as Chevron merged with its Caltex partner Texaco in 2002. Again, costs savings were spotlighted, with at least $1.2 billion to be achieved within a six to nine month timeframe. Conoco and Phillips merged in 2002, forming at the time the sixth-largest publicly held energy company in terms of crude oil and natural gas reserves, and the fifth-largest global refiner. The target was to achieve $1.25 billion in synergies by the end of 2003. Royal Dutch Shell's acquisitions, also in 2002, were modest compared to the other IOCs, consisting of Pennzoil-Quaker State, Enterprise Oil in the UK, and Texaco's downstream interest in the United States.

Decoupling the value chain

The decades-old tight integration of the big oil companies began to weaken with OPEC's evolution. Decoupling of previously interdependent business units became the order of the day. New realities demanded that each part of the value chain be accountable for its own return on investment. Some referred to this as the "atomization" of the industry, and it opened up opportunities for niche players in each part of the chain. Everyone was trying to improve return on investment in the upstream in this low crude-price environment. Companies like Royal Dutch Shell pruned their upstream assets, particularly in the United States. Figure 1.11 also shows the substantial shakeout among the independent producers that did not have the financial strength to weather the downturn.

With OPEC countries lacking downstream assets, trading activities grew in the early 1980s, buoyed by new market tools such as NYMEX's crude oil and gasoline futures contracts. During the 1990s, a new midstream part of the value chain formed in integrated companies involving trading, pipelines, and storage facilities. This intermediation function enabled the full decoupling of the core upstream and downstream businesses. Information technology was a major catalyst in this transition, and Enron took full advantage by developing risk management tools and innovative financial arrangements to become the dominant player in this part of the value chain. Unfortunately, Enron got caught up in the urgent need to continually improve financial performance on a scale that was out of proportion to its underlying energy trading competence, which in principle was a sound business model. Nevertheless, Enron demonstrated how a first-mover strategy can lead to a superior competitive position. The IOCs' mind sets were based on a long-cycle mentality, and they were slow to respond; however, today companies like BP, ConocoPhillips, Royal Dutch Shell, and investment banks have filled the trading gap left by the demise of Enron and the financial crises that plagued all of its energy trading imitators.

Downstream Unattractive Performer

Companies invested in the downstream refining part of the value chain only reluctantly, because the return on investment in this business segment was very poor. Refining and Marekting returns over the 25 year period of 1977 through 2003 average 6% which was half the average of the S&P industrials at 12% during the same period. The need for improved air quality drove significant expenditures in the industry; for example, in the United States between 1980 and 1998, petroleum refiners spent almost 83 billion dollars, which did reduce air emissions by 75%.[70] The heavier gravity in crude slates provided opportunities for building upgrading units such as cat-crackers, hydro-crackers, and cokers that break down the oil molecules into lighter products. Unfortunately for the industry, the herd instinct to capture the potential value from this "light-heavy differential" reduced the expected profitability.

The petrochemical business came under enormous pressure at the tail of the chain. It had largely become a commodity business requiring scale, operational efficiencies, and advantaged feedstock costs. Not envisioning relief in upstream and downstream profitability, which were considered core, several companies divested petrochemical operations. Texaco sold its chemical business, completing the final piece in March 1997. One rationale for these divestitures was that companies should get out of businesses where their parenting skills lacked impact. Shell exited several downstream chemical businesses. BP exited the olefins business in 2004. Chevron and Phillips merged their chemical businesses into a new company with each owning 50% in 2000.

Another competitive factor for the downstream is the feedstock cost advantage of the Middle East NOCs. This has driven a significant increase in their petrochemical capacity, from about 9% of world total in 2003 to potentially 20% in 2010.[71] The largest capacity still resides in the United States, where excess natural gas made ethane feedstock relatively cheap. However, this so-called gas bubble vanished around 2002/2003, and this change impaired American's global competitiveness in petrochemicals. ExxonMobil has clearly been an exception to this strategy and has maintained the premier petrochemical business in the energy industry. Its contrarian approach to business acquisition and divestiture has led to superior results. ExxonMobil also strove to capture the integration benefits between refining and petrochemicals at the same site, and achieved low-cost producer status.

Technology to the Rescue

Under the pressure of low oil prices and low refining margins, technology advanced to lower the cost of finding and developing oil, enabling new resources to be tapped, and more oil to be recovered from existing resources. Oilfield service and supply companies developed new competencies as oil companies outsourced more activities in the search for lower costs. The international oil companies moved much faster than the national oil companies, allowing them to present a compelling value proposition of lower cost, higher recovery allowing greater tax revenues to host governments, as the geopolitical winds shifted in the direction of privatization and opening resource opportunities in the 1990s.

Upstream exploration driven by information technology

As Michael Economides puts it, for oil and gas companies, "Finding new reservoirs is the name of the game."[72] Beyond a few geologic theories, drilling for oil in the early days was pretty much a guessing game. The guessers got a big boost, however with the 1924 discovery of oil beneath the Nash salt dome in Brazoria County, Texas, the first to be based on seismic data. This demonstrated the value of mapping subsurface structures by recording reflections from low frequency sound waves.

Seismic technology continued to advance, and the next big breakthrough occurred in the early 1960s with the advent of digital technology. This enabled processing of enormous amounts of data and provided dramatically improved knowledge of subsurface geology. By the late 1960s, massive data acquisition capabilities and faster computer processing led to three-dimensional (3-D) seismic, yielding not just a slice of the earth, but a chunk. The result today is 3-D visualization.

Modern 3D-seismic technology is applied to the entire range of exploration, development, and production operations to reduce uncertainties. Now, time-lapse or four-dimensional seismic, created by multiple surveys repeated over time, monitors reservoir performance and helps improve recovery efficiency.[73] The importance of visualization technology was not just its superior representation of the subsurface. It also led to different ways of working, enabling geophysicists, geologists, petroleum engineers, and even drillers to discuss together how to address the next prospect in real time, rather than as a sequential set of tasks.

Drilling and production high tech

Drilling and production technologies have made similarly spectacular advances. In the early 1920s, the Schlumberger brothers developed well logging to characterize the drilled rock's properties and pinpoint hydrocarbon zones. At about the same time, Erle Halliburton pioneered well casings to prevent commingling of zones and other well problems. Perhaps one of the most startling advances is the ability to drill wells horizontally. In the 1970s, engineers developed down-hole motors that could rotate the drill bit at the bottom of the well, allowing drillers to steer the bit and gradually guide the hole off vertical, eventually turning 90 degrees. Such wells are very productive for reservoirs that have flat, thin pay zones. Drilling technology's biggest contribution has been to increase petroleum recovery and reduce costs, which allow explorationists to purse more challenging prospects, such as those in ultra-deepwater in depths to 10,000 feet.

Equally remarkable are the floating production platforms and sub-sea well systems to extract oil and gas found in deepwater fields.

Getting the most out of a reservoir is an economic imperative, particularly when oil is in short supply. Enhanced recovery technology has provided impressive increases in the ultimate recovery of the total oil in place in a reservoir. For example, injecting CO_2 into old oil fields has in some cases doubled recoverability up to 60% of oil in place, further extending the field's life.

The incredible increase in computer workstation power has enabled reservoir engineers to use advanced simulation and visualization technology to better understand how to realize a field's full potential. Today, interest is strong in a systems approach to exploiting an oil and gas field by utilizing the tremendous advances in information technology. The industry has coined a number of terms for this system integration, such as intelligent oil fields, smart fields, e-fields, i-fields, and so forth. The great benefit here is that current recovery rates can be increased dramatically by adroitly managing reservoir pressures and well flow rates.

Service sector expands its competencies

From the early 1970s to the early 1990s, the large oil companies provided more than 80% of R&D expenditures in exploration and production.[74] Smaller oil companies relied on technology seepage from the large R&D programs. Are current R&D levels high enough? Economides and Oligney express doubt in their book *The Color of Oil*: "Why is it, then, that the petroleum industry, so technologically dependent, is the industry with the smallest R&D spending? The healthcare sector leads all industries with 11% of sales going to R&D; the electrical and electronics industry spends 5.5%, and the chemical industry spends 4.1%. In this light, the petroleum industry's R&D spending of less than 0.5% of sales is striking."[75] They argue that the petroleum industry can coast on existing production technologies for current oil plays, with the bigger payoffs coming from exploring for new elephant fields. Moreover, the payoff for new products is more attractive in the high margin pharmaceutical industry than in the low margin oil industry. These justifications for low R&D expenditures play well when the industry is in a down cycle, but they do not placate an angry, skeptical public in a time of rising energy prices and supply uncertainty.

As a result, energy company executives like Jeroen van der Veer, CEO of Royal Dutch Shell, suggest that "New technologies will have to be applied on an unprecedented scale and pace to increasingly demanding projects." Beyond making the most of fossil resources, he advocates that "We must reduce carbon dioxide emitted by fossil fuels, make alternative forms of energy economic, and increase energy efficiency."[76] With the emerging shift to new resources, technology expenditure in large oil companies is again on the rise.

Today, however, oilfield services companies are funding much of the industry's R&D work. In the early days, oil companies operated in a nearly self-contained manner when pursuing upstream opportunities. They owned their own drilling rigs, ran extensive R&D efforts, and tried to capitalize on proprietary technologies. This supported a strategy of providing fully integrated capabilities to host countries by marketing their differentiated technology advantages.

However, as these technologies matured, economies of scale created cost advantages for oilfield service companies like Schlumberger and Halliburton. Specialty niche technology services also evolved. This decoupling of technology from traditional Big Oil has enabled host countries to pursue strategic choices beyond the IOCs' value proposition.

Discontinuities Ahead

What are the fundamentals behind the higher crude oil prices in the 2000s up-cycle? Figure 1.12 provides some insight. OPEC crude oil production dropped from levels approaching 35 million barrels per day in the latter 1970s to less than half that by 1985. However since the mid-80s price collapse, OPEC production has climbed steadily and is now almost back to its earlier peak levels. The cartel's market share of world crude oil supply has risen to around 40%. But in 1998, OPEC experienced a tremor when prices collapsed again down to around $10 per barrel, in part because the impact of the Asian economic crisis was greater than anticipated. OPEC responded by cutting about three million barrels a day from its output, pushing prices to around $25 per barrel. Its ability to influence price levels was re-emerging. During the 1990s, prices had largely stayed in a band between $15 and $20 per barrel. In 2003, a new pricing band appeared in the upper $20 per barrel range, and it became clear that a new up-cycle was underway.

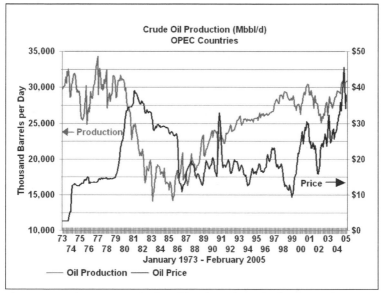

Figure 1.12: The 1990s price band has been breached as OPEC has approached capacity limits. *Courtesy of WTRG Economics: www.wtrg.com*

Terrorism and geopolitical disruptions

A discontinuity took place on September 11, 2001. The attacks on the World Trade Center and the Pentagon revealed a new geopolitical environment that would clearly have impacts on world energy markets. Crude prices rose sharply immediately after September 11, touching $30 per barrel over concerns about US retaliation in the Middle East.[77] The absence of a counterstrike and the expectation of weaker economies in the U.S. and Europe drove prices back down in the next few days. However, the threat of additional terrorism on American soil continues to shadow U. S. energy policy, renewing rhetoric about independence from Middle East oil supplies.

So the Middle East is once again at the center of the geopolitical stage, further enhancing OPEC's role in the global energy drama. In the OPEC meeting of September 25, 2001, the secretary general, Ali Rodriguez, said, "It's a very delicate time at the moment but we're all working to maintain price stability."[78] The discontinuity took on new dimensions after the U.S.-led invasion of Iraq. Oil markets have become very sensitive to world turmoil, whether it involves the Iranian nuclear program, resurgent resource nationalism in Venezuela and other Latin American countries, or the uprising among the disenfranchised inhabitants of Nigeria's main oil region in the Niger Delta.

Clearly, the war on terror has had far greater geopolitical consequences than the Arab oil embargo of 1973. It has raised urgent national and homeland security issues in the U.S., Europe, and elsewhere, and increased tension throughout the network of global economic and political relations. With the diplomatic, military, and commercial strings pulled taut, local and regional disturbances propagate instantly across the global network, and the effects are felt everywhere. Ultimately, this puts a risk premium on energy prices.

Where is the new oil production capacity coming from?

Another potential discontinuity has emerged involving the adequacy of future oil supply. Concern for future oil resources has surfaced many times in the past, as we have learned. The debate about peak oil production remains unresolved. What's clearer is that excess crude capacity declined from over five million barrels per day in 2002 to about one million or less in 2005. As a percentage of demand, surplus capacity has fallen from over 5% to perhaps 1% to 2%. No one really knows how much surplus capacity exists, and it is probably something that cannot be determined precisely.

With no supply cushion, pricing shifts from a commodity surplus regime of relative abundance to a scarcity regime with an inelastic supply curve for the short-term. When coupled with world tensions, the tight supply situation drove the benchmark West Texas Intermediate (WTI) crude oil price to over $70 per barrel.

So what kind of price regime can we expect in the future? Price forecasting rarely proves accurate because both suppliers and consumers react to prices by changing their behavior. This tends to dampen the price curve over time, whether it is heading either upward or downward. The authors believe this up-cycle will exhibit a good deal of price volatility; but a 10 to 15 year period of higher prices would fit previous cycles and would not be unexpected.

China the new country pacesetter

A third discontinuity has come from the demand side. We have focused on the United States during the 20th century because it drove the economic pace of the world and remains the largest consumer of primary energy at around 23%.[79] However, as a region, Asia Pacific has become the biggest energy consumer at over 31%.[80] But more than that, world demand growth is also being fueled by the Asia Pacific Region, and in particular China, which may become the dominant energy consuming country as the 21st century unfolds.

This demand discontinuity will likely trigger repercussions and realignments in the competitive landscape. Western IOCs pursuing hydrocarbons offer host countries with good prospects a value proposition based on providing money, technology, and management capabilities to find and develop potential reserves, while expecting a return based on private shareholder economics. Asian INOCs offer the financing, technology, and management plus a broad array of infrastructure support to developing countries in exchange for access to oil reserves; their returns expectations are based on their government's strategic political and economic objectives without free enterprise concerns for risk and profit.

From oil men to energy executives

As could be expected with higher energy prices, IOC leaders have been called to testify before U.S. congressional committees, where they are badgered with questions implying that their robust earnings must be the result of collusion and greed. The energy industry's image has bottomed out again, and this always invites punitive congressional action to appease the public's discontent. The war on terror and the reliance on Middle East oil have already prompted shifts in energy policy. In his 2006 state of the union address, President Bush announced a national goal of replacing more than 75% of our oil imports from the Middle East by 2025 to break the "addiction to foreign oil."[81]

A fourth discontinuity has also emerged. World leaders in both government and business are looking for ways to address global warming, perhaps the greatest discontinuity of all, with strong impact on the energy industry and national energy policy. The energy choices of the future will surely be judged on the basis of CO_2 emissions. Given this multitude of discontinuities interacting together, the forecast is for strong turbulence ahead.

While some history will repeat itself, the world is entering a new energy cycle with new economic, geopolitical, and societal drivers. We have learned that the dominant energy source is often not the pacesetter. Wood was not the pacesetter after the beginning of the 19th century, but coal did not take over the leading role until the mid-1880s. Oil began driving energy economics after World War I, but it was 1950 before oil became the dominant energy source. While oil delivers the largest portion of U.S. energy supply today, and will likely continue in that role for the next 20 to 30 years, it may not be the strategic energy source setting the pace during this period. We are in a world where change is accelerating and becoming less predictable. The need for highly capable leaders is always greatest in a time of significant change. Energy, as Nobel laureate Richard Smalley states "is the most important issue we face." The captains of this industry must transform themselves from oil men to energy executives to navigate through these un-charted waters. We will continue our voyage of discovery into Terra Incognita by analyzing the new competitive landscape, pondering the resource dilemma, and examining society's changing expectations.

1 Barbara Feese, 2003. *Coal: A Human History.* (US: Perseus Publishing), pp. 30–33.

2 Andrew Roberts. Society and Science History TimeLine <http://www.mdx.ac.uk/www/study/sshtim.htm> Middlesex University.

3 Malthus, 1798. Essay on Population, chapter 16.

4 Peter Landry, June 2006. *The Industrial Revolution: Myth and Realities.* http://www.blupete.com/Literature/Essays/BluePete/IndustRev.htm (accessed Aug 19,2006).

5 Samuel P. Hays, 1957. *The Response to Industrialism* 1885-1914. (Chicago: University of Chicago Press), p. 94.

6 Robert S. Woodbury, 1985. *History of the Lathe to 1985.* (Ohio: Society for the history of Technology), pp. 18,19.

7 Trevor I. William, 2000. *A History of Invention.* (US: Checkmark Book), pp. 176–179.

8 Feese, pp. 71–74.

9 William, p. 171.

10 Mary Bellis, 2005. *The Inventions of Thomas Edison.* http://inventor.com http://www.inventors.about.com/inventors/bledison.htm (accessed Aug 17, 2006)

11 EIA.com, 2006. http://www.eia.doe.gov/kids/history/timelines/index.html (accessed Aug 14, 2006)

12 Paul Halsall, Aug 1998. *Tables Illustrating the Spread of Industrialization.* http://www.fordham.edu/halsall/mod/indrevtabs1.html (accessed Aug 14, 2006), 1.

13 Adams, Sean. "US Coal Industry in the Nineteenth Century". EH.Net Encyclopedia, edited by Robert Whaples. January 24, 2003. URL http://eh.net/encyclopedia/article/adams.industry.coal.us. 6

14 Robert L. Galloway, 1969. A history of Coal Mining in Great Britain. (NY: Augustus M. Kelley Publishers), 62.

15 Feese, pp. 131–132.

16 Feese, pp. 131–132.

17 Feese, pp. 131–132.

18 Feese, pp. 133–136.

19 www.netstate.com, 2005. http://www.netstate.com/states/peop/people/pa_hcf.htm (accessed Aug 14,2006). 1–2

20 Schreiner, 1995. *Jr. Henry Clay Frick. The Gospel of Greed.* (NY: St. Martin's Press), p. 11

21 www.netstate.com, 2005. http://www.netstate.com/states/peop/people/pa_hcf.htm (accessed Aug 14, 2005), pp. 1–2.

22 Les Standiford. *Meet you in Hell.* (New York: Crown Publishers), pp. 278–281.

23 NETL, Secure & Reliable Energy Supplies-History of US Coal Use, http://www.netl.doe.gov/KeyIssues/history of caoluse.html (accesed Dec 7, 2006), p. 1.

24 A. Dudley Gardner and Verla R. Flores, 1989. *Forgotten Frontier*, (London: Westview Press), p 146.

25 Alexander Thompson, 1979. *U.S. Coal Industry.* (US: Garland Publishing), pp. 10–12.

26 Thompson, p. 29.

27 Thompson, p. 26.

28 Thompson, pp. 12–15.

29 Feese, pp. 140–141.

30 Feese , p. 35

31 AEI.1981, *The Clean Air Act Proposals for Revisions. Legislative Analysis No.30, 97th Congress.* (American Enterprise Institute for Public Policy Research), 2–3.

32 Feese, 147.

33 Daniel Yergin, 1991, The Prize: The Epic Quest for Oil, Money, & Power (US: Simon & Schuster), 78–80.

34 James A. Paul, Oct 2002. *Great Power Conflict over Iraqi Oil: the World War I Era* http://www.global_policy.org/security/oil/2002/1000history.htm (accessed Aug 8,2006) 1

35 Anthony Sampson, 1975. *The Seven Sisters: The Great Oil Companies & The World They Shaped,* (US:Batham Books), 25

36 Yergin, 52–53.

37 Yergin, 43–45.

38 Yergin, pp. 58–62.

39 Kendall Beaton, 1957. *Enterprise in Oil.* (NY: Appleton-Century-Crofts, Inc), p. 43.

40 Beaton, p. 54.

41 Yergin, pp. 71–72.

42 Sampson, pp. 52–57.

43 Austin Leigh Moore, 1948 *John Archbold and Early Development of Standard Oil,*(NY:Columbia University Press), pp. 213–214.

44 Yergin, p. 104.

45 Benjamin Shwadran, 1959 *The Middle Ease, Oil and the Great Powers.* (NY: Council for Middle Eastern Affairs Press) 225–260.

46 Sampson, p. 25.

47 Sampson, p. 90.

48 Yergin, p. 392.

49 Yergin, p. 334.

50 Yergin, p. 409.

51 Yergin, p. 436.

52 Sampson, p. 113.

53 Yergin, p. 392.

54 Yergin, p. 409.

55 Yergin, pp. 615–626.

56 Yergin, pp. 615–626.

57 Yergin, p. 702.

58 John Palfreman, 2005: *Why the French Like Nuclear Energy.* Pbs.org/ http://www.pbs.org/wgbh/ pages/frontline/shows/reaction/readings/french.html (accessed Aug 15, 2006). 2–3.

59 Langlois, McDonald, Rogner and Toth, Nuclear Power: Rising Expectations, 2005, Nuclear Power: Rising Expectations, (UK: International Power & Utilities Finance Review 2005/2006, Euromoney Institutional Investor), 17.

60 Herfindahl-Hirschman Index(HHI) is a measure of the size of firms in relationship to their industry. It also provides an indicator of the amount of competition among and effective market

control by firms. It is defined as the square of the market share of a firm represented on a scale of 0 to 10,000 where 10,000 would indicate a monopolistic producer. A 900 HHI indicates an effective 30% market share (30X30=900). If an oligopoly's combined membership holds 30% of the market or 30% oligopoly share, it can only affect full control over the sum of their shares when operating as a single entity (i.e. cohesiveness equal to 1).

61 Kenneth W. Costello, Daniel J. Duann, 2006: "Turning up the heat in the Natural Gas Industry". http://www.cato.org/pubs/regulation/reg19n1c.html (accessed Aug 15, 2006)

62 Robert Schnapp, *History of the U.S. Electric Power Industry*, 1882–1991 http://www.eia.doe.gov/cneaf/electricity/page/electric_kid/append_a.html (accessed Aug 20, 2006)

63 Tyson Slocum, 2001. "Energy Controversy" Bulletin of Science", Technology & Society, Vol. 21. No. 6.

64 Yergin, p. 774.

65 Thomas L. Friedman, 1999. *The World is Flat: A Brief History of the Twenty-first Century*, (NY: Farrar Straus Giroux).

66 Energy Information Administration, 2003, *Performance Profiles of Major Energy Producers 2001* (Washington DC: US Department of Energy) 8.

67 BP Press release of August 11, 1998.

68 TotalFina Press release of July, 5, 1999

69 Exxon Press release November 30, 1999

70 API,2006, *Environmental Compliance in Refining*, http://api-ep.api.org/industry/index.cfm?bitmask=002007004005000000 (accessed Feb. 2, 2006), p. 1.

71 Patrick Rooney, *Factors That Influence The Petrochemical Industry In The Middle East* Middle East Economic Survey Vol. XLVIII No. 23 June 6, 2005, http://www.mees.com/postearticles/oped/v48n23-50D01.htm (accessed Feb. 19), p. 1.

72 Michael Economides and Ronald Oligney, 2000. *The Color of Oil*. (TX: Round Oak Publishing) 93.

73 SPE, 2006 *Seismic Technology*, http://www.spe.org/spe/jsp/basic_pf/0..1104_1714_10004089.000.html, (Accessed Dec. 7, 2006), pp. 1–6.

74 Economides, p. 89.

75 Economides, p. 89.

76 Jeroen van der Veer, 2006. *Delivering technology-they key role of international energy companies*, CERA Week, Houston, Feb. 7, 2006

77 Eia.com. Oct 2, 2002. *This Week in Petroleum*, http://tonto.eia.doe.gov/oog/info/twip/twiparch/021002/twipprint.html (accessed Sep 12), 1.

78 Larry Elliot, Andrew Clark, Sep 25, 2001. *Oil Price tumble in fear of slump.* http://www.guardian.co.uk/waronterr/story/0,1361,557696,00.html (accessed Sep 12), 2.

79 BP Statistical energy data base.

80 BP Statistical energy data base.

81 Office of the Press Secretary, 2006. State of the Union http://www.whitehouse.gov/news/releases/2006/01/20060131-6.html (Accessed Sep 12).

82 Richard Smalley, 2005. Future Global Energy Prosperity: The Terawatt Challenge, MRS Bulletin, Volume 30, June 2005, p. 412.

2 New Global Competitors—Rise of National Oil Companies

In this chapter we will discuss how the competitive balance in the global oil market is shifting further against the traditional International Oil Companies (IOCs). Producer National Oil Companies (NOCs) are increasing their dominance in resource rich countries, while Internationalizing National Oil Companies (INOCs) are grabbing a greater share of the remaining available resources. The IOCs find themselves squeezed into narrowing confines where they use their expertise and capital to compete for favor in a world of shrinking prospects.

Driving Forces

In the decades of Big Oil's dominance, the IOCs saw themselves as independent global economic entities benefiting society by finding, developing, transforming, and distributing petroleum while also serving the interests of their shareholders. As described in chapter 1, they received an occasional helping hand from their home governments in securing access to resources in resource rich areas. From 1970 on, "the wind of change" of national self-determination in developing countries (see chapter 4) meant that developed country home governments were less likely to be helpful and led the IOCs to redefine their role in commercial rather than political terms. Their role was negotiating business terms with producing countries and supplying energy to consuming countries. Despite the strategic and political ramifications of their activities, the IOCs saw themselves primarily as commercial organizations initiating a constant flow of business transactions. The magnitude and scope of their operations made them the power brokers in the business of managing the worldwide energy enterprise. This mental model is still prevalent today in some of these companies, although renewed winds of change are blowing them further off their old course.

The many nations within which the IOCs operated as "foreign" developers or suppliers of petroleum saw the IOCs as instruments of their European or North American countries of origin. The United States, as the economic juggernaut of the 20th century, was most closely identified with IOC activities. The immense economic and national security impact of oil was apparent to all. Then as now, the governments of other consuming countries were uneasy with reliance on U.S.-based companies to supply such a critically important resource. Producing countries were also wary of entrusting the development of their natural resources to foreign companies, and over time became dissatisfied with their revenues from deals they had cut and with their lack of control over the pace of resource development.

Both producing and consuming governments recognized that their national laws were inadequate to force international companies to respond fully to their interests. After all, if these international companies did not like the conditions that governments imposed, they could always, in theory, walk away. Consumers would be left scrambling to make up the supply deficit, while producers would be abandoned with undeveloped resources, no markets for their oil production, and, worst of all, no revenues.

In reality this risk was small, since the companies all wanted to grow and, in practice, generally allocated oil volumes fairly across consuming countries during times of shortage. Further, if one company left a country, there would normally be others that would be pleased to pick up the relinquished market share. However, governments did not see it that way and, from the earliest days of the industry, suspected collusion among companies and monopolistic behaviors. Of course, the U.S. government fanned the flames of this incipient paranoia through the trust-busting activities of President Teddy Roosevelt between 1906 and 1910 (see chapter 1). If the U.S. government did not trust Standard Oil, how could any other government trust an "Anglo-Saxon" oil company?

Consuming countries thus sought a national champion whose primary mission was to assure secure supplies of oil to the nation in times of need. These companies generally could count on the support of the state in dealings with resource rich countries.

Producing countries took various paths. Several tried to build up state companies organically to compete with the international oil companies, to build a national capability to find and develop domestic oil resources, and to participate in discoveries made by IOCs. Beyond that, however, producing countries were driven by a sense that the IOCs were raking off excess profits and that their investment programs were not designed to maximize the host country's interests.

Underlying these suspicions was an important information imbalance. The IOCs had a better understanding than the governments did of the size and nature of the national resource base, of finding and development costs, of the value of the crude oils produced, and of market demands. The international oil companies also had access to the same information for other producing countries and could shift investments away from nations that imposed disadvantageous terms to countries with more generous terms or to those that were most aggressive in their demands. As governments tried to negotiate what they thought would be an equitable division of the potential economic rent, they often felt themselves severely disadvantaged.

So, producing countries demanded equity participation in the oil concessions they had previously granted and in many cases nationalized IOC assets outright. And they created NOCs as stewards of the state's equity interest in its oil and gas resources.

Over time, the NOCs have evolved. Most consumer NOCs have been partially or totally privatized and compete head to head with shareholder-owned IOCs for access to resources and markets. Producer NOCs have in some cases been encouraged to expand internationally, while others are limited to the role of custodians of the national patrimony and remain under full government control. We like to separate the current competitive arena into four groups (table 2.1).

Table 2.1: Categorization of Former and Current National Oil Companies

Provenance	Consumer Countries	Producer Countries		
Original Role	Access to international oil production	Play an important role in developing national hydrocarbon resources		
Current Role	Shareholder Value Creation	Dominant National Producer	Supervision of IOCs	International Expansion
Acronym	IOC (Today)	NOC	NSC	INOC
Examples	BP Total ENI Repsol Petro-Canada	Saudi Aramco NIOC KPC PDVSA Pemex	NNOC Sonangol ADNOC QP PetroEcuador	CNOOC ONGC India Petrobras Petronas Statoil

So, we are proposing the following categorization of competitors that we will use later in this chapter to discuss the current competitive reality:

1. National Oil Companies (NOCs) such as Saudi Aramco, that dominate production in their country;
2. National Supervisory Companies (NSCs) such as Nigerian National Petroleum Company, whose primary role is to own the government interest and to represent that government interest in joint ventures with IOCs;
3. International National Oil Companies (INOCs) such as Petrobras, that are strong competitors in their home countries and have international expansion goals;
4. International Oil Companies (IOCs) such as ExxonMobil, Shell, Chevron and the privatized former consumer national oil companies, that operate globally from bases in consuming countries;
5. Local Oil Companies (LOCs) such as TNK/BP in Russia and PD Oman, which is a joint venture between Shell and the Oman government.

We retain the "oil company" label because of its familiarity; however, we argue throughout this book that these organizations should be considered energy companies as they extend their interests and expertise from conventional oil into gas, alternative oils, power, and other energy forms.

Consumer Country National Oil Companies

The concept of a national oil company emerged from European colonialism and the perceived need of oil-thirsty consuming nations to ensure adequate supplies of an increasingly strategic commodity. It persisted among consuming nations surprisingly late into the 20th century before giving way to market realities in an era of abundant supplies. It has more recently re-emerged as an instrument for resource-poor nations to challenge IOCs as competition for petroleum supplies intensifies.

The Old Guard

The first consumer country National Oil Company was probably British Petroleum. In his foreword to Henry Longhurst's "Adventure in Oil" in 1959, former British Prime Minister Winston Churchill wrote:

I am glad that the history of the British Petroleum Company should now be written. I myself was closely associated with it in its early days in 1913 and 1914, and its development from the Anglo-Persian Oil Company is of much interest. This great enterprise has played a notable part in the history of the past 50 years, and has contributed to our national prosperity in peace and our safety in war. The pioneering of the vast oil industry of the Middle East is a story of vigor and adventure in the best traditions of the merchant venturers of Britain.

Churchill's sentiments may seem to belong to another era, with its mercantilist association of national and commercial interests. However, NOCs invariably owe their existence to perceived national interest. In the case of BP, Churchill had been responsible for the decision to transform the Royal Navy from dependence on coal as bunker fuel to the use of oil. He argued that "The advantages conferred by oil were inestimable."[1] He cited ship speed, acceleration, the ability to refuel at sea, and enhanced gun–power. But he recognized that Britain would have to import from abroad and sought a company able to provide secure supplies. In his words, "An unbroken series of consequences conducted us to the Anglo-Persian Oil Convention." In fact, he told Parliament that he chose the fledgling Anglo-Persian over the established Royal Dutch Shell as supplier because of Shell's "voraciousness."

The British government satisfied itself that Anglo-Persian had access to substantial resources in Iran, and after protracted negotiations led by the Chancellor of the Exchequer, Lloyd George, the government agreed to provide much needed capital in exchange for a guaranteed supply of oil for the Navy. The enabling legislation was laid before Parliament in May 1914. Churchill commented, "We knew that by our contract we should confer upon the Anglo-Persian an enormous advantage which, added to their concession, would enormously strengthen the company and increase the value of their property. If this consequence arose from the action of the State, why should not the state share in the advantage which we created?"

Although Anglo-Persian was a national oil company in that the government had a voting interest of over 50%, it differed from many subsequent NOCs. The government's intent was that it should remain an independent British company; the government was to appoint two *ex officio* directors with a limited power of veto; and apart from its contract to supply bunkers to the Royal Navy, it had no monopoly rights. It was, however, followed by similar companies created in other European consuming countries.

In 1920, following the first World War, the French government negotiated the San Remo agreement with Great Britain in which France would get 25% of Mesopotamian oil[2] by adopting the German share in Turkish Petroleum that had been seized by the British during the war. France was looking for other ways to enhance its oil position by forming its own state company as a national champion, and created the *Compagnie Francaise des Petroles* (CFP, Total's antecedent) in 1924 as a vehicle to take up the interests. This company was originally private, though with two government-appointed directors. In 1928, the French government introduced its system of organizing the oil market based on quotas to curtail the "Anglo-Saxon oil trusts" and build a domestic refining industry. The quotas were designed to favor French companies but to allow competition and to assure that CFP would embody French interests. The French government took a 25% interest in the company.

In the 1950s, CFP was heavily involved in the development of prolific Middle East fields and had no need and little appetite to explore for oil elsewhere in the French Empire. So the French government chartered a new set of state oil companies for that purpose. In 1956, *Regie Autonome des Petroles* (RAP) discovered oil in Algeria. In 1965, the French government consolidated its various petroleum entities outside of CFP into Elf-ERAP, a new company 100% owned by the French government. In hindsight, the relationship between Elf and the government proved to be corrosive, and three former Elf senior executives were jailed and fined in 2003 for making illegal payments to foreign politicians to secure access to resources, as well as to French politicians for domestic political advantage.

Italy followed the example of France by creating a national refining and marketing company, AGIP, which by the 1930s had secured approximately the same market share in Italy as Shell and Esso (Exxon). However, it did not initially have the same mandate to find and develop international oil resources. This changed when Enrico Mattei took over as leader of AGIP after the Second World War. Mattei presided over the consolidation of numerous state hydrocarbon entities in 1953 into ENI, which became a broad conglomerate with interests not only in oil and gas, but also in tankers, pipelines, engineering and construction as well as unrelated activities such as real estate and hotels. Mattei's attempt to secure a stake in Middle East resources through a deal with Iran in the late 1950s was not very successful, but proved to be disruptive to the established players' strategy of holding concession terms constant. The prevailing terms were a 50/50 split of net profits such that the IOCs' net income after taxes equaled the government's tax take. Mattei's deal with Iran allowed 75% of the net profits to accrue to the Iranian government with ENI only taking 25% in net income after taxes.

Privatization

During the 1980s and 1990s, when oil appeared to be in abundant supply and was increasingly controlled by NOCs, consumer country NOCs began to lose their *raison d'être*, and a wave of privatizations began. As described in *The Concise Encyclopedia of Economics* by Madsen Pirie:

> Governments all over the world were confronted in the seventies by the problems inherent in state ownership. Because state-owned companies have no profit motive, they lack the incentive that private companies have to produce goods that consumers want and to do so at low cost. An additional problem is that state companies often supply their products and services without direct charges to consumers. Therefore, even if they want to satisfy consumer demands, they have no way of knowing what consumers want, because consumers indicate their preferences most clearly by their purchases.

> The result is misallocation of resources. Management tends to respond to political, rather than to commercial, pressures. The capital assets of state businesses are often of poor quality because, it is claimed, it is always easier for governments to attend to more urgent claims on limited resources than the renewal of capital equipment. In the absence of any effective pressure from consumers whose money is taken in taxation, state industries tend to be dominated by producer interests.

Privatization of government-controlled companies became an international phenomenon after the early successes registered in the United Kingdom under the Conservative government of Margaret Thatcher. Over a period of 10 years, this government successfully privatized state interests in aerospace, automobiles, radio-chemicals, trucking, ports, telecommunications, steel, electricity, gas, water and oil. The government also substantially reduced marginal income tax rates on its citizens, sold public housing at discounted prices, changed attitudes toward labor unions and undertook substantial reforms of health and social services.[3] It is perhaps ironic that high oil prices of the early 1980s provided a favorable fiscal context for these reforms through high oil and gas tax and royalty receipts from the North Sea, which together with the receipts from privatizations combined to finance these moves.

The first privatization by the Thatcher government was to sell some holdings in British Petroleum to reduce its ownership below 50% in 1979. By the end of 1995, the government had sold its final 1.8% ownership interest in BP shares.

Perhaps because BP was clearly set up as a consumer country NOC, the Labor government of 1974 did not perceive BP as an adequate representative in North Sea exploration and production. Accordingly, shortly after assuming power, it set up the British National Oil Company (BNOC) in 1974 with the objective of taking a 51% interest in all North Sea fields. This proved unacceptable to the existing operators and the government settled for a purchase and sell-back arrangement at market prices. However, BNOC did secure substantial equity positions through the provisions of the fifth and sixth licensing rounds and was also granted several blocks directly outside the regular rounds.[4] In the pithy words of Frank Parra: "In 1982, BNOC's exploration and production assets were transferred to a new company, Britoil, whose shares were then sold of in stages. BNOC was left with only its trading arm, and it was finally closed down altogether in 1985 after incurring trading losses in the previous years. It had all been pretty much of a pointless exercise."

The government also spawned two new, successful, private upstream companies out of the oil and gas exploration and production assets of the national natural gas monopoly, British Gas. The new British Gas, now known as the BG Group (see Chapter 5 for a discussion of its origins and shareholder value proposition), inherited the natural gas production assets and has become one of the most successful growth companies in the international gas arena, and Enterprise Oil, which inherited the British Gas oil assets, was also quite successful in building an international business until it was purchased by Royal Dutch/Shell.

Other governments followed suit. The French government sold 14% of Elf/ERAP to the public in 1986, and by 1995, foreign investors owned 35% of Elf's shares. In due course, Elf would be acquired by Total, which had already acquired the Belgian company Petrofina. Italy sold 14.7% of ENI in 1995, and the company went on to simplify its businesses and focus on its core domestic and international oil and gas businesses. ENI has been much more successful in expanding its reserves since privatization than it ever was before. From its inception as a modern company after WWII, for 40 years through 1995, the company had accumulated oil and gas reserves of 4.3 billion barrels of oil equivalent (boe); in the 10 years from 1995 through 2004, the company added a further 2.9 billion barrels to reach a total of 7.2 billion.

Petro-Canada was formed in 1975 with the deliberate goal of providing Canadian policy makers with a better control and understanding of energy developments. Under its controversial chairman Bill Hopper, it made a number of high profile acquisitions of refining and marketing assets, introduced a national retail brand, and obtained exclusive exploration rights for certain Canadian properties. After a change in government in 1984, a privatization process began, and by 1995 the government share had been reduced to 20%.

The Spanish company Repsol was founded by the government in 1987 as a vehicle to manage a variety of government owned upstream and (primarily) downstream properties. The government then sold a 24% stake in 1987 and had reduced its holding to 10% by 1996.

These NOCs, mainly European, had served their purpose in the developed countries. They disrupted the status quo dominance of the national hydrocarbon sector by foreign firms and spawned a number of successful private IOCs, firmly domiciled in their countries of origin, that compete successfully with the large U.S. companies and with Shell. Most are listed on the New York Stock Exchange as well as in local stock markets and attract capital from worldwide sources.

Today, they are available to their home governments to explain what is going on in the global energy business. They provide jobs, societal contributions and tax income to their home countries. BNOC may have been "pretty much a waste of time," but the countries that developed and later privatized their national oil companies would be poorer without them. In most cases, it is doubtful whether private capital could have built up institutions of the scale of the former national oil companies. Indeed, Anglo-Persian could well have foundered for lack of capital resources without Churchill's initiative, in which case BP would not exist today.

The National Champions

Historically at the fringes, but more recently playing a more central role, are the national champions of successful developed countries such as Japan and Korea. We will discuss the actions of the "New Wave" of Chinese and Indian companies later in this chapter. Competition for resources from developed consumer country champions has again changed the oil market landscape. Depending on the price assumptions used, some IOCs believe that the most likely outcome on these winning bids will be a return below the cost of capital for shareholder-owned companies. Clarence Cazalot, CEO of Marathon, is concerned about the economics of the recent international bid rounds and the bidding philosophy of the INOCs: "To get a cost of capital return, forgetting about any risk adjustments, they have to be projecting $70–80/B crude prices for a long time to come. They may be right, but it's very hard for us to make significant long term commitments on that basis: if you're wrong, the company's results will be negatively impacted for quite some time."

In December 2005, the South Korean National Oil Company won an offshore lease in Nigeria by offering a high lease bonus, and also by benefiting from preemption rights granted in exchange for a South Korean commitment to invest in infrastructure projects. Two months earlier, it had been the Japanese's turn. On October 2, 2005, the results of the second round of international bidding for exploration and development rights in 26 Libyan oil zones were announced. More than 60 oil companies from around the world participated in a fiercely competitive environment. Asian companies won 10 of 23 contract areas. In this round, the "x-factor," which determines the amount of oil available for cost recovery and profits, declined to 13.2% from the previous round's 19.5%.[5] Japanese companies captured rights to six of the most promising oil zones. As one oil official told Reuters, "It was quite a tight race, it was cut-throat competition, but the Japanese were the big winners." The Japanese approach has always been distinctive, and relies on the Japanese private sector to implement government policy, aided by government incentives to mitigate exploration risk and finance development capital needs. Until recently, this assistance was provided through the Japanese National Oil Company (JNOC), but the government became frustrated that this entity consistently failed to be self-financing. The Japanese government has now reorganized its support of key resource capture under Japan Oil, Gas, and Metals National Corporation (JOGMEC). The official Web site describes its overall role: "Since Japan has few domestic natural resources, Japan depends on imports for almost all of its crude oil, natural gas, non-ferrous metals, and minerals. JOGMEC, an incorporated administrative agency established by Japanese government, will seek to maintain a stable supply of natural resources and energy to Japan, contribute to the industrial development of Japan and enrich the lives of all the Japanese people, through our various activities involving oil, natural gas, non-ferrous metals, and minerals."

Although the Japanese private companies are different in kind from the European consumer NOCs, the close collaboration with the government creates the same effect of transforming the international playing field for resource capture. The Asian Times report of the Libyan round is informative:

> But how is it that Japan could beat out all of the majors and other national companies to attain such an impressive share of the spoils?
>
> In the first round of bidding in January, Japanese efforts failed dismally. At that time, 15 zones were available, with 11 going to American companies. All Japanese bids were rejected. Some Japanese business leaders were shocked at these results as they had anticipated some degree of success. In wake of the Japanese

companies' failure, they reexamined their approach and found new ways forward.

In the January round the Japanese oil companies tended to work separately and offered independent bids. It is thought that by thus spreading themselves so thinly, they gave Libyan officials the impression that the financial resources behind each bid were rather small, posing that undercapitalized companies, if given the development rights, would be unable to produce results.

Also, the Japanese companies conveyed the impression that they were only interested in business deals and had no real long-term commitment to the prosperity of Libyan society. This was not conducive to creating an atmosphere of trust.

However, the Japanese companies in 2005 revised their strategies and found new approaches that worked much more effectively. Rather than operate independently, this time the Japanese organized group bids. Some of these partners were other Japanese and some were European companies that also failed in the first round. This gave the Japanese bids added strength.

Japanese oil companies also did a better job of assuring the Libyans that they were interested in more than just oil, and were committed to Libya's long-term development and prosperity. An early milestone in this campaign was the March visit of a group of Japanese oil executives headed by a former foreign minister, Kakizawa Koji. Kakizawa is noted for having championed Japanese business relations with Libya even before the UN sanctions were lifted in 2003. Meeting with senior Libyan officials in Tripoli, this delegation began the process of putting Japan-Libya relations back on track. This group met with Saif al-Islam Gaddafi, the leader's second son and possible successor.

On April 3, Saif al-Islam Gaddafi began a six-day visit to Japan. The young Gaddafi met privately with Prime Minister Junichiro Koizumi, Finance Minister Tanigaki Sadakazu, Economy, Trade and Industry Minister Nakagawa Shoichi, Communications Minister Aso Taro, Education Minister Nakayama Nariaki and House Speaker Kono Yohei.

Two days later, Saif al-Islam Gaddafi opened an art exhibition in Tokyo called "The Desert Is Not Silent," which included both archaeological artifacts from Libya and the young Gaddafi's own paintings. Present at this opening were Prince Mikasa Tomohito, a former prime minister, Mori Yoshiro (who is involved in Japan-Africa relations), and Environment Minister Koike Yuriko (a fluent speaker of Arabic). Finally, the young Gaddafi also participated in the Aichi Expo where Libya had set up a national pavilion that was drawing 12,000 visitors daily. The national day events were held on April 6.

The Japanese oil companies were involved in some of this diplomacy. For example, "The Desert Is Not Silent" exhibition, which was held in Akasaka, was 75% financed by contributions from these oil companies. Despite the young Gaddafi's busy schedule, the oil companies also had some face time booked with him. While Gaddafi holds no official position in the Libyan government, the importance of reaching understandings with him was widely recognized.

The first official high-level exchange soon followed. In June, Senior Vice Foreign Minister Aisawa Ichiro traveled to Tripoli and met with Libyan leader Muammar Gaddafi. Aisawa carried with him an invitation from Koizumi for the elder Gaddafi to visit Japan some time in the future.

All of these efforts formed the backdrop to the startling Japanese success on October 2.

The oil-development rights gained in Libya were shared by five Japanese companies: Inpex Corporation, Japan Petroleum Exploration Company, Mitsubishi Corporation, Nippon Oil Corporation, and Teikoku Oil. For some of these companies, this was their first successful oil bid in the Arab world. If all goes well, actual oil production at these sites should begin in 2011 or 2012.

It is worth reflecting on what this means for the competitive position of the traditional IOCs. New rivals are unfolding new paradigms of doing business. The traditional IOC value proposition to bring money, technology, and management in an integrated package is being outflanked. IOCs will need to modify their value propositions, seek conventional resources in non-traditional places, or develop unconventional resources where their traditional value proposition is still valid. INOCs and their government sponsors would do well to reflect on whether their actions will actually add to their security of supply, or will just raise the costs and reduce the returns on resource development projects.

Producer Country National Oil Companies

As shown in Chapter 1, NOCs as resource stewards of resource rich producing nations first emerged in Russia after the revolution and in Latin America in the 1920s and 30s. Major producing nations in other regions began to establish NOCs several decades later, leading eventually to the founding of the Organization of Petroleum Exporting Countries (OPEC), surely the most famous cartel the world has ever seen. Several of these NOCs have expanded beyond their national borders and now compete globally with IOCs for access to undeveloped resources. While most NOCs have delivered some of the benefits envisioned by their founders, they have often been trapped by ambiguity in roles between the NOC and the government and have exposed their nations to economic hazards.

The Underlying Dilemmas

Natural resources have been sought after because they are believed to automatically confer wealth on their finders and developers. History, however, tells a different story. The experience of the Spanish State with its colonial haul of gold created echoes that are still being felt throughout the oil producing world. "During the first century [between 1492 and 1600] approximately two billion pesos' worth of gold and silver was shipped from the colonies to the mother country. This was at least three times the entire European supply of these metals before the discovery of America. . . . Because of the enormous influx of gold and silver from the colonies, the Spaniards at home became unwilling to work hard to accumulate a competence, but would far rather by hook or crook get their share of the incoming wealth. As a consequence they permitted their industrial enterprises and even their agriculture to deteriorate."[6]

There is a natural desire to control the development of the national patrimony, and to assure that the benefits flow to the nation. Unfortunately, it is frequently distorted such that the profits are siphoned off into the hands of the privileged few in the governments and their favored entrepreneurial collaborators, and society comes to believe that wealth flows from the ground rather than being created by human endeavor.

Even when the country is modern in its institutions, hydrocarbon resources can create economic problems as well as benefits. One example is often referred to as the "Dutch Disease." Following discovery and development of the giant Groningen gas field in Holland by Exxon and Shell, the value of the Dutch guilder rose on international currency markets, to the significant disadvantage of Dutch industrialists, who found that the cost of their inputs rose compared to their international competitors when measured in terms of international currencies. Conversely, the cost of imports declined when measured in guilders. This was not because of mismanagement of the economy, or irrational behavior by industrialists. It merely reflected the fact that currency traders saw that Dutch bonds had lower default risk compared to alternative international investments, and they priced them accordingly. The flow of funds into Holland drove up the value of the currency.

In addition to the "Dutch Disease," the ability of a government to spend money without needing to raise taxation from its citizens creates what economists call a "moral hazard" and seems to encourage a paternalistic rather than participative relationship between those in power and the citizens. A further complication is caused by volatility in oil prices. A national budget that is balanced at oil prices of $30 per barrel will be severely in deficit at $20 per barrel, and substantially in surplus at $50 per barrel. A number of producing governments have recognized this and have chosen to institutionalize the transfer of oil revenues above a threshold amount into specified uses. The state of Texas endows its permanent fund for higher education with oil severance taxes and royalties. The state of Alaska cuts a check for each citizen each year on grounds that the citizens can better direct the funds than their government. Kuwait and the United Arab Emirates have established large funds that are invested outside the local petroleum sector and held in trust for their citizens. Venezuela adopted a similar policy, but it is widely believed that the funds have been raided and redistributed into domestic programs supported by the current administration. Opponents believe their purpose is to "buy" votes in forthcoming elections.

In times of low prices, there tends to be extreme stress in the relationship between the management of the national oil company and its government. The government is the shareholder and in principle should look for the success of its investment. But at the same time, the government is the taxing agency looking to maximize its receipts from the NOC. A similar conflict exists on environmental performance. The government sets and enforces the rules that should lead to respect for environmental standards. Yet the government is also the shareholder that has to fund the investments and higher operating costs needed to comply with the rules. The government has to play the roles of both gamekeeper and poacher, and experience suggests that the gamekeeper frequently turns a blind eye to the poacher's misdeeds to the detriment of the environment.

The government also faces a dilemma when deciding how to use the NOC in policy development and execution and in infrastructure development. In several producing countries, the NOC houses a disproportionate share of the nation's capabilities. Investment funds and capable people are often easier to access than they are in the government. Consequently the national oil company may be called upon to devise and execute policy and to invest in roads, ports, schools, and hospitals. In several countries, national oil company personnel are the experts in petroleum economics and are best placed to design fiscal regimes for competitive bid rounds for exploratory or development acreage. But that should be the government's job.

Mixing roles can create animosity and conflicts of interest. Similarly, the NOC may be much more experienced in major project design and execution than the government. However, diverting those talents from hydrocarbon projects to civil projects comes at a cost of reduced capacity to undertake revenue-creating investments. Finally, the government may seek to influence NOC personnel policy. This can range from demands that relatives of government members are provided jobs, to a mandate, as at one time in Algeria, that *moudjahiddin* who had fought in the war of independence from France be provided employment, to a general decree limiting management's ability to reduce staff.

A further bone of contention is the pricing of petroleum products into domestic markets. Citizens in oil rich countries tend to believe that they should be able to access petroleum products at low prices. In reality, these low prices involve subsidies since the products could be sold for more on international markets. The subsidies encourage inefficient use and result in lower income to the national oil company than it could receive selling at international parity prices.

Indonesia is a poster child for the risk that a policy of subsidies entails, as they have recently moved from being a net exporter to a net importer of petroleum products. They now have to import products at international prices and fund their resale at lower domestic prices. In response, the government has been obliged to raise the controlled price level, to the dissatisfaction of the citizens. When Carlos Andres Perez of Venezuela tried to eliminate subsidies in 1988, the resulting violent protests forced him to withdraw the proposal and to retreat from an important platform of his economic reform program. Indirectly, it also led to the failed coup led by current president Hugo Chavez, which provided the name recognition that enabled his presidential bid.

Finally, if an NOC takes on international obligations such as investments in refineries, it can expect government officials to be fickle allies. Government's institutional memory is short, and the political conditions that made the investments desirable at one time may not be present at a later time. Venezuelan officials have been particularly inclined to second guess the refining deals made in the 1980s and 1990s to provide outlets for heavy oil. These deals included netback pricing of crude oils into the foreign affiliate, which at times of low refining margins can appear disadvantageous to the producer. At times of high refining margins, such as in 2005, the deals are advantageous to the producer and second guessing subsides; but the suspicion remains that the negotiators of the original deals could have done better.

These conflicts are not insuperable, but they add a complexity to the management of NOCs that is not generally present in shareholder owned oil companies. The latter, in their dealings with the former, need to recognize the pressures that their potential partners are subjected to and try to help find solutions.

Saudi Aramco governance

Abdallah S. Jum'ah, CEO of Saudi Aramco, believes that his country has defined the roles of government and NOC with sufficient clarity that the conflicts that have been seen in other countries will not be an issue in Saudi Arabia. First, he cites the operational independence of Saudi Aramco:

> I have spoken at many forums about the peculiarity of Saudi Aramco. You know, we are an international oil company, yes. We are a national oil company, yes. But Saudi Aramco is different in many ways: let me just look at the difference between Saudi Aramco and its sister national oil companies.
>
> When we moved from being Arabian American oil company to the Saudi Arabian oil company, Saudi Aramco, it was the wisdom of the government at that time to say, "We don't want to see a change in this organization—it's working well. If it ain't broken, don't fix it. It's [to] continue to operate." And therefore our employees—nobody felt a change. There was no fanfare. No big headlines in the papers. People didn't feel it. It was a seamless transition. And the reason is, the bylaws of the new company were written [to assure] that this is an organization that will continue to operate as a commercial entity and for profit. And we will continue to deal with the government exactly like our predecessor company, the Delaware corporation, on a tax and royalty regime. And we deal with the government as a shareholder. When we have enough money after satisfying all our requirements, if we have enough cash, we declare a dividend. So we are operating like your normal commercial company.
>
> The Minister of Finance and the Minister of Petroleum do not pick up the phone and tell me, 'Do this, and do that. Or give me money, or transfer money from here to there . . .' The lines of communication between the Minister of Petroleum, or the government as a whole, and Saudi Aramco are drawn—very clear, very crisp.

Now above the board is the Supreme Council for Petroleum and Minerals, which is headed by the king. Membership [includes] the Minister of Petroleum. I am a member. And a few other people. It is a small council, which represents the stockholders.

They appoint the president of the company. They look at the annual financials. They approve the external auditor and his remuneration. They approve an annual report of operations. All the activities that are vested in a general meeting of the stockholders of a company, are vested in this council. So in a sense, we are operating like any other company.

[Production policy] of course, happens at the Ministry of Petroleum, but promulgated through the Supreme Council. The Supreme Council is not solely a shareholder of Saudi Aramco. It also deals with minerals. It also deals with oil policy.

[Interference by the ministry in Saudi Aramco operations] does not happen. I'll tell you also, we are happy the ex-CEO of Saudi Aramco is now the Minister of Petroleum. So he understands.

The First Wave—Latin American Origins

National oil companies for oil rich countries first took shape in Latin America, where ancient royal prerogatives and Spanish colonial influence established the concept of state ownership of resources. Nationalist and socialist influences in Latin America also encouraged the formation of national oil companies charged with the responsibility of developing known hydrocarbons resources for the benefit of the nation.

YPF (Yacimientos Petroliferos Fiscales) was formed in 1922 as a government-owned monopoly to develop Argentina's hydrocarbon resources. Under its director general, General Enrique Mosconi, YPF grew into a large corporation capable of providing a defense against the pressures of Royal Dutch/Shell and Standard Oil. Nevertheless, in 1928, General Mosconi said, "Without a petroleum monopoly it is difficult, or more exactly, impossible for a state organization to win in commercial competition with private capital organizations." Accordingly, on August 1, 1929, YPF took over regulation of the Argentine domestic market to "set prices equitably and thus defend the consumer and as well, defend the national economy against other products of foreign origin. . . . From this day the companies have seen a reduction in their excessive profits." He then retired to Europe, as had the leader of Argentine independence, General San Martin, before him.

YPF coexisted with Royal Dutch/Shell and Standard Oil's successor company ExxonMobil (known in Latin America mainly under the Esso brand) in the refining and marketing sector in an environment that was quite comfortable for the IOCs. Since they were obliged to sell at the same prices as YPF and their volumetric sales were predetermined, their revenues were preordained. All they had to do was to be more efficient than YPF and agree not to rock the boat by competing too aggressively. YPF also allowed IOCs to operate small producing assets in Commodoro Rivedaria (Amoco) and Tierra de Fuego (Total), and negotiated several service contracts with Argentine firms to operate YPF owned fields in the more prolific Neuquin area. These Argentine firms formed the nucleus of a dynamic Argentine private oil sector once the YPF monopoly was lifted.

The YPF precedent moved steadily north. ANCAP was formed in Uruguay in 1932, YPF Bolivia in 1935, and CNP Brazil in 1938. But the most dramatic event happened in Mexico. Mexican oil production had fallen dramatically in the 1920s as Venezuelan production ramped up. The Mexican government "blamed the foreign companies exclusively, rather than acknowledging the effects of a depressed international market and of domestic conditions that were decidedly inhospitable to foreign investment." President Lazaro Cardenas was a leftward-leaning nationalist who was committed to increasing government control of the oil industry. The industry was still operating within a colonialist mindset. The resident manager of Mexican Eagle, the largest Mexican producer controlled by Shell, said of Royal Dutch/Shell's CEO Henri Deterding that he was "incapable of conceiving of Mexico as anything but a colonial government to which you simply dictated orders." Eugene Holman, head of Standard Oil of New Jersey's production department, was also intransigent: "the company would prefer to lose everything it had in Mexico rather than acquiesce in a partnership that might be regarded as partial expropriation." To do otherwise might set an unfavorable precedent. Cardenas' position was that the companies should transfer managerial control to the unions, and this was totally unacceptable to the companies. In March 1938, Cardenas signed the expropriation order, and Pemex was formed to operate the nationalized assets. Later, the companies were compensated monetarily, and government ownership of hydrocarbons resources was written into the Mexican constitution.

Pemex as an institution has a lot to be proud of in its development of Mexican oil resources. However, it suffers from all the constraints described previously. In particular, it has difficulty financing its development projects because government and company compete for oil revenues, political interference has led to excessive bureaucracy, and the power of the unions has led to substantial overstaffing.

Brazil has always been different from the rest of Latin America. It was colonized by the Portuguese, not the Spaniards. It was not perceived initially to be rich in minerals, so it was developed as an agricultural rather than mining economy. It received a large influx of African slaves to work in the sugar plantations of the North. It secured independence without significant bloodshed through a surreal set of royal maneuvers during which the Portuguese king John VI fled Portugal for Brazil as Napoleon threatened and his son Dom Pedro declared Brazilian independence from Portugal and became Emperor of Brazil. It became a strong industrial force behind the dynamic center of Sao Paulo. And its national oil company, Petrobras, was formed in advance of the discovery of any significant hydrocarbons in the country and adopted a mission that more closely resembles those of other consumer country NOCs, including participation in international as well as domestic exploration and production. Consequently, we will discuss Petrobras later in the section on internationalizing national oil companies.

The Second Wave—OPEC Producers

During and after the Second World War, there was a hiatus in the formation of national oil companies. As described in chapter 1, the focal point of the international oil industry shifted to the Middle East, OPEC was formed in response to a world of surplus oil, and a process of participation and nationalization began in the 1970s with the 1971 Algerian takeover of 51% of French producing interests and Libya's expropriation of BP.

By the end of the 1970s, the IOCs had been fully nationalized in Algeria, Venezuela, Saudi Arabia, Kuwait, Qatar, Iran, and Iraq. They retained reduced interests in Libya, Nigeria, the United Arab Emirates, and Indonesia among the OPEC countries, and in Gabon, Angola, and Ecuador (Ecuador and Gabon were OPEC members in the 1970s, but left the organization in 1992 and 1995 respectively) among the non-OPEC countries. Those countries that fully nationalized in most cases developed highly capable NOCs that tended to reflect the character of the IOCs that had gone before. Saudi Aramco (Saudi Arabia) and PDVSA (Venezuela) in particular built on the capabilities of their predecessor IOCs, applied modern technologies to the further development of the national resource, and even mounted substantial research programs.

Each national oil company has a hierarchy of objectives, ranging from a modest supervision of the IOCs to an ambitious intent to compete with them. By way of illustration, here is how we characterize the current strategic intent of three major producing NOCs.

Saudi Aramco: Fulfill the Kingdom's policies to maintain a production capacity cushion of two MMBD, selectively secure outlets for heavy oil through local investment in preference to international JVs. Grow employment opportunities by investment in hydrocarbon value chain extensions through joint ventures with established players having access to markets, with a focus on Asia.

Kuwait Petroleum Company: Maximize the value of Kuwait's hydrocarbon resource through investments in the hydrocarbon value chain domestically and internationally, pragmatically involving IOCs where they have specific capabilities to offer, and add value by investing in international downstream facilities.

Qatar Petroleum: Create the dominant international natural gas industrial center, leveraging partnerships with major oil companies to capture international markets.

Saudi Aramco enjoys a preeminent place among producer NOCs by virtue of the enormous resource base to which it has access and its overall competence. There is no doubt that Saudi Aramco is a highly capable company, technically, and managerially. It is fully competent to develop the Saudi Arabian oil resource in alignment with Saudi Arabia's perceived national interest. This interest includes:

- Taking a long term view, and extending the life of the oil fields for the benefit of future generations
- Assuring maximum recovery of the oil in place
- Responding to the world's need for more oil by expanding production, but within the constraints defined by the previous objectives
- Maintaining a cushion of spare capacity to be able to dampen global price spikes caused by episodic supply crises

Commercial companies would find these policies to be anathema. The imperatives of discounted cash flow drive them to maximize early production consistent with good reservoir practices, and worry about ultimate recovery only later when new technologies may be available. And they would resist investing in capacity that might not be used. The fact is that Saudi Arabia does not need IOCs at the present time to develop its conventional oil resources.

Saudi Aramco CEO Abdallah S. Jum'ah notes three major differences between his company and the IOCs:

> The IOCs in general, when I compare to Saudi Aramco, have a shorter time horizon, simply because they are under the scrutiny of the financial markets. So they have to produce results for the next quarter, and so on. Our horizon is much longer. It is much longer because, first by virtue of having roughly one quarter of the world's

reserves, we want to maximize the value of that. When we go to a field, we don't want just to try to make our high teens, or mid teens, or low teens, returns whatever it is, in the shortest time possible, and we leave. Our management is way long term because we want to maximize the value of that field, extract that last possible drop from it, you know, in the most efficient manner. We are not coming to a place and [then leaving]. These fields are going to be with us forever, and we want to maximize their value.

The second difference is Saudi Aramco's ability to maintain a cushion of spare capacity:

> By government directive, we have a mandate to keep between one and a half to two million barrels of extra capacity for "just in case." And this just in case did happen many times. Saudi Arabia wants to have a stable oil market, it doesn't want hiccups, wants to keep that capacity cushion, so that the world doesn't see a problem. I give you an example. In April 2003, we had the war in Iraq. We had problems in Venezuela, problems in Nigeria, a very cold spell in the eastern United States. We elevated our supply to the U.S. from a normal 1.5 million barrels per day to 2.2 million barrels per day. Think if we were not able to do that, what would have happened to the gas prices or the U.S. economy. Likewise it happened during the 1990s, when Iraq production was out, Kuwaiti production was out, we came to the fore and supplied that shortage. There is no IOC whose shareholders would ever agree to have that capacity and spend that money to maintain it for "just in case."

He also noted Saudi Aramco's role as an agent of change in Saudi Arabia:

> So the government would like Saudi Aramco to be the transferer of technology; be a model for the business community in Saudi Arabia; be a responsible citizen in terms of supplying energy to power production, to water [desalination], providing feedstock to the national industry, training, providing an example for proper training and development; and, of course, encouraging local industry. And we have a lot of activities that are going on now: Saudizing the work force, trying to create jobs, and so on.

> At the end of the day, these are not costs on us. We look at this in two ways. One, it is our social responsibility. And second, actually, it gives us dividends at the end of the day. It may not be

that dividend coming really soon, but in the future it comes. For instance, if you create local industry, then you are not at the vagaries of the [international] market. If you need supplies from the local market, it is very close to you. We are very far from Japan and the U.S. and Italy and so on. So if you create that industry, help the industry, and help to create that manufacturing in the Kingdom, then it helps us. If we help contractors and develop Saudi work for their projects, we are not awaiting whether there are visas for the Filipinos, or visas for the Koreans. At the end of the day, we get a return for this investment we are putting in.

These comments describe what host governments want oil companies to do in the way of "extracurricular" activities, and IOCs need to heed them if they wish to compete with NOCs.

Other producer NOCs are at various stages of development. Some, such as Nigeria's NNPC, have grown up in a supervisory role, with a primary duty of representing the national interest in joint ventures with IOCs. Many of these are working to enhance their capabilities as operators so that they can take a more direct leadership role in important hydrocarbon development projects. Others, such as Algeria's SONATRACH, are evolving with the objective of being competitive with IOCs, not only in their home country but internationally as well.

Our view is that the evolution of SONATRACH is a natural step for a country with a maturing resource base. Incremental production in a mature province is higher cost and creates less economic rent for the government. Further, a competitive approach applying diverse mental models from a wider range of companies will likely be more creative and will discover reserves that might otherwise not be discovered. Incremental production provides economic benefits in hard currency earnings and work for local contractors. At some stage the government and the NOC will decide that the national interest is best served by opening up the resource to international companies.

This was the case in Venezuela in the early 1990s, when Luis Giusti was planning coordinator and recognized the opportunity to expand production but could not see any way to finance the required investment through PDVSA. Accordingly, he embarked on a strategy of first opening up the Orinoco tar sands, since there was a political consensus that this massive resource might otherwise never be monetized. He went on to auction off marginal fields that had good potential value, but were at the tail of the PDVSA investment opportunity queue. Finally, when CEO of PDVSA, he opened up exploratory acreage that PDVSA did not think was very prospective.

It turned out PDVSA was right, and IOCs spent a billion dollars to access acreage that contained very few economic prospects. The roll-back of this policy by the Chavez administration will prove to be a mistake, as PDVSA has lost the technical capability to sustain oil production, and the government's social and political priorities will continue to demand higher expenditures that can only be met if oil prices continue to rise. However, the Chavez administration has certainly been quick to appreciate the changing competitive reality, and it is actively seeking to diversify its investor base by inviting new INOC entrance into the Venezuelan petroleum sector.

The Third Wave—Internationalizing National Oil Companies

A new phenomenon has been the expansion of NOCs whose original purpose was to develop national resources into the international arena. This expansion of NOCs has challenged the IOCs, and is exemplified by the attempt of Chinese National Offshore Oil Company (CNOOC) to buy Unocal.

Unocal had in the early 1980s been one of the larger second tier U.S. oil companies with a national refining and marketing network as well as North American and international exploration and production. However, the company came under attack from Boone Pickens, who argued that the company was squandering shareholder value by continuing to invest at a time when oil prices were declining. In order to fend off a potential acquisition by Mr. Pickens, the company recapitalized in 1985 and took on substantial new debt. This defense against acquisition was similar to that employed by Phillips Petroleum, and both companies found themselves hamstrung by high debt service payments as oil prices continued to decline. Over time, Unocal sold off its refining and marketing assets and focused on exploration and production in North America and the Far East.

Chevron prized Unocal's businesses in the Far East, in the deepwater Gulf of Mexico and in the Caspian. Unocal's exploration and production businesses in Myanmar, Thailand, and Indonesia provided Chevron with long-lived assets that fit well with its existing strength in Asia, which had been reinforced by the merger with Texaco and the consolidation of Caltex under a single owner. Unocal's deepwater assets also fit well with Chevron's successful Gulf of Mexico business, and Unocal's Caspian business complemented Chevron's strong position in Kazakhstan. After informal discussions with Unocal management, on February 26, 2005, Chevron offered to purchase Unocal in an all-stock deal, exchanging 0.94 Chevron shares for each share of Unocal stock. This offer was rejected by the Unocal Board on March 1,

2005, indicating that it would be necessary to include some cash in the deal. On April 4, 2005, the Unocal Board accepted a revised offer by Chevron of 0.7725 shares of Chevron stock plus $16.25 per Unocal share. Chevron shares promptly declined, lowering the value of the bid to Unocal shareholders.

China has three national oil companies, formed out of the previous Ministry of Energy, each of which has some international investors. China National Petroleum Corporation (CNPC) has primary responsibility for Chinese domestic onshore exploration and production; China Petrochemical Corporation (SINOPEC) is primarily a refining and marketing company; and CNOOC is responsible for exploration and production offshore China. The primary areas of responsibility have blurred over time, and each company now has activities in the areas of focus of the others. In particular, each company has international exploration and production investments.

CNOOC, by virtue of its close relationship with western companies in developing offshore oil and gas fields, is the Chinese company that most closely resembles a western oil company in organization and performance. CNOOC has recently spearheaded plans to import liquefied natural gas (LNG) into China and has leveraged its access to the Chinese market to farm-in to exploration and production assets operated by IOCs. CNOOC's management saw Unocal as an extension of this strategy that would increase its Asian natural gas reserves and also provide deepwater expertise, and they decided to submit a bid to buy the U.S. company.

Then fate took an interesting twist. CNOOC's outside directors became concerned that the proposed bid would not create shareholder value. Unocal was financially a huge acquisition for CNOOC that would be financed by loans of $2.5 billion for two years at zero interest and $4.5 billion for 30 years at 3.5% interest from its parent company (essentially the Chinese government). The resulting debt load would be proportionately higher than for other large oil companies, albeit at very low interest rates. As directors of a public company, even one with a majority government ownership, they might be liable to shareholder action if it were deemed that CNOOC had overpaid for Unocal, in essence putting the security of supply concerns of the Chinese government above the financial interests of private investors. The external directors commissioned an independent report[9] to test the recommendations of CNOOC's investment banking and legal advisors. Following delivery of that report, the external directors determined that they could support the CNOOC bid, and on June 22, 2005, CNOOC submitted an official all-cash bid of $18.5 billion, or $67 per share. This bid trumped the Chevron bid that was worth approximately $60.50 per share at the Chevron stock price prevailing at the end of June.

Chevron cried "foul," complaining that the CNOOC offer was enabled by subsidized loans from the Chinese government. Chevron's complaint echoed similar arguments by Boeing with regard to subsidies paid by the EU to Airbus Industries to develop new jets. So what started as a commercial competition rapidly became a geo-political issue. Chevron's CEO, Dave O'Reilly told us:

> The industry has always been at the intersection of economics, technology and geopolitics—always. I don't see that changing. It's a business that's just too strategic. It's been that way since 1879 and I don't see how that can go away.
>
> The entry of new Chinese and Indian competitors is not necessarily bad. You can look at it as a glass half empty or half full. It's such an essential business that plays such an essential role in the economic well-being of the globe, and the quality of life that people enjoy or aspire to enjoy, that it is necessarily going to be political. The idea that it might become a purely commercial business like widgets is just not realistic. We have gone through a whole history of experiences that suggest that there may be times of tension around the business as a whole, but somehow the combination of technology and economics seems to balance the geopolitical dimension and the industry works its way through it.

He went on to say with specific reference to the Unocal transaction:

> This was a very unusual set of circumstances. When the transaction was well along on its way, in comes a late hostile bid from a Chinese company. We dealt with that as a company by first of all increasing our bid, but also pointing out to the policy makers that there's an issue here that needs to be addressed around a level playing field. If a small company can suddenly show up with $20 billion and bid in a market environment where we have to go to the financial markets and pay commercial rates for debt and equity, where does that lead if it isn't addressed now? We did get thoughtful people engaged in that debate. It was helpful that the Chinese government said this was purely commercial. It was helpful because the data showed, when it had to be disclosed in the Securities and Exchange Commission filings that it was not a purely commercial bid.
>
> Now what was regrettable in a sense, was (and this is a matter of judgment that you would have to ask China about) that they put

this in at a time when there were all sorts of other issues. When you have people coming out of the extremes of left and right that normally do not support oil companies but because it was an opportunity to promote their anti-China agenda, they used this as an opportunity to promote their position. That was an unfortunate consequence of the timing of this and the fact that it was China. Our argument was that it doesn't matter that it was China; this could be any government-financed acquisition. It was taking away an opportunity from a commercial enterprise. It was a trade and investment issue. But unfortunately it got blown up with a number of other issues around trade and currency, so you had the pro-labor people jumping on this because of job concerns, you had the protectionist people jumping on it.

In the end, Chevron's revised bid of $64 per share was supported by institutional investors, prior to its consideration August 10 at the scheduled Unocal shareholder meeting, on grounds that the lower bid carried less risk. Seeing the lay of the land, CNOOC withdrew its bid on August 2 in a public announcement:

Reference is made to the Company's announcement dated 22 June 2005 concerning the Company's proposed merger with Unocal Corporation ("Unocal"), under which the Company offered U.S.$67 in cash per Unocal share (the "Merger Proposal"). The offer valued Unocal at approximately U.S.$18.5 billion. The Company wishes to announce that it has withdrawn its Merger Proposal, and has notified Unocal accordingly.

The Company has given active consideration to further improving the terms of its offer, and would have done so but for the political environment in the United States. In deciding to withdraw its offer, the Company took into account the political opposition amongst some members of the United States Congress that followed the announcement of its Merger Proposal as well as the Company's overall commercial objectives and the best interests of its shareholders.

Undeterred by the Unocal experience, China has gone on to acquire the Canadian company PetroKazakhstan through CNPC for $4.5 billion and is claiming preemptive rights to buy from Nelson Resources Ltd., a Bermuda-based company, half the North Buzachi field in Kazakhstan. The Russian private company Lukoil has acquired a

65% stake in Nelson, and Nelson claims that "no such preemptive right exists in the current circumstances."

Oil and geo-politics are again inextricably intertwined in the restructuring of the Russian petroleum sector, a work in progress that will undoubtedly be the subject of many books before it is finished. For our purposes, it is sufficient to note that Lukoil is a private company, with a minority ownership by ConocoPhillips. TNK/BP is also private and is owned 50/50 by Russian interests and BP. There are several other Russian private companies, but the whole industry operates under the giant shadow of Gazprom. Gazprom is controlled by the Russian state and owns most of the country's enormous gas reserves and the entire gas infrastructure. It has recently added oil production through a Byzantine set of maneuvers involving the prosecution of Mikhail Khodorkovsky and dismemberment of Yukos. It has ambitions to compete internationally with the super majors. Gazprom is the dominant force in the Russian oil industry and intends to leverage this dominance to become a global influence.

The Chinese and Russian NOCs are not the only ones with ambitions to expand internationally. NOCs of other countries have also been active in building an international portfolio of development opportunities. Statoil, the Norwegian NOC remains over two thirds owned by the Norwegian State after a partial listing in 2001. Statoil recently shocked incumbents by its successful bid of $2.2 billion to capture the deepwater Gulf of Mexico assets of EnCana. Statoil already has significant producing assets in Algeria, Angola and Venezuela, with important development projects in Iran and Azerbaijan. The private ownership share and listing on the NYSE provides important discipline to the company, which is seeking to carve out a differentiated position as a state-controlled company. It promotes a special understanding of the needs of other state-owned companies in resource rich countries while also noting that it is managed to the highest international standards of efficiency and effectiveness. Its strategy includes the statement: "Cooperating with former and existing national oil companies over the use of core technological expertise to achieve international growth targets will also remain important."

Helge Lund, Statoil CEO, gave his view on Statoil's positioning with respect to IOCs and NOCs:

> I started in Statoil on 8th of March, 2004. And then I spent the next few months trying to understand the company. I went quite deeply into what the company called at that time the NOC-NOC strategy to partner to access resources.
>
> I thought it was a quite nice story on the conceptual level, and that the question was, does this really give us an advantage in terms of access to resources and winning projects outside

Norway. And I guess in the beginning I was quite skeptical as to whether this idea was really worthwhile to entertain. But I must say that, having been here for a couple of years, and been around representing Statoil all over the world in business negotiations and opportunity qualifications, I feel that the legacy of Statoil is part of the competitive advantage, or edge.

Lund's comments tell us how the new competitive reality might unfold. He went on to describe how he felt that Statoil's experience has led to the potential for competitive advantage in accessing resources:

I like to think about Statoil's position in three layers. The first is our background; I think most people regard the Norwegian model as maybe one of the most successful ways of organizing oil and gas activities to benefit not only the general public and government finances, but also to build a strong oil and gas company and a strong service industry. So I think that has a certain attraction to most of the nations today that have significant oil and gas resources that they want developed.

The second leg is our competitive positioning related to the way we work. I think, generally, Statoil is regarded as a company with extremely high standards related to HSE and ethical standards, and being quite open with our technology. In the sense I think many people consider us as an equal and good partner, so they are not afraid of partnering with Statoil. They are not afraid of being dominated by Statoil because I think we have, as a corporate culture, a quite open, sharing way of working. And I don't think that Norwegians have had any imperial ambitions since the Viking age!

And the third leg, which in the end has really made the difference, is the technology and the professional experience that we have. We are relatively, compared to most other oil and gas companies, short on history; but I think in many ways we are long on technological achievements because the operating environments that we have been operating in for the last 40 years have been extremely demanding and there has been an attitude in the company, of embracing technology progress.

He described how National Oil Companies are coming together to share experiences:

> I think Statoil, maybe Petrobras, and maybe Petronas have a leg in both camps, both in terms of being an international oil company with those kinds of aspirations and ambitions, as well as being a national oil company. And Statoil, Petronas, and Petrobras are the only companies that are regarded also as international companies that are participating in this national oil company forum. But we really sit together in Saudi Arabia and in New York with all the other truly national oil companies to discuss opportunities and learn from each other. So in a way, I hesitate to categorize Statoil as a national oil company or as an international oil company. I think in a way we have both aspects included in our business proposition.

Lund has noted the tendency of other INOCs to offer a broad portfolio of economic development assistance, beyond the normal skill set of oil companies, but reflects the view of a partially privatized company when he says:

> I know some of the offerings that we have seen recently go way beyond what I think is meaningful for an international oil and gas company to enroll themselves in. And clearly these guys are driven also by strategic considerations and intents that go way beyond the individual deal. It is tough. But I am quite pragmatic as to what kind of value you are offering as long as you can see total value. I will not exclude that at some point we can engage in a downstream or midstream project to qualify for upstream opportunities. But that must be seen in the same context of value creation.

In Brazil, the oil industry was for many years under tight national control. The Web site of the Brazilian NOC Petrobras states:

> In the 1930s, there began a nationalization trend in Brazil of the subsoil resources. In 1938, all oil activities were by law to be obligatorily carried out by Brazilians. That same year, the National Petroleum Council (CNP) was created to assess the requests for research and mining of oil deposits. The decree creating the CNP also stated as a public utility the national oil supply and regulated imports, exports, transportation, distribution and trading of petroleum and byproducts, and the running of the refining industry.

> Moreover, the oil reserves, although not yet located, were now considered the property of the central government.[10]

Petrobras succeeded the CNP. Its status is similar to that of Statoil, with the Brazilian federal government now owning 56% of the common shares after a partial privatization in 2000. Petrobras is recognized as a global leader in deepwater exploration and development and is justly proud of its accomplishments:

> 1974 was a major landmark in the successful history of Petrobras—the location of the Garoupa field, the first discovery in Campos Basin, off the coast of Rio de Janeiro state. Later, in the mid-1980s, Petrobras directed its exploration activities especially toward the deepwater regions in the Campos Basin, culminating in discoveries of giant fields, such as Marlim, Albacora, Barracuda and Roncador. Today, the Campos Basin is the largest oil-producing province in Brazil and one of the largest deepwater oil producing provinces in the world.

In the mid-1980s when Petrobras was making its major deepwater discoveries, oil prices were collapsing and international oil companies were drastically retrenching and narrowing their focus primarily to the North Sea and Gulf of Mexico. It is doubtful that, even if Brazil had been open to international oil company investment, the IOCs would have been interested in investing in an area with historically disappointing exploration results in a search for resources requiring for their development a set of technologies that did not yet exist. By the year 2000, Petrobras had established the Brazilian deep water as a major oil province, and the country was ready to introduce some competition. A new regulatory agency, the ANP, was formed to regulate the activities of a competitive industry; Petrobras was partially privatized and obliged to relinquish a number of exploration blocks that were opened to bids from international oil companies; and several gas and power utilities were sold to international investment consortia. Petrobras still controls the Brazilian refining industry and has increased its international operations, with a primary focus on Latin America and West Africa. It has also built an interesting position in the U.S. Gulf of Mexico.

Petronas, the Malaysian NOC, has also developed a growing international business and has ambitions to be a major global player in liquefied natural gas (LNG). Petronas is still fully owned by the Malaysian State. It was conceived in the 1970s, a time of growing nationalism worldwide:

> During the Israeli war in 1973, several Arab oil-producing countries decided to stop oil shipments to certain countries. As a result, there was a mad rush for oil, prices shot up overnight and the

world experienced its first energy crisis. The 1973 oil embargo also made oil producing countries of the world realize the importance of controlling their own petroleum resources.[11]

In Malaysia, it led to the promulgation of the Petroleum Development Act in 1974 and the formation of a national oil company to ensure that the nation's petroleum resources could be developed in line with the needs and aspirations of the nation. Article 2 of the Act vests the entire ownership in, and the exclusive rights, power, liberties, and privileges of exploring, exploiting, winning, and obtaining petroleum whether onshore or offshore of Malaysia, in PETRONAS.

In practice, Petronas has partnered with IOCs in developing its Malaysian hydrocarbons resources. The company is very clear in laying out its vision statement, *"A Leading Oil and Gas Multinational of Choice,"* that it intends to build a company that will compete effectively with IOCs, even while remaining fully owned by the Malaysian government. The elements of the vision statement are elaborated on the Petronas web site in terms that would do credit to any IOC.

Steve Lowden, who has started a new international energy company with private capital, called Sun Energy Resources (SER), recalls his introduction to Petronas when he was with Premier Oil, a small UK independent:

> Well, a gentleman I'm extremely fond of, his name is Tan Sri Hassan Marican, the president of Petronas, one of the most intelligent, most calm characters I've ever met. But he's also one of the most creative energy leaders. Petronas is the NOC everyone looks up to. Most like an IOC. All the benefits of an NOC. I mean, it is a dream machine. And I remember when he first started looking to go international, he came alongside Premier Oil[12], tiny little 40,000 barrel a day Premier Oil. He came alongside us, and he let us very happily build a multi-billion dollar project. And he learned. He was prepared to go through the learning process. And in the end, he bought the entire asset. It was a business model that he and I put together, and the fact he was obviously very hungry to learn, and he wasn't somebody that needed to necessarily hobnob with Lee Raymond. Perfectly happy to go with whomever would teach him.

The Petronas web site also expresses clearly its core values:

Loyalty: Loyal to nation and corporation

Professionalism: Committed, innovative, and proactive and always striving for excellence

Integrity: Honest and upright

Cohesiveness: United in purpose and fellowship

In many national oil companies the simultaneous commitment of loyalty to nation and corporation has presented challenges when the interests of the nation, as defined by the government, have diverged from the interests of the national oil company as perceived by its management.

There is, however, a "dark side" to the new competition coming from the INOCs. It was one thing in 1959 for Winston Churchill to laud "the merchant venturers of Britain," but we should remember that those merchants often relied on the crown for political and even military support. We have described in this chapter the contest between CNOOC and Chevron to acquire Unocal, which Chevron CEO Dave O'Reilly argued was not conducted on a level playing field. Chinese and Indian national oil companies have also bid aggressively for assets in the Caspian, Latin America, and Africa. They have picked up assets in countries under boycott by developed countries because of human rights violations, such as Sudan and Myanmar. And the Chinese government has thwarted attempts by Western powers to bring these governments to account within the United Nations through imposition of sanctions.

We describe in chapter 4 the societal influences on energy companies. It will be quite unfortunate if IOCs, which by and large are taking the high road on relations with resource rich governments and NSCs, are displaced because some INOCs are prepared to take the low road of condoning human rights abuse and corruption.

The New Competitive Reality

Each year, the weekly energy newsletter Petroleum Intelligence Weekly publishes a ranking of the world's top 50 oil companies, based on total oil and gas production and reserves, refinery capacity, product sales, and employee count. The ranking provides an interesting perspective on the current competitive scene. Using the categorization proposed in table 2.1, and including all shareholder owned companies as IOCs (including former national oil companies as well as other international oil and gas companies), the current top ten includes four NOCs (Saudi Aramco, PDVSA, NIOC (National Iranian Oil Company, and Pemex), five IOCs (ExxonMobil, BP, Shell, Total, and Chevron), and one emerging INOC (PetroChina). The top 50 includes eighteen IOCs, nine NSCs, seven NOCs, ten INOCs, five LOCs, and Gazprom, which defies categorization.

The top 50 companies account for 80% of the world's oil and 70% of the world's gas production and sales:

Table 2.2: Top 50 company share of world oil and gas

	Liquids Output (MBD)	Gas Output (MMCFD)	Liquids Reserves (MMB)	Gas Reserves (BCF)	R/P Oil	R/P Gas
IOC	17,264	62,095	76,296	332,083	12.1	14.7
NSC	6,858	16,303	126,291	1,038,660	50.5	174.5
NOC	25,149	23,830	696,285	1,679,120	75.9	193.0
INOC	9,062	24,563	50,009	347,797	15.1	38.8
LOC	4,784	3,315	31,817	44,142	18.2	36.5
Gazprom	240	52,574	14,372	1,140,000	164.1	59.4
Total Top Fifty	63,357	182,680	995,070	4,581,802	43.0	68.7
Total World	80,260	258,800	1,188,600	6,337,400	40.6	67.1
Top Fifty Share	78.9%	70.6%	83.7%	72.3%		

NOCs as a group control the largest share of global oil production, followed by IOCs. While IOCs as a group control significant production, they have a much lower share of reserves (fig. 2.1). Consequently, IOCs operate at a much lower reserves to production ratio than NOCs. The four NOCs in the top 10 and KPC, Iraq National Oil Company, and Rosneft control 70% of the world's oil reserves. Access to these reserves is difficult or impossible for IOCs. NSCs allow access by IOCs, but control output to maintain a reserve to production ratio closer to NOCs than IOCs. The INOCs as a group operate at reserve to production ratios close to those of the IOCs and are competing aggressively with IOCs for access to resources.

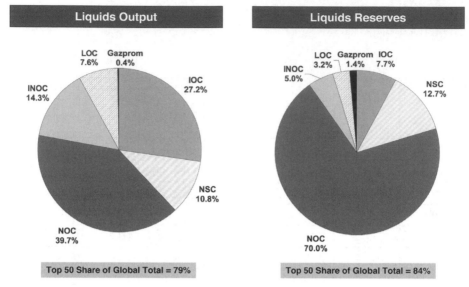

Figure 2.1: Top 50 companies' shares of oil production and reserves. *Source: Petroleum Intelligence Weekly—PIW's Top 50: How the Firms Stack Up (www.energyintel.com)*

IOCs have the greatest share of refining capacity and oil product sales (fig. 2.2), with 60% of total top 50 product sales and 50% of refining capacity. (Note that the top 50 only control 60% of global refining capacity, which is a more fragmented segment of the industry.) This reflects the historical complementary roles of NOCs and IOCs, whereby NOCs control the resource and IOCs control the markets. INOCs as a group have production equal to their refining capacity, while IOCs have refining capacity equivalent to 150% of their production and NOCs have refining capacity equivalent to 40% of their production. However, NOCs are steadily increasing their involvement in international refining, especially those with low-value crude oils and with special attention to the fast growing markets of Asia.

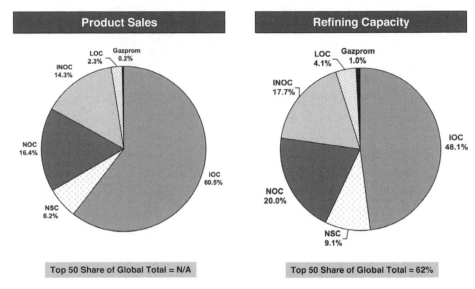

Figure 2.2: Top 50 companies' shares of oil product sales and refining capacity.
Source: Petroleum Intelligence Weekly—PIW's Top 50: How the Firms Stack Up (www. energyintel.com)

Abdallah S. Jum'ah, CEO of Saudi Aramco, sees his firm expanding its international downstream from its strong upstream base in the future:

> When you look at where is our strength, our strength is in our reserves. We have one-quarter of the world's reserves of crude. And we are adding to our reserves every year no less than what we produced in the previous year. We are looking at a Saudi Arabia that is under-explored. So the potential for additional crude is there. We are today the world's fourth largest owner of natural gas reserves. Our exploration program and development program is big, and we expect to find more. Today our target is to add to our reserve five trillion cubic feet of gas [per year]. So every plan, we put in 25 trillion cubic feet [of reserve additions]. And we have been able to be successful in that. Looking to the future, we will be bigger in downstream. Because now at this time, with good margins, we are expanding our refining. We are expanding our joint ventures. We are expanding our refining and our joint ventures outside the Kingdom, you saw that about Motiva: Now we are moving at Port Arthur. The Koreans are seriously looking at doubling the capacity; we have a 35% interest in Korea. We went there when they were

producing 90,000 barrels per day. Today it is a 520,000 barrels per day refinery. We are looking at the complex in China, in Fujian, with Exxon and Sinopec. So I believe we will be a bigger player in the future than we are today.

The reserves to production ratio is over 50% higher than for oil in aggregate, and natural gas resources have only recently been targeted for exploration as a global commercial trade system develops (fig. 2.3). Gazprom is the largest global gas producer and resource holder, but IOCs have the highest share of global gas production apart from Gazprom, especially in terms of natural gas marketed to third parties rather than reinjected or consumed in production. In many cases, NOCs and NSCs are allowing IOCs access to their natural gas reserves, which require integrated supply chains into consumer markets in order to monetize the resources. International gas has become a growth segment for IOCs, and the INOCs have generally been prepared to partner rather than compete with the IOCs.

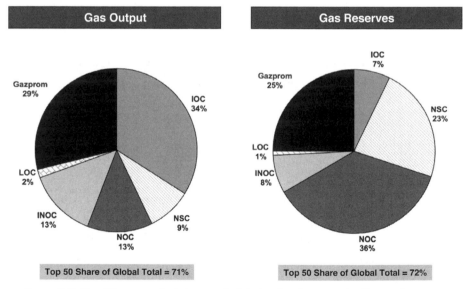

Figure 2.3: Top 50 companies' shares of global natural gas production and reserves

In the 1990s, it seemed likely that the large resource holders would open up to IOC investment. Their fiscal positions were tenuous at low oil prices, and there was perhaps excessive optimism that technology would rein in demand growth and stimulate further production growth such that low oil prices would persist. The industry was mature or aging in several countries, such that incremental production growth was becoming more capital intensive and incurring higher operating costs.

NOCs with constrained capital budgets were able to address only the best few of the investment opportunities, and production seemed more likely to decline than expand.

At higher prices, resource owners' fiscal conditions have improved markedly, and NOC capital budgets can be increased without jeopardizing other government priorities. The logic for opening up resource opportunities for IOC participation is therefore less compelling, and the arguments for conserving the resource for future generations gain weight over those advocating increased production. The NOCs can with some justification quote the Brundtland Commission on Sustainable Development (see chapter 4): "Sustainable development is development that meets the needs of the present without compromising the ability of future generations to meet their own needs."

Michael Porter's Five Forces model provides clear insight into changing petroleum industry conditions. (We describe the Five Forces model in more detail in chapter 6.) This model analyzes the competitive intensity of an industry by assessing the relative bargaining power of suppliers and of customers, the threat of new entrants, the possibility that technology will lead to substitutes, and the natural internal rivalry between existing firms. It is clear from this model why competitive intensity has escalated for the upstream petroleum industry, which is best understood for our purposes as converting prospective resource acreage to saleable hydrocarbons:

- The suppliers of prospective acreage have enormous bargaining power, and the NOCs, which control 70% of remaining oil reserves, control access to and production from their reserves very tightly.
- The INOCs represent a group of new entrants that in some cases may be driven by national security concerns more than economics.
- Technology is expanding the supply of substitute acreage only slowly, opening project opportunities in deep water, oil sands, and stranded gas that are highly capital intensive and have long lead times.
- Contrary to public perceptions, rivalry within the industry for prospective acreage has always been intense.
- The only good news from the industry's point of view is that customers for the hydrocarbons have virtually no bargaining power. This is only partially good news, however, since it reflects the reality that consumer government attempts to influence resource owners to expand access as well as production have generally been ineffectual.

Arguably, it is the role of the IOCs to resolve this situation by expanding the resource base. In the 1970s, they did that by discovering resources in the North Sea and Alaska and developing them with new offshore and Arctic technologies. They are currently exploiting new resources in deep water, oil sands, and stranded gas, but as we see in the next chapter, their efforts seem unlikely to match trend line petroleum demand projections, and we will be surprised if the NOCs expand their production to fill the shortfall. Current IOC leaders in many cases were seared by the experience of the 1980s, when their predecessors had blown through money like casino gamblers and expanded their payrolls so their successors had to downsize painfully. They are a cautious group of people with regard to organic growth, and they don't want to repeat the mistakes or the previous generation by over-expanding. We believe, however, that their caution has become excessive.

The INOC's aggressive competitive behavior is analogous to that of ENI under Mattei—more likely to alter the terms of trade and reduce the returns on investment for all competitors in favor of resource rich governments than to add significant hydrocarbon resources. In many places, INOCs have a competitive advantage over IOCs by virtue of their access to low cost government financing and the willingness of their government/owners to offer a broad package of economic assistance. By contrast, IOCs tend to have adversarial relationships with their own host governments. It seems likely that IOCs will continue to lose market share in established oil producing regions if these conditions persist, and inevitably their production of conventional oil will begin to decline. Their challenge will be to make value-enhancing acquisitions and to develop resources in unconventional resources and non-traditional places.

Conclusions

So what have we learned from our study of NOCs and INOCs? We have seen a series of transitions occurring over the past thirty years. IOC assets were nationalized in resource rich countries in the 1970s, and the companies lost their lucrative positions. National oil companies in consuming countries were privatized in the 1980s, and several made the jump to IOC status. Resource rich countries opened some of their least prospective acreage for IOC access in the 1990s. And a few NOCs upgraded to emerge as INOCs capable of entering global competition with the IOCs in the 21st century. Thus was the playing field transformed.

Are national oil companies a good idea? On the positive side, it is easier to align NOCs than IOCs with government policy, when the desired policy outcomes differ from the results that would be delivered by the natural economic drivers of shareholder owned IOCs (e.g., Saudi Aramco). IOCs can be ordered to develop currently uneconomic resources (e.g., Petrobras); they can compile detailed technical and economic knowledge to help governments structure bid rounds (e.g., PDVSA and Petrobras); and they can become flagship corporations, providing fulfilling careers for nationals and a vehicle to expand from exploitation of local resources to the pursuit of international prospects (e.g., Statoil, Petronas, Petrobras and CNOOC).

On the negative side, they can become bloated bureaucracies with unions that block efficiency improvements (e.g., YPF prior to privatization, Pemex); they can be subject to political interference on personnel selection and corporate strategy, frustrating efforts at rational decision making (e.g., PDVSA); they can be starved of necessary capital by the funneling of cash from industrial investment to social programs or general government expenses (e.g., PDVSA and Pemex); they can be obliged to provide subsidized fuel into the national market (e.g., NNPC, Pertamina, PDVSA, and many others); and they can adopt very low environmental, health, and safety standards if they can persuade the government that the alternative will require them to retain more income.

We will admit to bias towards privatization of national oil companies and opening of natural resource access to a variety of local and international companies, with the role of government being to issue and enforce the rules of the game. However, history has shown that NOCs can be a useful evolutionary step when there are no reliable, indigenous private sector firms capable of taking the financial and technical risks involved in the oil business, or when the IOCs cannot be persuaded to align with national priorities. In the long term, nations and their NOCs benefit from competition for resources; and introducing a range of ideas, technologies, and capabilities will almost always result in more discoveries and higher production. In that competition, NOCs can still play a useful and even dominant role, as PDVSA did in the 1990s, and as Saudi Aramco plans to do even if the gas initiative is extended in the future.

Helge Lund of Statoil agrees that a mixed economic model is likely to be the most successful:

> I look at the different countries, those that are succeeding in developing their oil and gas resources, and those that are not succeeding. If you look, for instance, at Iran and Qatar, I think Qatar has been extremely successful in the sense that they have been open to foreign investments, open to technology transfer, invited people in; and they have had a beautiful journey. [By contrast] the Iranians that are working on the same field have been much more skeptical. They do not see win-win situations, so they are more skeptical to taking on international experience and technology. They are failing.
>
> And I think if you look at the development of the Norwegian oil and gas sector, I think also there has been a history of a country being open for all the international players in the service sector as well as in terms of the oil and gas companies, while at the same time

building up our own skills and competencies. And I can't see why they don't see that it is impossible for any country today to attack all the challenges in this industry unless they are open to taking the best players in the industry, and the best technology.

Interestingly, he makes an exception for Saudi Aramco: "We have a close relationship with Saudi Aramco. And I must say, based on what I see, I am really impressed with the technology standards and level they have in their organization. So maybe that is the one exception."

It has been common practice to view the IOCs as the primary suppliers of oil worldwide. It is time to recognize that the eighteen IOCs in the top 50 global oil companies control less than one quarter of global oil and gas production, own about 30% of worldwide refining, and account only for about 6% of global oil and gas reserves. Therefore, they have limited influence, let alone control, over global oil commodity prices. With 70% of conventional oil reserves controlled by NOCs, consuming countries need to change their mental models of the international oil market, which tend to be frozen in the paradigms of the 1960s or even the 1890s. On their part, the IOCs can help by explaining the new reality more clearly to stakeholders in consuming countries.

The changing competitive landscape is challenging the IOCs' traditional business model. They have been "solution integrators" bringing money, technology, and management as a package to resource rich nations. This model may not be sufficient in today's world to satisfy developing countries' needs, particularly when INOC competition brings offerings beyond those dictated by the economics of shareholder return on investment. It may also not be fulfilling to consuming countries expecting affordable and sufficient energy supply. And it may not satisfy shareholders, as the pipeline of conventional investment opportunities created in the 1990s dries up in the 2010s. We believe there will be a competitive shakeout in this new petroleum up cycle, with a number of different strategies unfolding. Some IOCs may reunify the value chain through strategic partnerships and combinations, while others focus on new resources and advanced technology plays. This is the subject of the chapter 7 on setting the course.

---

1 Wintson S. Churchill, 1959. *The World Crisis*, quoted in Henry Longhurst, 1959, *Adventure in Oil: The Story of British Petroleum*. London: Sidgwick and Jackson Ltd.

2 Daniel Yergin, 1991. *The Prize*. New York: Simon & Schuster.

3 Norman Fowler, 1991 *Ministers Decide*. 1991. London: Chapmans Publishers Ltd.

4 Francisco Parra, 2004. *Oil Politics: A Modern History of Petroleum*. London: I.B. Taurus & Co. Ltd.

5 *Oil & Gas Journal*, October 24, 2005.

6 John A. Crow, 1992. *The Epic of Latin America*. Berkeley and Los Angeles: University of California Press

7 In reality the argument tends to be on the validity of the transfer price between two wholly owned affiliates. The consequences of shifting earnings from domestic production to foreign refining are in where and how much taxes are paid. Tax rates on production earnings are generally much higher than those on refining earnings, so a transfer of income from production to refining is in the interest of the national oil company (and the refining department) and against the interest of the government (and the production department). This imbalance is compounded if the refining earnings are abroad, since the taxes on refining earnings are now paid to a foreign government. However, there is very limited room to play with transfer prices, since the IRS has strict rules that require that transfers be at market prices. This fact does not seem to affect the emotional intensity of the arguments.

8 Daniel Yergin, 1991. *The Prize*. New York: Simon & Schuster.

9 Financial Times, 17 June, 2005, "Rothschild – alongside consultants CRA International and the law firm Skadden, Arps, Slate, Meagher, Flom – was hired by CONOOC's four non-executive directors following their rejection of the first bid for Unocal just days before Chevron's offer."

10 Petrobras web site: http://www2.petrobras.com.br/EspacoConhecer/ing/HistoriaPetroleo/conselhonacpetro.asp.

11 Petronas Website: http://www.petronas.com.my/internet/corp/centralrep2.nsf/frameset_corp?OpenFrameset.

12 Petronas actually bought a 25% interest in Premier, and later exchanged this ownership interest in the company for specific assets in Myanmar and Indonesia.

13 *Petroleum Intelligence Weekly*, December 12, 2005.

③ The Resource Dilemma

We shall not cease from exploration
and the end of our exploring will be to arrive where we started
and to know the place for the first time.[1]

This chapter and the next present the forward-looking aspects of a strategic assessment. They identify the trade winds that can propel companies forward and the headwinds and doldrums that can make even the best strategy extremely difficult to accomplish. In our experience, there are three overriding contributors to success in the energy business:

- positioning a company down wind of important economic and social trends;
- determining what the company is not going to do as well as what it is going to focus on, and being disciplined in removing distractions;
- building the competencies needed to execute the strategies selected.

The Nature of the Dilemma

The energy industry has a poor record forecasting the future, particularly crude oil prices, whether in an up-cycle or a down. What one can do, however, is identify important trends. Understanding these trends provides a backdrop for developing strategies that offer a persuasive value proposition for stakeholders. Sometimes trends suggest that the world can continue on its present course even though discontinuities are looming. Other times the trends just do not look stable. We believe that historical energy trends have now become unsustainable, and that energy industries face a discontinuity similar to the massive changes in the 1970s. As then, this discontinuity creates threats and opportunities, and the successful companies will be those that manage the former and capitalize on the latter.

This is why we call this chapter "The Resource Dilemma." *Webster's* dictionary defines a dilemma as *a choice between evenly balanced alternatives or a predicament defying satisfactory solution.* That accurately describes the forthcoming energy scene. The world needs more energy to allow more people to prosper. We cannot in all conscience decline to provide it, but increasing energy supplies has undesirable consequences. There is no obvious right answer at the moment. Fossil fuels may be the easiest to access but are messy and depleting; renewables are not yet ready for prime time; and many people are still frightened by nuclear energy. The bottom line is we are in a sorting-out period to identify the strategic energy source over the next quarter- to half-century.

From Cassandra to Chicken Little, there have always been prophets of doom, foretelling the end to civilization as we know it; and there have been perennial optimists, like Voltaire's Dr. Pangloss, who remain convinced despite all evidence to the contrary that "all is for the best in this best of all possible worlds." What's different today is the sheer magnitude of the numbers we are dealing with. It all starts with the global population, which is now about 6.5 billion people heading, according to the United Nations, towards 9 billion by 2050. The more developed regions will have approximately the same population in 2050 as in 2005, at about 1.2 billion, so all the net population growth will be in the less developed regions. The World Bank estimated in 2001 that roughly 2.7 billion people lived on less than $2 per day and 1.1 billion on less than $1 per day; 2.4 billion people relied on traditional biomass fuels for cooking, with dire effects on the environment in the form of emissions and deforestation; and 1.6 billion lacked access to electricity.

The good news is that most regions seem to be on a path to reduced poverty levels (fig. 3.1).

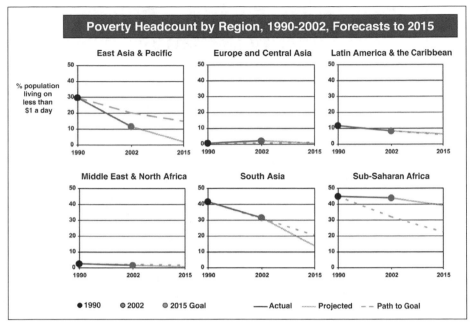

Figure 3.1: Poverty reduction is ahead of Millennium Development Goals in Asia but behind goals in Africa. *Source: Global Monitoring Report 2006.*

There is, however, considerable debate among the experts about measuring progress in reducing poverty. Certainly, the income level of $1 per day defines a grim poverty. But at what income level does poverty cease? Further, the data is interpreted differently by different analysts. The World Bank tends to focus on the absolute number of people living in poverty, and notes the tremendous progress in Asia in reducing the absolute numbers in extreme poverty, particularly in China and India. They further note the even greater improvement as a proportion of the growing overall population. For example, the proportion of Chinese living in extreme poverty has declined from 63.8% in 1981 to 16.6% in 2001, while in India the proportion has declined from 54.4% in 1981 to 34.7% in 2001. Other observers remain concerned at the proportion of the population living in poverty in Latin America and the Caribbean, while noting the increase in Sub-Saharan Africa from 41.6% in 1981 to 46.9% in 2001.[2] They are outraged by the increase in inequality, as the rich get richer, and the poor remain trapped at a bare subsistence level. The data are the same but the emphasis is quite different.

Developing Countries Driving Energy Demand

The corollary to the good news is that continued progress in this direction will require energy and lots of it. Long-range energy forecasts range between a doubling to a tripling of energy demand by 2050. Jamal Saghir of the World Bank notes: "In modern times no country has managed to substantially reduce poverty without greatly increasing the use of energy. Modern energy has the biggest effect on poverty by boosting poor people's productivity and thus their income. It also reduces poverty in other ways. By powering lights and modern equipment, electricity helps improve health care and education for poor people and makes it more likely that women will read and children will attend school regardless of their income level."[3]

The UN also recognizes the connection between poverty reduction and availability of commercial energy sources.[4] The United Nations Development Program collaboration with the World Bank identified four major challenges:

1. Widening access to energy services for the poor as a means of supporting development overall;
2. Enhancing environmental performance of energy supply and consumption;
3. Mobilizing financial resources to expand energy investments and services;
4. Linking energy planning to goals and priorities in other sectors and sustaining political commitment to sound energy sector management and governance.

So, there is a general sense that the world has been making and should continue to make progress in reducing poverty, and that access to commercial energy is essential for socio-economic development. But there is a vocal group of *civil society* representatives who believe strongly that current rates of progress are not fast enough, and that the inequalities between rich and poor are unacceptable. This group does not have a policy platform that would achieve its desired results, other than transfer payments from rich to poor. Despite the failure of socialist economies to provide sustainable welfare to constituents, transfer payment and populist rhetoric have obvious appeal to people who believe it is the existing system that is responsible for their poverty. This appeal will retard poverty reduction efforts in countries that succumb to it.

As Jamal Saghir notes: "Clearly energy for the sake of energy is not useful. Its utility lies in facilitating human development." The energy needs of the poor are focused on electricity for lighting and kerosene and LPG for cooking. Providing commercial energy to the poor increases demand for electricity, kerosene and LPG. However, the more substantial driver of increased energy demand is the commercial and industrial development required to provide jobs and income that will raise the poor from subsistence to a broader participation in economic life, moving them up Maslow's hierarchy of needs. This again requires electricity, and also transportation. Improved transportation systems are required for farmers to get their surplus products to market and move from subsistence to income generation; they are necessary also to make products available for purchase, and they are further necessary to allow specialization and the collection and assembly of components into products for export. Finally, the desire for personal mobility seems to be extremely powerful, and a private vehicle becomes one of the first uses for discretionary income.

Desire for Personal Mobility

In "Mobility 2030: Meeting the Challenges to Sustainability," the World Business Council on Sustainable Development observes: "One factor has continued to impress us throughout our assessment—the strength of people's desire for enhanced mobility. Mobility is almost universally acknowledged to be one of the most important prerequisites to achieving improved standards of living. Enhanced personal mobility increases access to essential services as well as to services that serve to make life more enjoyable. It increases the choices open to individuals about where they live and the lifestyles they can enjoy there. It increases the range of careers that individuals can choose and the working environment in which they can pursue these chosen careers. Enhanced goods mobility provides consumers with a greatly widened range of products and services at more affordable prices. It does this by enabling people to market the products they grow or manufacture over a much wider geographic area and by reducing the cost of inputs they must use. The vast expansion in the number of automobiles and trucks over the last one hundred years has been one of the most important manifestations of this desire for enhanced personal and goods mobility."[5]

Unfortunately, there are two complications. The first is the growing belief that man-made emissions of greenhouse gases are contributing to global warming and to other societal penalties. The WBCSD report continues its assessment of sustainable mobility: "These vehicles have provided their users with unprecedented flexibility in terms of where they can go and when they can go there. But people are increasingly aware that their enhanced mobility has come at a price. This price has included the financial outlay that mobility users must make to providers of mobility systems and services to permit them to supply such systems and services. But it has gone beyond this. Enhanced mobility has tended to be associated with increased pollution, emission of greenhouse gases, congestion, risk of death and serious injury, noise, and disruption of communities and ecosystems." This complication will be addressed in the next chapter.

Recognizing that most transportation today is fueled from the refining of crude oil into gasoline and diesel, the second complication is the belief that the world's oil supplies may become limited sooner rather than later. The question today, as it might be posed by the more pessimistic among us, is how can we satisfy the socio-economic aspirations of a large and growing global population without running out of fossil fuels, and oil in particular? Put in a less calamitous way, there are serious problems with projections of energy supply and demand that assume a continuation of past trends. We will explore the problems with these projections in this chapter, dealing first with the determinants of demand, then supply. The focus will then shift to the supply options for the future and the implications for different business segments of the energy industry.

Energy Consumption Drivers

Let's take a closer look at the impact of the developing world's drive toward economic growth on the demand side of the energy equation.

The link to economic growth

Energy demand and economic development go hand in hand. There are no examples of countries or regions that have increased their standards of living without increasing their consumption of energy products. However, as countries and regions increase their prosperity, the amount of energy it takes to produce an incremental dollar of GDP tends to decrease (fig. 3.2).

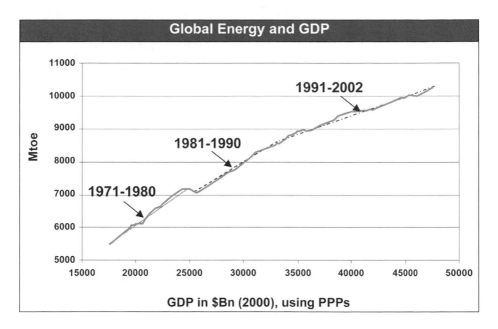

Figure 3.2: Energy requirements for incremental economic growth decrease as economies develop. *Source: World Energy Outlook © OECD/IEA, 2004, Figure 1.1, p. 42*

"The so-called income elasticity of energy demand—the increase in demand relative to GDP—fell from 0.7 in the 1970s to 0.4 from 1991 to 2000."[6]

Energy demand forecasts

There are three main sources for intermediate term energy demand projections that are accessible to the public: the International Energy Agency in Paris, the Energy Intelligence Agency (part of the DOE) in Washington, D.C., and ExxonMobil. We include these forecasts not as accurate predictors, but to highlight trends and identify issues with potential for discontinuities. The intermediate term forecasts are done annually, and normally the change from year to year is not significant. Longer term forecasts to 2050 project energy demand doubling to tripling over that extended time frame driven by the population growth and economic development of countries like China and India. The three intermediate-term forecasts arrive at similar conclusions: that energy demand will continue to grow; that most of the growth will be in the developing countries; that natural gas will have the highest growth rate among fossil fuels; and that renewable fuels will grow most rapidly, but from such a small base that they will still be barely material by 2030 (table 3.1).

Table 3.1: Future energy demand projections (million tons oil equivalent)

	2004	2030 Projections			
Source	WEO 2006[7]	IEA WEO 2004[8]	IEA WEO 2006	EIA IEO 2006[9]	Exxon 2005[10]
Coal	2773	3601	4441	4839	4000
Oil	3940	5766	5575	5918	5250
Gas	2302	4130	3869	4700	3900
Nuclear	714	764	861	859	1000
Hydro	242	365	408	n/a	400
Biomass and Waste	1176	1605	1645	n/a	1400
Solar and Wind	57	256	296	n/a	150
Total	11204	16487	17095	17861	16100
Energy Growth (% pa)		1.7%	1.6%	2.0%	1.6%
GDP Growth (% pa)		3.2%	3.4%	3.8%	

The major difference among the three sources of these projections is that the IEA expresses concern that its Reference Scenario may be undermined by lack of investment in the Middle East and North Africa (MENA) oil production capacity, and thus presents an alternative "Deferred Investment Scenario." The EIA is the most optimistic on the supply of fossil fuels. ExxonMobil presents its outlook as feasible, but is silent on the subjects of sustainability and energy prices. To the extent that the ExxonMobil projections show lower global economic growth and strongly emphasize the role of conservation, one might conclude that ExxonMobil has a higher implicit price assumption than the IEA and EIA. However, ExxonMobil is adamant that it does not make price assumptions in its forecasts.

After describing its Deferred Investment Scenario in WEO2005, the IEA submitted: "In no sense, however, can either vision of the energy future be considered sustainable. G8 leaders, meeting with the leaders of several major developing states at Gleneagles in July, 2005, acknowledged as much when they called for stronger action to combat rising consumption of fossil fuels and related green-house gas emissions." In 2006, the tone is even more severe: "Current trends in energy consumption are neither secure nor sustainable—economically, environmentally or socially." The IEA responds by proposing an alternative path in its World Alternative Policy Scenario (table 3.2), in which consumer governments enact a series of policies already under consideration to reduce demand for fossil fuels.

Table 3.2: IEA WEO demand scenarios (million tons oil equivalent)

	2004	2030 WEO Projections		
	WEO actual	Reference	Deferred investment	Alternative policy
Coal	2773	4441	3551	3512
Oil	3940	5575	5068	4955
Gas	2302	3869	3639	3370
Nuclear	714	861	772	1070
Hydro	242	408	369	422
Biomass and waste	1176	1645	1690	1703
Solar and wind	57	296	278	373
Total	11204	17095	15367	15405
Energy growth (% pa)		1.6%	1.3%	1.3%

The IEA Alternative Policy scenario is clearly designed as a signal to consumer governments that they can have their cakes and eat them too. By adopting not very radical fuel efficiency targets, they can reduce their fossil fuel consumption and lower oil prices at the same time. The IEA claims, "In aggregate, the new policies and measures analyzed yield financial savings that far exceed the initial extra investment cost for consumers." The IEA claims, "In aggregate, the new policies and measures analyzed yield financial savings that far exceed the initial extra investment cost for consumers." Of course, democratic governments also have to address the winners and losers of the new policies and decide where the votes will lie. But the alternative is to continue with current policies until supplies fail to meet demand and prices rise so that demand and supply find a new equilibrium; then they would have to face an angry electorate demanding to know why they did nothing. At that point, politicians and the press will generally blame the energy companies; and the "invisible hand" of the market will cause the investments to be made that had previously been recommended. Where the "invisible hand" may fall short is in pricing global commons issues such as a potential future shortfall in supply, "externalities" such as pollution costs, and in particular the long-term cost of climate change.

Most of the demand growth is in the power generation and transportation sectors, as displayed in table 3.3 below.

Table 3.3: Energy demand by sector (million tons oil equivalent)

			History			IEA WEO 2006 reference scenario		
			1971–2002	Growth			2004–30	Growth
	1971	2004	% pa	Share	2030		% pa	Share
Transport	856	1969	2.6%	20%	3111		1.8%	20%
Power Gen	1211	4133	3.8%	52%	6926		2.0%	47%
All other	3469	5102	1.2%	28%	7058		1.3%	33%
Total primary	5536	11204	2.2%		5668	17095	1.6%	5891

Between 1971 and 2002, primary fuels consumed in power generation tripled and transportation demand nearly doubled, while all other uses (industrial, commercial, and residential) of primary fuels increased by less than 50%. This reflected the rapid electrification of these three sectors, which reduced their relative consumption of primary fuels, and the growing importance of transportation in economies as they develop. Broadly, these trends are expected to continue in the future so that two-thirds of the growth in global energy demand will come from the transportation and power sectors.

Choices for the Power Sector

Energy is increasingly being consumed in the form of electricity. We are becoming more of an "electricity world." In a period of transition and uncertainty in primary energy sources, electricity offers flexibility through diversified options. It can be generated from all the fossil fuels, as well as from nuclear, hydro, geothermal, wind, and solar energy. Current constraints on the further use of electricity are storage and distribution.

The IEA expects global power demand to grow steadily, with natural gas and coal providing most of the primary fuels increase (fig. 3.3); however, this is a decline from its historical growth rate. Coal and natural gas are expected by the IEA to increase market share, taking share from oil and nuclear. The EIA, noting higher prices for oil and natural gas, shows higher growth rates for coal than the IEA. Renewable fuels will grow rapidly but still have only a less than 10% share in 2030.

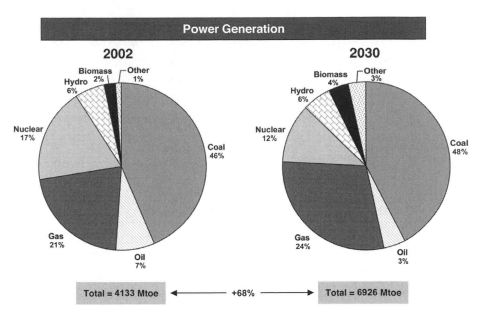

Figure 3.3: Coal and natural gas are competing for power growth. *Source: World Energy Outlook © OECD/IEA, 2004, Annex A pp. 430-431 as modified by the authors*

Underlying these projections are some fundamental economic drivers. The first is the importance of electricity to economic development, which drives the overall demand for power. The second is the relative economics of different technologies for generating power. Coal is the cheapest fossil fuel because it is abundantly available and can be mined inexpensively, particularly when the deposits are close to the surface and can be produced through strip mining using a highly mechanized approach with low labor intensity. Such approaches are prevalent in the Western U.S.A and Australia. Coal is also abundant and cheap in China and India, where it is the major energy source, and in South Africa, where during the apartheid years of isolation it was not only burned for power, but was also converted into transportation fuels.

John Wilder, CEO of TXU in Dallas, Texas, is not impressed by the technology choices made by the U.S. power industry while it was fully regulated: "I can't find one technology choice that was right. Not one. There was a massive nuclear build-out to the tune of about $250 billion, and it generated a net present value of the assets of maybe $2 to $3 billion, becoming a huge loser. There was a big coal build-out in the '70s, same thing with a huge loss on an NPV basis. Then the choice was, 'We've got to go to clean fuels, go to natural gas.' That was about a $150 billion build-out. Right now that asset class across America is probably worth about $50 billion. The good news about the last technology decision was that a lot of it became based on the equivalent of deregulated markets. So capital took a hit rather than consumers."

Wilder is hopeful that a market solution will provide better technology choices for the next generation of new-builds. However, he is open-eyed about the risk inherent in the decision: "We've modeled out each of these different technology classes to build a probabilistic distribution. The range of values is just enormous. Depending on the technology and other factors, a billion dollars of capital could be worth minus two billion to positive five billion."

Coal Addresses Environmental Problems

Coal has one major drawback: burning it releases a number of undesirable chemicals that are damaging to the environment and to human health. In particular, sulfur and nitrogen oxides (SO_x and NO_x) are believed to cause acid rain, which damages forests. Together with particulate emissions (soot), they also cause respiratory problems in humans and animals. In addition, airborne emissions can also contain mercury, a toxic element that can build up throughout the food chain, and has been found to accumulate in fish.

There are two main approaches to reducing these emissions: to clean up the gases produced when the coal is burned using flue gas desulphurization (stack gas scrubbing), or to extract the impurities prior to burning using a gasification technology. The second is more technologically effective.

Of current clean-up technologies, most power companies prefer stack gas scrubbing, which is perceived as less expensive than the alternatives. The overall process is called pulverized coal (PC) because the coal is first pulverized before it is burned using a fluidized bed and "Low NO_x" burners to limit nitrogen oxide emissions. Then the flue gases are scrubbed using a "wet" flue gas desulphurization technology in which the gases are reacted with limestone (calcium carbonate) to produce gypsum (calcium sulphate), which can be used as a building material. In addition, small particles can be removed by using an electrostatic precipitator or by passing the flue gases through a fabric filter. Some, but not all of the mercury is removed through these processes. The heat from coal combustion is used to produce steam at supercritical temperatures that powers turbines. The overall heat efficiency from coal to electricity is around 34%.

That is the technology chosen by John Wilder, CEO of TXU, based for the most part on its lower cost. TXU operates in the deregulated ERCOT market of Texas, and has analyzed the different technologies carefully under different assumptions for fuel prices, capital costs, regulatory, and permitting factors, environmental impact, and the potential for greenhouse gas controls. Priorities were to satisfy customers' wishes for reliability and lower prices and to select technology that was environmentally clean. After those priorities were assured, the most important consideration for Wilder was the ability of his team to engineer costs out of the construction program: "We announced a big coal construction program, the largest in the United States, $10 billion, in fact. It rivals the China coal build-out program. We've had Bechtel engineers, Fluor engineers, TXU engineers, TXU traders—a big integrated team—working this for about nine months. And I can show you chart after chart of where we've broken down the construction process. We began with the regulated coal plant model with an off-the-shelf coal configuration at $1,500 per KW cost. Right now we are targeting $1,100 per KW cost and doing everything we can to push the cost lower. But most importantly, in terms of the commodity cycle, the time to production was conventionally 60 months. We're down to 36 months."

Wilder is vehement about the superiority of a free market to a regulated market model. He takes the coal plant example further, asserting that in a regulated environment there is less incentive to minimize capital costs: "A regulated coal plant builder might get an agreement that, 'It's going to be $1,600 a KW costing $2 billion to build this plant.' Their incentive then is just not to exceed $2 billion because any excess is called a disallowance. There'll be hearings for years about why they spent $2.1 billion. Now what we want to say is, 'Let's put together a high-performance capital allocation construction plan, and if we assume it's $1.2 billion, and we come in at $1.3 billion, we are still much better off than if we would have assumed $2 billion and come in at $2 billion.' But in the regulated model, you would never do that because the probability of disallowance on an underestimation is almost 100%."

An alternative technology is called Integrated Gasification and Combined Cycle (IGCC). This process removes impurities from the coal chemically before combustion by gasifying the coal. It was widely used prior to the commercial availability of natural gas to produce town gas. In IGCC, coal is reacted with water to produce a "syngas" of hydrogen and carbon monoxide. The syngas can then be burned in a highly efficient combined-cycle turbine using low NO_x burners. The advantage of this process is that SO_x can be separated by absorption and converted to elemental sulfur using established chemical technologies, and the mercury remains as a solid in the slag with other solid impurities and can be disposed of safely, as outlined in figure 3.4. Combined-cycle turbines have an energy efficiency of over 60%, and the overall energy efficiency of IGCC plants is estimated at around 40%.

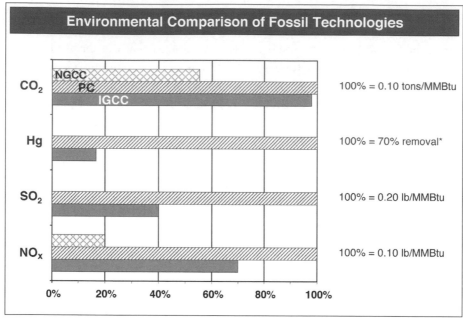

Figure 3.4: IGCC has superior environmental performance with regard to sulfur and mercury relative to PC. *Source: CRA Estimate based on industry & trade data*

The further advantage of IGCC is that the combustion products are water and a pure stream of carbon dioxide that can be captured and sequestered, if greenhouse gas controls are implemented. The disadvantage of IGCC is that the initial capital cost is higher, and existing plants have not been as reliable as conventional fluidized bed combustion. Some estimates suggest that the cost of electricity from IGCC could be 20% higher than from pulverized coal.

The Electric Power Research Institute (EPRI) has studied the costs of different technologies extensively. They have concluded that current technologies for 2010 commissioning favor coal and nuclear energy ahead of other power sources if there are no penalties for emitting CO_2, confirming the conclusion reached by John Wilder of TXU (fig. 3.5). IGCC is approximately 20% more expensive than PC. However, if there is a penalty of $50/T for emitting CO_2 then coal fired generation becomes the most expensive power source and natural gas at $6/mcf is preferred over coal. Because there is a wide spread between the costs of the different power sources, and the relationship between the sources changes substantially depending on the level of penalty for emitting CO_2, the investment decision for new power plants is particularly difficult and risky.

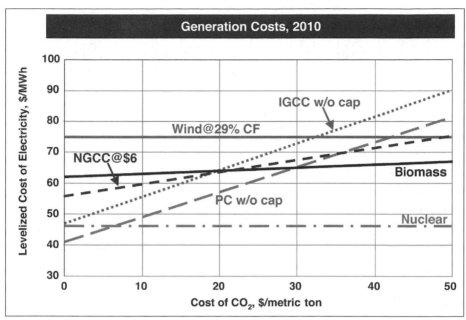

Figure 3.5: Comparative costs of generating options in 2010. *Source: Electric Power Research Association* © *EPRI*

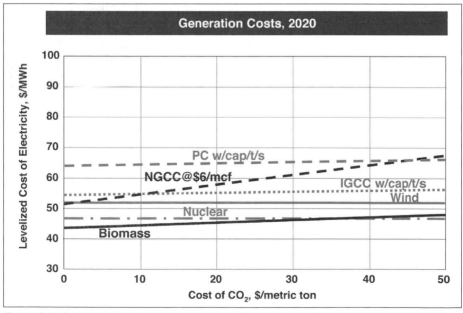

Figure 3.6: Comparative generation costs in 2020. *Source: Electric Power Research Association* © *EPRI*

EPRI further estimates the potential impact of technological advancements for power plants to be commissioned in 2020 (fig. 3.6), and concludes that new technologies will narrow the spread between the costs of power from different sources. EPRI notes that IGCC coal technology will be favored over PC if CO_2 emissions must be captured and sequestered. Further, EPRI expects wind and biomass to be competitive with fossil fuels by virtue of their lack of carbon emissions and further improvements in costs from technology advances. This could allow growth of renewable power sources to accelerate even faster.

Nuclear power plants have been particularly prone to massive cost overruns in the past, and none have been completed in the U.S. for more than 10 years, so current cost estimates do not benefit from recent experience. In EPRI's estimation, nuclear energy is both cost effective and carbon neutral, but only time will tell whether the public can overcome its fear of nuclear energy to the extent that politicians will allow permanent and safe storage of nuclear waste.

Natural Gas Gaining Supply Alternatives

As an international commodity, natural gas is some 30 years behind crude oil. Until recently, natural gas has been a regional commodity. North American gas markets have been independent of those in Europe, Algeria, and Russia, and both Atlantic markets have been independent of the Pacific market of Japan, Korea, and Taiwan (JKT), with Southeast Asian and the Middle East suppliers locked in by long-term contracts. This situation is changing rapidly, for reasons similar to those that drove oil market changes in the 1960s and 1970s when North America became a significant and rapidly growing import market. By 2015, inter-regional trade may account for 30% of global gas supplies, and it is likely that an active global spot market for liquefied natural gas (LNG) will develop over the next decade (fig. 3.7).

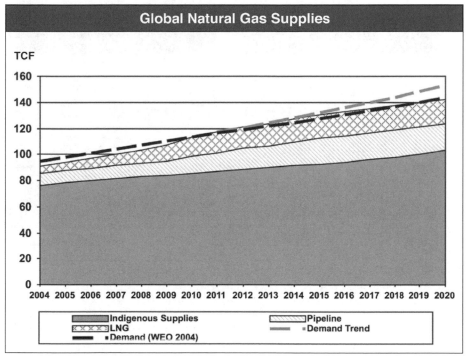

Figure 3.7: By 2015, inter-regional natural gas trade will account for 30% of supplies.
Source: CRA Global Gas Model

The natural demand trend for natural gas has shown strong growth in the U.S. and in Europe due to its environmental benefits and competitive prices, enabled by growing indigenous supplies from the Gulf of Mexico and Canada for North America and from the North Sea for Europe.

Both North American and European indigenous supplies appear to be peaking; thus, incremental demand must be met by imports (fig. 3.8).

The European and North American markets are of similar size. However, Europe is currently dependent on imports for just over 50% of its supplies, while imports to North America have only recently begun. By 2015, North American import dependence may reach about 15%, while European import dependence may reach over 70%.

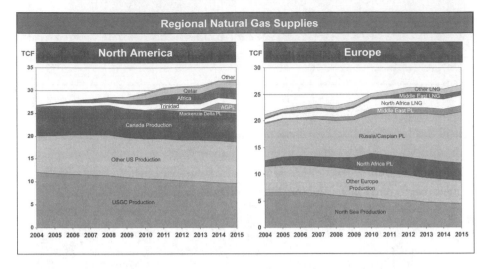

Figure 3.8: LNG and new pipelines are needed to supply European and North American natural gas demand growth. *Source: CRA Global Gas Model*

The internationalizing of European and North American gas markets raises two primary geo-political issues. First, Europe and North America will be competing for international supplies of LNG, so surging demand in one region will result in higher global prices and thereby impact the other region. And second, Europe will continue to depend on Russian supplies, an uncomfortable situation which was highlighted in early 2006 by the acrid dispute between the Russian national gas company, Gazprom, and Ukraine over gas supply and pricing. Russian-European trade was a tri-lateral bone of contention between Europe, Russia and the U.S. during the Cold War. It is now a bi-lateral issue, related to the desire of Gazprom to be a monopolist in Russia while becoming an international oil and gas giant.

Another growing geo-political issue is the emergence of China and India as potentially significant LNG importers. Both countries are heavily dependent on coal for domestic power production. Both would benefit environmentally from increasing gas imports. Both were successful in securing initial long-term, low-price import contracts with Middle East, targeted at competitiveness with coal ($3–3.50/ mcf, regasified). However, both found out that subsequent international price levels could escalate rapidly, and Chinese companies lost out to Japanese companies in securing purchases from the huge Gorgon project being developed on Barrow Island to commercialize Australian Northwest Shelf gas reserves. China and Japan are also at odds over a disputed area of the East China Sea, where China appears to be making preemptive development moves. Since Japan's motivation to expand in Asia during

World War II was in part motivated by the desire to secure access to oil resources, the coincident need of Japan and China for gas must be a cause for ongoing concern.

Steve Lowden, CEO of Suntera, believes that "dirty coal" is becoming a less feasible option for China and India as the middle class grows in each country and their populations rise up the Maslow hierarchy of needs and wants:

> I guess another very important trend is gas-to-power, particularly in the developing world. We've had gas-to-power in the developed world. We are now going to have gas-to-power in the developing world. And so that is going to drive up the value of gas in those domestic markets, even where they are competing with coal.

> Why? Because I think the coal equation has been unfairly portrayed in terms of economics, which don't take account of the environmental issues. And the developing countries are as affected by the environmental issues as the developed countries. The pollution in Delhi was horrible until it was solved by going to CNG (Compressed Natural Gas) vehicles. The anti-pollution gains being made in the developing world are much more impressive than those being made in the developed world because they are the bottom end of the curve. Twenty million people are added to the middle class in India every year. They all care about the environment. They have to go to hospital with respiratory problems and lose money when they don't go to work.

> The reason why coal works in India and China is that it is actually subsidized. It's protected and subsidized. It is not sustainable. I looked at a coal-fired power plant in India recently. It currently doesn't even match gas. Even at gas that's $5, $6 a million BTUs, it's not competitive. But it's subsidized. And is it going to be forever subsidized? No, absolutely not. The subsidy is going to come off because the Indian government is rapidly going to get to the conclusion, once they've dealt with this energy security issue, they will no longer subsidize coal. And they are going to deal with energy security because they are highly motivated right now to do so.

Lowden may overly discount the potential impact of technology in lowering the cost and improving the environmental performance of coal, but he is certainly wise in avoiding investments that are dependant on government subsidies for their economic success.

New LNG Business Model

The energy industry has jumped on the need for increased imports by investing heavily in LNG projects worldwide. In doing this, they have transformed the LNG business model. The explanation starts with understanding LNG fundamentals.

LNG is natural gas, primarily methane, which has been cooled to its liquid state at –260°F (–162.2°C). This reduces the space it occupies by more than 600 times to a volume and density that make practical to store and transport it in vessels. LNG (the liquid itself) is neither flammable nor explosive. The vapor (methane) is colorless, odorless and non-toxic, but can become an asphyxiant when it displaces the amount of oxygen that humans need for breathing. LNG vapor (methane) typically appears as a visible white cloud when it contacts air since its cold temperature causes humidity in the air to condense. Cold LNG vapor (methane) is flammable when it occurs in a 5% to 15% concentration in air; however, at higher concentrations there is too little air to provide enough oxygen to sustain a flame, while at lower concentrations the gas is too dilute for ignition. The safety record for LNG has been excellent, since LNG vapor is not explosive in an unconfined environment; and after LNG vapors become warmer than –160°F (–106.7°C), they become lighter than air and will rise and disperse rather than collect near the ground.

LNG can be delivered into the U.S. and other importing markets at competitive costs. LNG project economics are highly dependent on the capital costs, which tend to be higher in regions with large resources but the least existing infrastructure. They are also sensitive to the distance to market. Table 3.4 shows the approximate capital costs for a single 5-million-ton per year (tpy) liquefaction train, with an LNG tanker fleet and regasification terminal sized to match the liquefaction plant. Some recent trains being developed in Qatar have 7.5 million tpy capacity, and therefore had higher capital costs but lower per-unit cost than this example. Upstream capital costs for exploration, production and gas gathering will be additional, so the full supply chain for a 4–5 million tpy plant can cost from $3–5 billion depending on the project. The two liquefaction trains for Qatargas 2, each with a nominal capacity of 7.8 million tpy, which are capable of delivering approximately 0.4 TCF per year of regasified natural gas, when combined with upstream, tanker and regasification capital, may cost a total of around $14 billion.

Table 3.4: LNG netback values to liquefaction plant input

	Delivered to:					
($/mcf)	U.S. Gulf Coast			South Europe		
	Trinidad	W.Africa	M. East	N.Africa	W.Africa	M.East
Price	5.00	5.00	5.00	5.00	5.00	5.00
Regasification	(0.40)	(0.40)	(0.40)	(0.40)	(0.40)	(0.40)
LNG Tanker	(0.50)	(1.10)	(1.75)	(0.30)	(0.80)	(0.95)
Liquefaction	(1.15)	(1.40)	(1.15)	(1.25)	(1.40)	(1.15)
Upstream Value	2.95	2.10	1.70	3.05	2.40	2.50
Full Value Chain Capex ($ Bn)	2.20	3.15	4.00	2.30	2.80	3.75

As noted earlier, natural gas is competitive with clean coal in power plants at about $5/mcf. If prices reflect that competitive equilibrium, then investors in LNG supply chains will receive netbacks to their liquefaction plants of $1.50–3.00/mcf, a value that compares favorably with the price received by U.S. Gulf of Mexico producers for their gas during the 1990s. The cost of new liquefaction and regasification plants has risen due to inflation in the prices of special steels and strengthening engineering and construction margins. On the other hand, the scale of new plants has been rising such that new projects are still economically attractive.

The critical mass of the combination of LNG plants in place coupled with the new projects underway creates enough global supply to give credibility for building regasification facilities in supply-short consuming regions with some optionality beyond fixed contracts. This triggers opportunity for additional spot capacity from debottlenecking and adding on cost-effective trains to existing LNG Plants. The LNG business model will look more and more like the crude-oil-to-refinery chain as destination flexibility is built into new supply contracts and more cargoes are traded on spot markets. The new business model will allow LNG to flow to the markets with the greatest needs. The prior model typical of the 20th century of inflexible point-to-point long-term contracts with formula prices may well become obsolete. By default, the U.S. Henry Hub spot-gas prices may become the global marker just as forward Brent has become the global crude oil marker.

Renewed Nuclear Interest

Nuclear energy has been very unpopular since the accidents at Three Mile Island and Chernobyl, and unfavorable public sentiment in several countries was cemented by the movie *The China Syndrome*. Though there has been a renewal of interest since nuclear plants produce no greenhouse gases, the cost of power from new greenfield nuclear plants in the developed world is uncertain and estimated to be higher than the cost of power from pulverized coal plants.

However, that could change with new technologies and if standardized designs allow cost reduction. New-generation nuclear plants operate with more "passive" safety features that rely on gravity and natural convection. They either require no active controls or operational intervention to avoid accidents in the event of major malfunction, or at least allow considerable time for intervention. If additional regulation makes CO_2 emissions expensive, and if public disquiet over the disposal of nuclear waste can be allayed, then nuclear power could be heading for a new dawn.

One big impediment, particularly in the United States, is the lengthy regulatory process, which can consume a decade before final approval is granted by the Nuclear Regulatory Commission. Elapsed time becomes a critical capital return issue for the deregulated market place but can flip-flop into a great cost for the consumers in regulated regimes, as John Wilder describes: "Because nuclear involves a big rate base and long construction program, you get this weird incentive in the regulated business called interest in periods of construction. Regulated companies get a return on the interest that they've built up. So the dream asset class for a regulated business is a nuclear plant."

Much longer term, nuclear fusion, which combines atoms rather than splitting them, may play an important role. France has secured the site for a 10-billion euro nuclear fusion research project that, if expectations are fulfilled, would provide a cleaner, safer, and significant nuclear energy source. After completion of construction around 2015, the plan calls for operation and experimentation of the 500-megawatt ITER reactor that could go on for 20 years. At the moment, projections put the first commercial fusion plant in operation around mid-century.

Renewables: high growth, low share

Finally in the power sector, we have noted that none of the published projections show renewables gaining a substantial share of the global power market.

Although most renewable energy forms are economic or close to economic (fig. 3.9), they all are replete with problems. Wind farms attract opposition on aesthetic grounds from nearby property owners and from environmentalist trying to protect birds. Hydro draws fire from environmentalists for flooding existing ecosystems and affecting salmon spawning in the Pacific Northwest of the U.S. Commercially attractive opportunities for geothermal power are quite limited. Wind and hydro can be intermittent and depend on weather patterns. Solar technologies are not yet economically competitive for base-load uses, but can be economic for rural areas, where the cost of transmission from central power stations may be prohibitive. Nevertheless, solar and wind are projected to grow at over 10% annually in the IEA WEO 2006 Reference Case, which is a higher growth rate than any other global primary energy source. In certain regions, renewables should achieve extended double-digit growth rates, and the authors concur with EPRI that biomass has substantial potential.

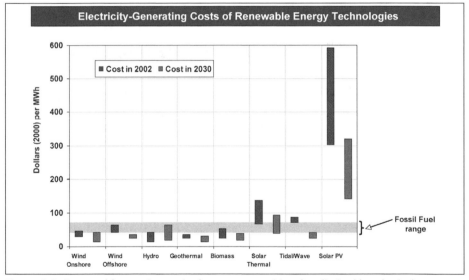

Figure 3.9: Renewables are increasingly cost competitive, but limited by physical constraints. *Source: World Energy Outlook © OECD/IEA, 2004, Figure 7.9, p. 233 modified by the authors*

Bob Lukefahr, President of BP Power Americas, is optimistic about wind power, and BP has announced aggressive plans to grow its business in this area. He points out that the potential exists for 900 GW of wind power in North America.

There are a large number of potential sites with attractive wind characteristics, and the impact on migrating birds has been overblown. The overall power system could accommodate using wind power up to 25% of dispatched electricity. In 2005, we had 9 GW of wind power installed capacity, so we have 891 GW to go! The issue is economics. Wind is economic compared to gas fired combined cycle plants at gas prices above \$7/mcf so it is quite close to economic now even without a production tax credit. If greenhouse gases were controlled such that CO_2 emissions credits, or emissions taxes, were worth \$30/ton, then wind would break even with gas at \$4/mcf or coal at \$40/ton. So if we are concerned about global warming, and more and more Americans seem to be, then wind will be economic. And then when you consider that it is a domestic resource, it becomes very attractive indeed. The other issue is transmission. The transmission system was not designed for the way the power generation supply growth seems to be going, and that needs to be addressed.

Less Choice for Transportation

Oil dominates the global transportation sector; it is the preferred fuel for ship bunkers (residual fuel), rail (diesel), road haulage (diesel), and personal vehicles (gasoline and diesel), as figure 3.10 indicates, and is expected to remain the preferred fuel.

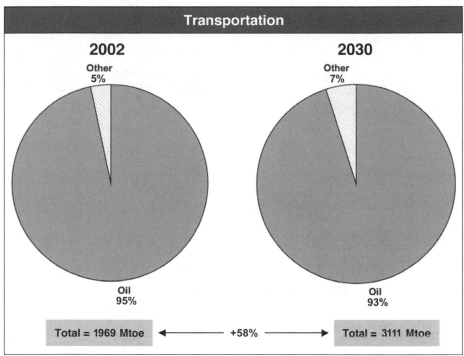

Figure 3.10: Oil has essentially no competition in the transportation sector. *Source: World Energy Outlook © OECD/IEA, 2004, Annex A pp. 430-431 as modified by the authors*

Fuel cells have received considerable attention as a potential substitute for the internal combustion engine in future power trains; however, the cost is still prohibitive, and the infrastructure required to provide the hydrogen for the fuel cells would have to be built. All public projections include only slow penetration of fuel cell vehicles over the next 25 years, but they could certainly play an important role during the following quarter century. More likely is some erosion in the market share of oil by bio-fuels and possibly "plug-in hybrids."

As has been described previously, transportation is central to a modern economy, and demand for transportation fuels rises with increasing economic activity. The IEA WEO 2004 Reference Case modeled a continuation of these trends (fig. 3.11).

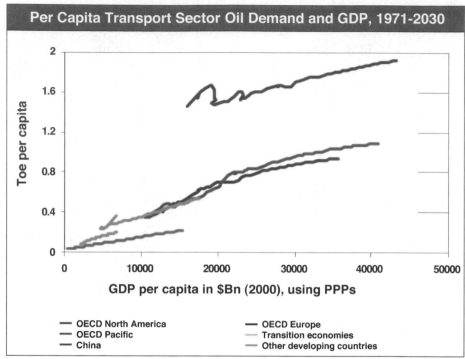

Per Capita Transport Sector Oil Demand and GDP, 1971-2030

GDP per capita in $Bn (2000), using PPPs

— OECD North America   — OECD Europe
— OECD Pacific   — Transition economies
— China   — Other developing countries

Figure 3.11: Transportation fuel demand follows prosperity. *Source: World Energy Outlook © OECD/IEA, 2004, Figure 3.6, p. 87*

It is noteworthy that North American use of transportation fuel is considerably higher per unit of GDP than other regions. This is because of the large distances between cities, the relative lack of public transportation, suburban sprawl, and personal preferences for large, heavy vehicles. All of this is underpinned by relatively low transportation fuel prices in North America, at least historically. European and Asian OECD countries tax motor fuels highly, both as a source of government revenue and as part of a policy to encourage conservation. In the U.S., proposals for raising gasoline taxes are considered political suicide, and they rarely become part of any proposed legislation even though they may be good public policy for conservation. Also noteworthy is the apparent IEA assumption that China will follow a new and different pathway from all other countries, which will result in a much lower use of transportation fuels as the economy grows than all other regions of the world. This assumption may be wishful thinking, in which case trend line oil demand will be higher than expressed in the public forecasts.

The most comprehensive publicly available study of future transportation fuels, emissions and safety was undertaken by a large team convened by the World Business Council on Sustainable Development for its Sustainable Mobility Project (SMP). This study commissioned the IEA to expand and further develop its model of the world transportation fleet, separating out the different modes of transportation by region and assessing vehicle ownership levels, utilization rates and likely energy efficiencies. Their report, "Sustainable Mobility 2030: Meeting the Challenges to Sustainability," concludes that "the present system of mobility is not sustainable, nor is it likely to become so if present trends continue. Not all the indicators point to a worsening of the situation. But enough do for the SMP to conclude that societies need to act to alter their direction. This is true, in particular, if mobility is to be made sustainable in the developing world." The final report is signed by the CEOs of General Motors, Toyota, Shell, BP, DaimlerChrysler, Ford, Honda, Michelin, Nissan, Norsk Hydro, Renault, and Volkswagen. These companies would not call for a change of direction that could require major technological and strategic shifts without very careful consideration.

Conservation still a challenge

If such a change in direction is not forthcoming, current trends will continue until market forces change them. Current trends include a tendency to add features to vehicles that increase their weight even in markets such as Europe where transport fuels are highly taxed (fig. 3.12).

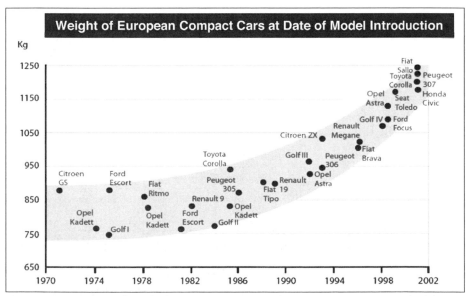

Figure 3.12: Tendency for vehicles to increase in weight. *Source: WBCSD Sustainable Mobility Report, used with permission from European Aluminum Association*

The increasing weight has reduced the gains in efficiency through improved power-train technologies. Nevertheless, the SMP report assumes that technology will continue to deliver improving overall energy efficiency in the future. Even with these improvements, the SMP forecasts that strong growth in transportation fuels demand will continue to such an extent that the result would be unsustainable.

The model developed for the SMP is available to the public. Embedded in the model are projected trends in vehicle ownership, scrappage and use, as well as expected energy efficiencies. The user is able to change assumptions on the composition of the fleet in different regions by changing assumptions about new vehicle sales in each year. The model then calculates the composition of the total fleet in each region over time, fuel use, average fuel efficiencies and emissions.

CRA International decided to test what would happen to transportation fuels demand if there were a strong move towards hybrid gasoline-electric vehicles as well as a further shift to diesel engines in North America, and compared this scenario with the SMP project reference case (fig. 3.13).

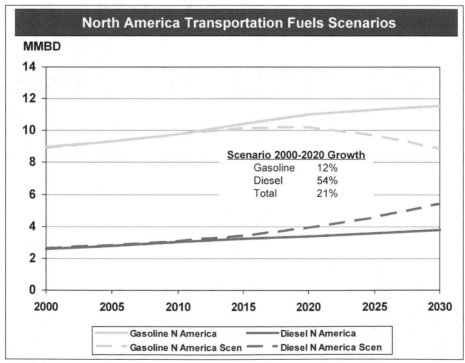

Figure 3.13: Vehicle technology will likely alter the demand barrel more than change aggregate growth. *Source: CRA International*

The study assumed that North American light-duty vehicle fleet sales would steadily change in favor of more fuel efficient vehicles such that by 2030 they would be: 35% conventional, 25% gasoline hybrid, 25% diesel, 15% diesel hybrid. CRA International did not change the assumptions for heavy-duty vehicles. The study found that the premised changes would have a significant effect on the demand barrel, increasing diesel demand and decreasing gasoline demand, but a much lesser effect on total fuel consumption. In earlier oil price spikes, there have been reductions in transportation fuel demand. However, much of the reduction was due to economic downturns that reduced the demand for freight haulage and lowered employment, and with it work-related travel, as well as discretionary travel by out-of-work individuals.

The study went on to make similar assumptions for the rest of the world (fig. 3.14).

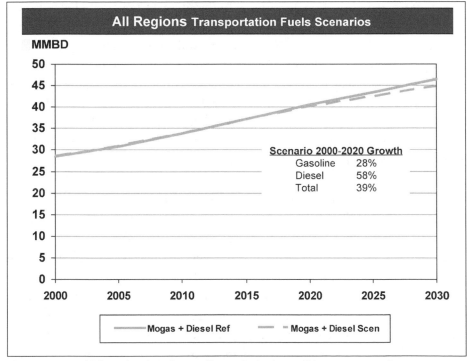

Figure 3.14: Vehicle technology has a disappointingly small impact on total transportation fuels demand. *Source: CRA International*

Because the installed base fleet is much smaller in the developing countries, the effect of increased efficiency is attenuated by the fact that there is not a large, inefficient fleet to be replaced. After adding the North American and other regional components together, the study found that the move towards greater fuel efficiency through a shift in power-train technologies had a disappointingly small impact. The study concluded that governments will need to take quite drastic action, or a lower economic growth rate will be needed to change the trend line.

Biofuels part of the solution

Measures recommended by the SMP to reduce greenhouse gas emissions include further improvement in the energy efficiency of transport vehicles; the development of hydrogen as a major transport energy carrier; the development of advanced biofuels; and planning and construction of new fuel infrastructures. To preserve and enhance mobility opportunities available to the general population, the SMP recommends consideration of increased utilization of buses for urban areas. Above all, the SMP concludes that "moving towards sustainable mobility will involve paying as much attention to institutional frameworks as to the inherent potential of any vehicle technology or the theoretical effectiveness or ineffectiveness of any particular policy or lever." This is because the policy changes that will be required to alter the fuel use growth trend will in many cases be unpopular and create winners and losers.

U.S. federal policy is moving in the direction of mandating significant use of fuels derived from biomass. The Energy Policy Act of 2005 requires 500,000 bpd of biomass by 2012, an attainable goal in light of 2005 production of 4 billion gallons, or 260,0000 bpd, and then maintaining its ratio to gasoline thereafter. A recent Oak Ridge study showed the potential for cellulose ethanol to make up 10% of transport fuels by 2020.

Brazil is a leader in this field, and produces ethanol from sugar cane at a cost that is competitive at recent oil prices. Brazil launched its ethanol program in 1973 in response to increasing international oil prices (at that time, Brazilian oil production was 170,000 bpd, one-tenth its current level), and production rose rapidly to about 200,000 bpd. However, lower oil prices in the 1980s undermined the economics of ethanol, and production fell back to 20,000 bpd by 1990. With increasing oil prices, production of anhydrous ethanol had again risen to about 130,000 bpd in the 2003-04 seasons (half the 2005 U.S. output). Flexfuel vehicles, which can run on pure ethanol, pure gasoline or any mixture of the two, accounted for 60% of Brazilian new car sales in 2005, so ethanol has become an important part of the nation's transportation fuel mix.

The European Union has chosen a tentative target of 20% biofuels by 2020, but has not yet defined any implementation plans. India and China are also developing biofuels programs. Biofuels are economic without subsidies at recent oil prices in Brazil, but until recently barely economic in the U.S. Economics would be enhanced by greenhouse gas controls, since growing the grain or sugar used as feedstock absorbs much of the CO_2 that is later emitted when the biofuel is consumed in transportation

Don Paul, Chevron's chief technology officer, thinks that bio-fuels will eventually be significant, but considers that corn-based ethanol has serious limitations:

> You can't use an expensive feedstock, and ultimately the political issues about competing corn in food supplies against corn in fuel will be a problem. But the biomass, the waste, is vastly larger. It is orders of magnitude bigger than the food stock: paper, pulp, garbage, sewage. These are enormous carbon, hydrogen, oxygen, nitrogen concentrations. But the biggest one by far is in the United States and Canada. The only other country that is in a surplus position is Brazil. That's the other thing about these unconventional things, they are even more regionalized than oil and gas. But the interesting news, they are all in the open countries.

> We get back to this old key issue: synthesize fuel instead of distill it. But it's early in the game. And it's still a cottage industry at this point. We don't even have standards for ethanol, and biodiesel is worse. It comes down to molecular manipulation. These technologies all have great potential. The question is timing. At the moment, the resources being devoted to them are inadequate for there to be any reasonable expectation of a material impact over the next fifteen years. However, a period of sustained high oil and gas prices will attract additional funding and accelerate the arrival of new solutions.

Continued Primacy of Fossil Fuels

Whenever oil demand outpaces supply capacity, causing prices to rise dramatically, the discussion inevitably includes the ultimate question: are we running out? Oil is a depletable resource, so the question is entirely fair and has been one of the underpinnings of the movement toward sustainability.

Are we running out?

A three-way debate, often acrimonious, is under way about energy supply. The participants include traditional economists allied with petroleum engineers; economists and environmentalists aligned with the late Julian Simon; and environmentalists aligned with conservationists.

Traditional economists and petroleum engineers. Traditional economists tend to relish the definition of economics as "the dismal science" and follow in the footsteps of Thomas Malthus and David Ricardo, who were both deeply concerned about the ability of the world's natural resources to keep up with the growing population. Their concern centered at the time on food production, and they reached their pessimistic conclusions by extrapolating farm yields using existing technology applied to a limited area of arable land, and comparing this with the needs of a rising population. Naturally, whenever an exponentially rising demand confronts a fixed supply, a crisis is inevitable.

However, the assumptions that both Malthus and Ricardo used were wrong, and the world did not deliver the reality they predicted. Instead, farms were mechanized, new lands were opened, and fertilizers, pesticides, and genetic engineering improved yields so much that countries traditionally subject to famine, such as India, became net exporters of grains.

But the Malthusian math exerts a strange attraction on researchers, and in 1970 the Club of Rome foretold the end of civilization as we know it, as exponential demand growth for minerals, including oil, collided with fixed resources. In this case, the math was much more complex, using Jay Forrester's systems dynamics approach, but the results still failed to describe the future accurately. Petroleum was discovered in the North Sea, in Alaska, in new parts of Africa, and offshore Brazil, and demand shrank dramatically as oil priced itself out of the power generation market. However, we should recognize that though civilization as we know it did not end, a significant discontinuity did occur involving where oil came from and how it was used.

And the fight continues. There is no urgent concern (yet) about the available resources of coal or natural gas (fig. 3.15); but concern over oil reserves is growing, with geologists and petroleum engineers taking over from economists as the prophets of doom. Proven reserves of oil are only sufficient to supply 40 years of demand at 2004 levels. By contrast, proven gas reserves will last for 65 years and coal reserves for 125 years at 2004 production levels.

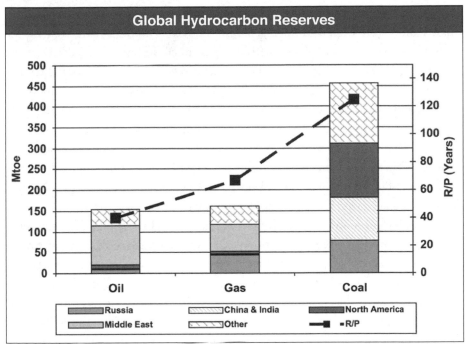

Figure 3.15: The reserve life of oil is much less than gas and coal. *Source: CRA, Data from BP Statistical Review of the World Oil Industry*

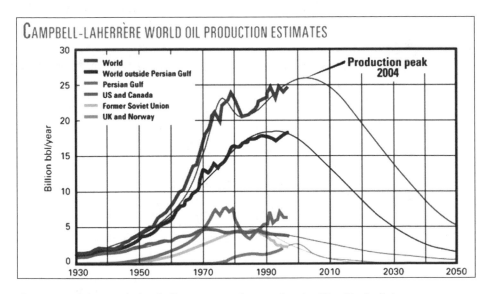

Figure 3.16: The pessimists believe we are close to "peak oil" with declining production imminent. *Source: U.S. Energy Information Administration*

Several authors have rediscovered "Hubbert's Peak" (Fig. 3.16), named after a famous and highly respected Shell geologist, M. King Hubbert. In 1956, he quite accurately described the peaking in 1970 and eventual decline of U.S. Lower 48 oil production. Others now apply the same approach to global oil reserves estimates and conclude that global oil production will have peaked some time between the years of 2000 and 2010. Kenneth Deffeyes was a technologist in Shell Oil working for Hubbert and is now professor emeritus at Princeton University. The title of his book, *Hubbert's Peak—The Impending World Oil Shortage*, provides a strong clue about his position. Other proponents of this point of view include Colin Campbell, former chief geologist of Amoco. The outlook for oil production, as seen by the pessimists, is pictured as a bell curve with a decline starting in this decade.

Julian Simon economists and environmentalists. However, there is a sub-set of economists who subscribe to the position established most famously by Julian Simon in his book, *The Ultimate Resource*. Simon's thesis is that there is never a shortage of resources, only a temporary shortage of human imagination. Given a clear economic signal, technology will rise to the occasion and deliver what is needed. One result of this process is that the price of resources will continue the declining trend that has characterized most commodities for most of the time since the early 19th century.

This position exasperates the geologists and petroleum engineers, who with some justification want to know where this oil is going to come from. They understand only too well the petroleum systems of the world, the biology and physics of deposition and migration, and the difficulty of discovering and producing oil resources. Nevertheless, history tends to favor the Simon advocates, and resources may again not be the problem.

ExxonMobil describes the situation somewhat artfully, but in the end reasonably, in their outlook (fig. 3.17).

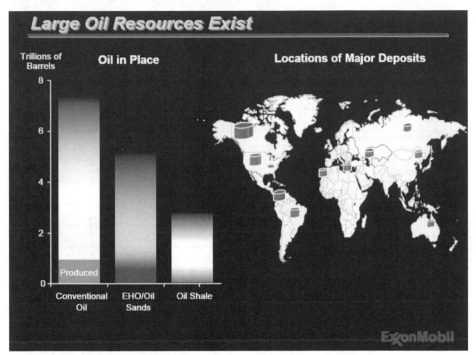

Figure 3.17: Oil reserves can be enhanced with higher recovery and unconventional sources—with time, technology, and capital

ExxonMobil's point is that conventional reserves can be augmented by increased recovery rates, while vast resources trapped in oil sands and shales can be liberated with new technology. However, ExxonMobil is silent on the price that will be required to trigger development and implementation of these technologies.

Environmentalists and conservationists. The third axis of our triangle is populated by environmentalists who much prefer that we reduce our consumption of fossil fuels, whether the resources are there or not. Indeed, it is not just environmentalists who think that way. One oil company executive told us that he believes that "we will run out of the planet's carrying capacity for greenhouse gases before we run out of hydrocarbon resources."

So we have one group of people convinced that we will imminently run out of oil; another that says resource limits are irrelevant, given time technology and capital; and a third that says we should reduce demand, not increase supply. "We live in interesting times," as the Chinese curse goes.

Where is the oil coming from?

Let's return to the public energy supply and demand projections that we reviewed earlier and examine their oil supply projections (table 3.5).

Table 3.5: Oil supplies (MBD)

	2005	2030 Projections			
	WEO	WEO2004	WEO2006	EIA2006	Exxon 12/05
OECD countries	18.9	12.7	14.1	20.7	
Transition economies	11.9	15.9	17.0	19.1	
Developing countries	16.0	14.8	19.2	21.4	
OPEC crude oil	29.1	64.8*	45.7	45.3*	47
OPEC NGLs	4.3		9.0		
Non-conventional	1.6	10.1	8.9	11.5	
Total world	81.7	118.3	113.8	118.0	115

* Includes OPEC NGLs

The IEA lowered its oil demand projection from 2004 to 2006 due to higher assumed prices, which also resulted in a slight increase in non-OPEC production. EIA has also raised non-OPEC production as a result of higher price assumptions. Overall, the IEA, EIA, and ExxonMobil project OPEC crude oil output of around 45 million bpd, a substantial reduction from projections made in 2004. History would suggest otherwise. When OPEC crude oil production reached 30 million bpd, in 1973, again in 1979, and once again in 2004, oil prices rose sharply. This led in due course to an increase in non-OPEC supply and a reduction in demand. OPEC production did not increase.

Why should we expect it to be different this time? The IEA WEO 2005 and WEO 2006 developed a "deferred investment scenario," in which the Middle East and North Africa (MENA) countries decide to limit their investment in oil production capacity to the same share of GDP as their average for the past decade. The deferred investment scenario results in significantly lower production from MENA and, therefore, from OPEC (table 3.6), though projected supply is some 10 million bpd, higher than in 2005.

Table 3.6: IEA WEO supply OPEC production scenarios (MBD)

	2005	2030 WEO 2005 Projections	
	WEO	Reference case	Deferred investment (est.)
Iran	3.9	5.2	
Iraq		6.0	
Kuwait		4.0	
Qatar		0.5	
Saudi Arabia	9.1	14.6	
UAE	2.5	3.8	
Algeria	1.3	0.7	
Libya	1.6	2.7	
Other OPEC	6.0	8.2	
Total OPEC crude oil	29.1	45.7	
OPEC NGL and Non-Con	4.5	10.5	
Total OPEC	33.6	56.2	45.0 (Est)
Average price (2005 $/B	~$50/B	~$55/B	~$75/B

Although revenues to OPEC would be lower as a result of the lower production, OPEC countries would have saved the investment required for growth and, at the end of the period, MENA countries would still have about 65 billion barrels in the ground that they could produce later. Also, their aggregate proven reserves to production ratio, which was 79 years at the end of 2004, would still be 37 years at the end of 2030 in the deferred investment scenario, even if no further reserves were added. By contrast, the R/P ratio would have fallen to 22 years in the reference case if no further reserves were added, which may not be a politically desirable outcome from the point of view of future generations.

There is also an issue of the validity of the reserves stated by OPEC members. Though there is no simple formula for determining production quotas within OPEC, various factors are considered. These include historic production rates, population, and reserves, and there has been a tendency for countries to claim high reserves to bolster their negotiating position with respect to quota rights. The issue of transparency in reserve statements has been taken up by Matthew Simmons, who also has serious misgivings about the future production potential of Saudi Arabia,[12] based upon an intensive review of petroleum engineering papers published by Saudi Aramco technical staff. Even if the major producing countries do in fact have even larger reserves than currently estimated, it would seem that Saudi Arabia's economic interest would lie with producing more slowly for a longer time rather than producing quickly for a shorter time.

What is in the best interest of producer countries?

And there lies the important strategic question. A commercial oil company with a project internal rate return of 15% and using a flat oil price assumption will almost always find that the best economic scenario is to produce the resource as fast as possible. However, a government may perform a different calculation. In a time of high prices when oil revenues exceed its immediate revenue needs, the value of incremental revenues is the returns on its incremental investments in, for example, U.S. treasury bills. They might, therefore, use a low discount rate to value the surplus revenues. Further, they may take the view that oil prices will be higher later than they are today. If they assume that oil prices will inflate faster than the rate on a U.S. T-bill, then they will be better advised economically to leave the oil in the ground and produce it later at a higher price.

As indicated in the discussion of oil price formation in chapter 1, whenever producer governments with national oil companies are running a financial deficit, the NOC's cost of capital can be extremely high because it has to compete with vital social programs for funding for its capital programs. In this environment, there is a strong tendency to maximize short-term production and even to break the NOC's monopoly by inviting foreign capital into the oil sector. This can result in tension around OPEC quota adherence and sometimes to a price war, as in 1998. Thus, there is a tendency to conserve when prices are high and perceived to be rising, but to produce at the maximum when prices are low.

There is also a popular saying in producing country circles, attributed by some to Robert Mabro of the Oxford Institute for Energy Studies: "The Stone Age did not end for lack of stones." The implication is that the Oil Age may end because of changes in technology, such as greatly improved hydrogen or electric powered vehicles, rather than because of shortages of supply. In this case, oil left in the ground after, say, 2030 may well have lower value than it does today, and the sensible course would be to produce it sooner rather than later. Also, informal conversations with senior Kuwaiti executives clearly indicate that they believe Western political support for the continued existence of their country requires them to be more expansive in their plans than they might otherwise be. So there are countervailing pressures that guide resource rich country production policy. Still, it is important to recognize that production levels will be set by policy and not left to chance.

Abdallah S. Jum'ah, CEO of Saudi Aramco, described for us in our interview the process by which his company makes investment decisions in production capacity, which he characterized as decision making amid great uncertainty:

> Our planning follows a five-year cycle. And the first thing we look at is the business environment in the whole world. We are one of the biggest world players, and therefore we want to see what is happening around us in terms of the global economy. But it's not five years for all our activities. When we look at the increment, we look at longer terms, like a 10-year horizon. But the first thing is to look at the business environment and what are the effects of supply and demand. We concentrate on the demand side of the business. Basically it is really moved by the economy. We look at non-OPEC supply. We look at OPEC supply. We look at the other sources of energy that are also taking share: gas, coal, nuclear, renewables, and so on. And out of this we formulate a position as to what is needed.

That covers the quantitative analysis, but he also is very conscious of the huge uncertainties:

> But that's not the end of it because it's not a straight line. You have uncertainties on the supply side; and you have uncertainties on the demand side. On the demand side, is business going to be as usual, is business going to change, is the economy is going to collapse? You really don't know. The supply side uncertainties, of course, are also many. You have the degree of maturity in OPEC, the degree to which what is said about what will be built actually happens. What is happening in non-conventional resources: whether the production is going to be higher or lower?
>
> Geopolitics, of course, creates a lot of uncertainty; 10 years ago, five years ago, three years ago, we said Iraq would be producing three million barrels per day. Today they are producing one million per day. Looking into the future today, we think that Iraq isn't going to add more than 100,000 barrels per day for the coming five years. So these are uncertainties. And out of this messy situation, we look at a reasonable capital program.

The quantitative analysis is married with an assessment of the uncertainties and government policy to develop a capital expenditure plan:

> When we plan, we keep between one-and-a-half and two million barrels of extra capacity over what we think is the call on

us. So by 2009, when we have 12 million barrels per day capacity, the call on us is going to be between 9.5 maybe 10. The 15 million barrels per day capacity is a scenario for "just in case." But we don't see today when that call on us is going to materialize.

We have seen forecasts of higher demand for OPEC and Saudi Arabian oil. We have numbers from 10–20 million barrels per day. But, these numbers are soft numbers. Nobody knows. We could well be producing only eight million barrels.

So the producers have a dilemma that cannot be resolved easily by quantitative analysis. On balance, it seems likely that the very large resource owners—Saudi Arabia, Iran, Iraq, Kuwait, and the UAE—will expand cautiously, more in line with the deferred investment scenario than the reference case. Even this could be on the high side since Iran is still at loggerheads with the U.S. and increasingly with the EU over its nuclear ambitions and support of terrorism. The recovery of Iraq's oil sector is highly uncertain. Further, there is little precedent for these countries to increase production capacity significantly; only Saudi Arabia and the UAE are producing above their historical maximums. Kuwait's maximum annual rate was 3.3 million bpd in 1972, Iran's was 6.1 million bpd in 1974, and Iraq's was 3.9 million bpd in 1979; current production in each country is lower by virtue of geology, policy, or politics. Seen in that context, even the major producer projections of the deferred investment scenario seem optimistic.

Balancing Oil Supply and Demand

So how will supply and demand come into balance? Economists tell us that price is the answer. In fact, the fundamental question for consumers is the price of energy, not whether conventional oil production is peaking. As ExxonMobil's analysis suggests, supplies could be further augmented by harnessing lower quality resources such as oil sands, by expanding gas-to-liquids (GTL) capacity at a faster rate than assumed in the base projections, by rediscovering an old technology for converting coal to liquids, and by exploiting shale oil deposits. With the exception of shale oil, which is so far proving difficult to produce in an environmentally acceptable way, the technologies for developing these resources are well established. CRA International estimates that synthetic oil can be produced from natural gas at a cost of $35-45/B, and from coal at a similar cost, depending on whether there is a penalty for emitting CO_2. Canadian oil sands were growing profitably when oil was priced at $20/B, though local cost inflation has driven up the cost. So in principle, there should be abundant supplies in the long term at 2005 prices. Demand, however, is less elastic. In the transportation sector, much of the consumer fuel price is sales taxes, which is why gasoline can cost $6 per gallon in parts of Europe and only half that in the United States. However, as noted, demand for oil used for heat and power may be more elastic.

Quite simply, if the world is provided with less oil supply than expected from major producers, then energy prices will be higher than expected. The higher price should induce companies to invest in new resources, and will encourage consumers to conserve and replace oil with other fuels where possible. Because of the nature of commodity cycles, the transition will not be smooth. In the short term, prices can increase to very high levels since consumers have few choices to reduce consumption and it takes time to bring new supplies on stream. In prior cycles, producers have eventually over-invested to bring new supplies on stream, just as consumers have made the tough decisions to change behaviors and replace fuel-inefficient vehicles and plants with more efficient equipment. So boom turns to bust. This is the specter that haunts energy company leaders, as we have noted previously. It is always dangerous to say, "It's different this time." But in fact, this time it may be.

Moreover, as Abdallah S. Jum'ah, CEO of Aramco, cautions, the global economy can suffer a major downtown, which is what happened during the 1970's up-cycle. Stagflation became a challenge for economic policy makers. While the relationship of energy prices to GDP is less direct today, commodity prices are emitting warning signals that could trigger flashbacks to the early 1980s. Clearly, the journey ahead will be rough as volatile energy prices deliver their jolts. These minor cycles should not distract us, however, from looking at the overall direction and broader trend lines.

Supply's response sluggish

Four problems are slowing the supply response to high prices. First, Western oil companies are denied economic access to resources, particularly in those areas controlled by the NOCs. Second, energy leaders are exhibiting an excessive caution that inhibits them from making necessary investments, although this is changing. Third, there is a critical near-term shortage of the capabilities required to undertake multiple international major projects simultaneously. And fourth, the time lag between identifying an opportunity and completing a major project can range from seven to 20 years depending on the technical, economic and political complexity.

The industry has been caught short by the sudden disappearance of spare capacity at every step along the oil and gas supply chain, from exploration through development and production to refining and product transport. Companies have been downsizing and "capturing synergies," a euphemism for lay-offs following mergers, for

20 years since the price collapse of 1986, and are now for the most part ill-prepared for a new, robust growth cycle. We come back to this theme in chapters 5 and 6.

So supply may be constrained not by lack of conventional physical resources, but more by a dearth of human resources and the need to shift to different physical resources for several years. During this period, the world must look for a demand solution. It is arguable that oil prices have not gone up by much (yet), when compared with spending power, or when adjusted for the decline in value of the U.S. dollar. This argument would draw an analogy to the first international oil crisis, when prices rose by a factor of about five between 1972 and 1974. Prices then stabilized until 1979, when they nearly tripled in dollars of the day (fig. 3.18).

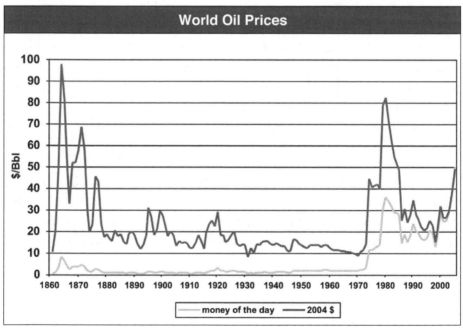

Figure 3.18: Have oil prices risen high enough yet? *Source: Data from BP Statistical Review of the World Oil Industry*

In 2006, prices made a strong move to return to the real levels of 1973, but they remain short of 1980 levels.

Demand response conditioned by prosperity

Economists expect rising prices to drive down demand through the operation of four distinct effects:[13]

1. An income effect
2. An efficiency effect
3. A fuel-switching effect
4. A longer term capital replacement effect

The income effect suggests that when people have to pay more for a basic commodity like energy, then they will spend less on something else and economies will slow down. Economies in the developed world were severely impacted by the 1973 price rise, but have remained robust in 2005 and 2006. Explanations for this could include the fact that OECD per capita income has approximately doubled over the same period so that consumers can afford to pay more for oil; that central banks have inflation under control; and that China's rapid economic growth is providing countervailing deflationary pressure to the rest of the global economy, and that is balancing the inflation of energy costs. Whatever the causes, the global economy does not seem to be slowing down at these higher prices.

The efficiency effect is the result in short-term changes in behavior. People may change their thermostat settings, or choose to shift travel to their most efficient vehicle; companies will reinvigorate their energy efficiency programs, sealing leaks and advancing maintenance and replacement of inefficient machinery. This may be happening, but it is not visible at the time of writing. Fuel switching and long-term capital replacement tend to require major capital expenditure, and thus only happen when the incentives are very compelling.

The last up-cycle taught us lessons about the fuel-switching effect. In 1973, at the beginning of the last oil price shock, world demand (excluding the former Soviet Union) for gasoline and distillate fuels stood at 27.3 million bpd, while fuel oil stood at 14 million bpd. Ten years later, after two major increases in crude oil prices, gasoline and distillate demand had grown to 31 million bpd, while fuel oil demand had declined to 10 million bpd, resulting in no change in overall demand for petroleum products over the period.

Three decades later in 2003, worldwide demand for gasoline and distillate had almost doubled to stand at just over 50 million bpd, and fuel oil actually declined further to 9 million bpd (fig. 3.19).

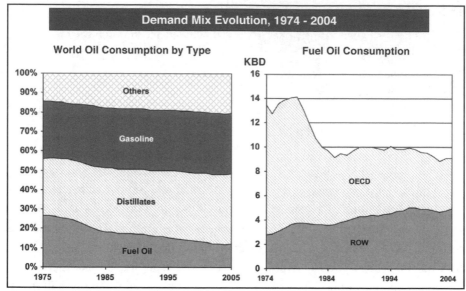

Figure 3.19: Residual fuel oil demand was displaced from 1974–2004. *Source: Data from BP Statistical Review of the World Oil Industry*

However, while OECD use of residual fuel oil continued to decline, this was not the case for the rest of the world. The non-OECD focus now needs to shift toward oil alternatives, particularly for power generation, since oil cannot be economic in that end-use sector at today's prices.

What is evident from history is that the demand for petroleum used for transportation is highly inelastic because there are no viable alternatives (yet), while the opposite is true for hydrocarbons used to generate heat and power. It is a common misconception that government policies directed toward lessening gasoline and diesel use were instrumental in flattening demand after the first oil shock in 1973. In fact, the loss of power plant demand and conversion of heavy fuel oil to transport fuels were the major drivers, as heavy fuel oil could not compete in the power sector with other less expensive alternatives.

There is still room for some reduction in residual fuel oil demand in response to higher oil prices. Expansion of natural gas use in Asia will help, but substitution will take time. There is also room for electrification and natural gas reticulation systems to displace LPG and kerosene use by domestic and commercial users. This again will take time. Switching energy sources requires very large investments, which must be based on considerable confidence that the economics supporting the decision will be robust. If there is a suspicion that the oil price increase may be transitory, then the decisions may require government support, which in turn requires political will.

Lead times to bring on additional energy supplies involve five- to 10-year time horizons and more. The demand side replacement cycle also has a built in lag. People don't just buy a car today and scrap it the next. In Houston, for example, where large pick-up trucks and SUVs roam the freeways in herds, empirical evidence indicates that prices are not yet high enough to cause significant change in behaviors.

A Different Future?

We believe that the technologies exist to resolve this resource dilemma. Indeed, they have existed for decades, but the economic motivators were not present. Higher energy prices will be providing the incentive, coupled with continuing pressures for environmental improvements, particularly with respect to global warming. New, more advanced technologies will emerge with time that will make existing technologies obsolete. In the meantime, the existing technologies could establish a new era of convergence between oil, gas, and coal in the quest to meet the world's needs for clean, sustainable mobility and power. Biotechnology should contribute much more suitable feedstocks for ethanol and biodiesel than corn, sugar, or rape-seed oil, further reducing requirements for crude oil.

Technologies drive demand reduction

Most importantly, there is tremendous potential for demand reduction. Steve Shapiro, former CFO of Burlington Resources, put it this way: "The only major new source of supply is conservation." Solving the transportation sector's supply/demand equation looms as the most significant problem with the current reliance on crude oil. Here, the primary avenue is slowing demand growth through improved efficiency. The principal technologies in this sector will be weight reduction through new materials and smaller vehicles, adoption of hybrid gasoline/electric engines, further utilization of diesel, and incremental improvements in conventional internal combustion engines.

In the power sector it comes down to customer choice and technology. John Wilder, Chairman and CEO of TXU, is quite excited about the potential in electricity markets: "In our business, the customers have never had the opportunity to express what they really want because our government has never given them a chance. Now that some governments, like in the Texas market, have started allowing a choice, we are out testing what the customers really want." Wilder explains the process TXU has gone through: "We've had focus group after focus group after focus group. And what's amazing to the evolution of that process is that three years ago the customers couldn't even describe what they wanted. They just kind of wanted it to be there, and cheap. Now they are becoming much more knowledgeable of their options and preferences.

Wilder points out that new technology enables variable pricing:

> Now what if I came to you and said, 'Give me the right to turn your thermostat up three degrees. And give me the right to do that 20 times a year. And if you do that, I'll reduce the price of your product 15%.' And you might say, 'Gee that could be kind of interesting.' The waste reduction potential is a result of the proper price signal to the customer, the right tools to the customer. And the beautiful thing about it, with the convergence of information technology and communications technology, you can almost make it painless. The point is that it is very expensive to provide every customer with full service at all times; but unlike other markets, there is no signal that relates the cost of the service on the margin to the customer. All customers pay for the expensive reserve power, and that hurts the less advantaged customers most. We have 20 gigawatts of production capacity, and over half of our production capacity runs less than 5% of the time. So there is huge capital investment in the entire electric network in the U.S., waiting there as safety stock; the government's social contract is that every customer needs all the call options. And those call options are enormously expensive for the physical option writer, and everyone has to pay for it in the socialist process. But for our more disadvantaged customers, we'll be able to find tools over time that will allow us to get resources to them that are a tenth of the cost that they pay today.

Supply side technologies already exist

On the supply side, the technologies and resources already exist to meet demand for many decades out into the future, even as the world's population swells and more and more people move from poverty into the global economy. The solution will, however, require change, and it is most important to allow markets to function in this process so that the best technological solutions will rise to the top. It is not presently possible to foresee which solutions will be the winners, and in the next decade, we expect convergence of multiple fuels providing clean power and clean mobility. The linking technologies for fossil fuels are gasification and Fischer Tropsch synthesis (fig. 3.20), enhanced by sequestration of carbon dioxide (see the next chapter for a discussion of greenhouse gas concerns).

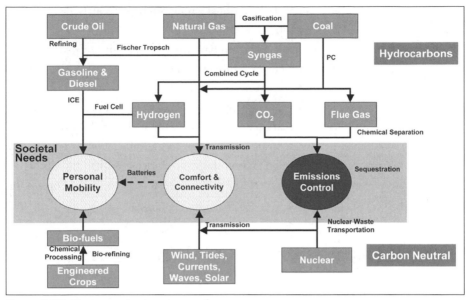

Figure 3.20: There may be a new era of convergence among energy sources

Fischer Tropsch technology was developed in Germany prior to World War II and was used to convert coal into gasoline in support of the German military in the war. It was later taken up by Sasol in South Africa during the nation's years of isolation, when its access to international crude oil supplies was severely constrained. Using these technologies, natural gas and coal can be converted to synthesis gas, the components of which can then be combined to form long paraffinic hydrocarbon chains, or wax. These chains can then be hydro-cracked to produce naphtha, a gasoline feedstock, low sulfur diesel fuel, and lubricants.

Synthesis gas is also a good source of hydrogen, which can be used in fuel cells to provide clean mobility, or as fuel for combined cycle gas turbines. In both cases, the only by-product is water. Further, the CO_2 produced in the synthesis process can be captured and sequestered, as explained in chapter 4.

The issue, therefore, is not technology, but economics and politics. Legislated market inefficiencies have been erected at all stages of the energy value chain designed to protect incumbents. These should be relaxed to encourage better technology solutions.

Overall, the world will face an oil supply challenge for at least the next decade, threatening the availability of transportation fuels. Natural gas and coal to meet future power demand are abundant. Markets responded to the crises of the 1970s and 1980s by eliminating the use of oil in power generation. The response to the current situation may well be to extend the same play, by transforming fuels previously destined for power generation into transportation use.

We have touched on renewable fuels and have noted that many are close to being economic. Have the public forecasters underestimated the potential for change? Possibly, although the intermittent nature or wind and solar limits their utility until a sufficient energy storage medium evolves. However once the initial capital outlay is made, maintenance costs are marginal compared to the high variable cost of fossil fuels. The incremental supply of an installed base of renewables could significantly impact energy prices; as we pointed out in chapter 1, prices escalate rapidly when supply capacity is limited; but the marginal addition of renewables could create the necessary cushion to buffer prices.

Will Disruptive Technologies Evolve?

In President Bush's 2003 State of the Union Address, he announced "a $1.2 billion Hydrogen Fuel Initiative to reverse America's growing dependence on foreign oil by developing the technology needed for commercially viable hydrogen-powered fuel cells—a way to power cars, trucks, homes and businesses that produces no pollution and no greenhouse gases." A controversy soon arose about whether the initiative represents a realistic vision for the future. Certainly over the next quarter century, a hydrogen-based economy seems very unlikely to develop. Today, natural gas offers the most cost-effective means of producing hydrogen. In the longer term, however, large-scale renewable ventures could produce hydrogen through electrolysis, with the hydrogen serving as an energy storage medium. If nuclear fusion became commercially successful, it could power electrolysis to produce hydrogen. With a public mandate for significant investment, particularly in infrastructure, a new paradigm could emerge. In Las Vegas, however, it would be called a wild card. If it were to happen, this would be a massive discontinuity altering many underlying energy business models.

There are other emerging technologies that have the potential to change the whole energy equation in equally disruptive ways. Biotechnology could develop plants that can more easily and more economically be transformed into liquid fuels for transportation, and into feedstocks for gasification and subsequent combined-cycle power generation. This could be a very attractive solution for energy consuming countries because a home-grown fuel source would give them greater control over their energy security. Biotechnology could also develop neural networks with lower energy requirements for computational power that would allow Moore's Law to continue operating indefinitely.

Smart highways could organize traffic in ways that reduce traffic jams and increase fuel efficiency. The future electricity grid could become a super highway for transporting energy. Superconductive wires and armchair quantum wires (carbon nanotubes), both currently under development, could substantially reduce long-distance transmission losses, enhancing electricity as a preferred choice of energy transportation. Beyond transmission, nanotechnology may have a pervasive impact in miniaturizing devices that might reduce energy demand in some sectors by similar orders of magnitude to those associated with the semiconductor/diode improvement. The nanotechnology revolution has just begun.

Strategic Implications

The accelerating speed of change and uncertainty about its direction carry significant risk for businesses that are locked into historical paradigms. Pricing economics from oil supply and demand trends, if continued, will bring on a new energy supply mix and consumption patterns sooner than the oil industry is anticipating. Public projections are misleading in suggesting that supply and demand will balance at moderate prices, when it is not in the economic interest of major producing countries to increase output by the large increments needed to match demand trends at low prices.

Up-cycles always initiate transitions as higher prices induce energy source substitutions and provide incentives for new and enhanced technologies affecting both supply and demand. Transitions create strategic opportunities for those companies willing to look beyond their current business models. However, the authors believe that the current situation is more than an up-cycle; it signals a "phase-change" during which the global energy complex will be retooled. The reason is quite simply that the transportation system and supporting energy supply system that has been constructed to satisfy mobility needs in the OECD countries cannot be simply scaled up to meet the needs of China and India, let alone the rest of the world. Mobility needs must be met using more efficient forms of transportation on the demand side and a broader array of fuel sources on the supply side, along with a more complex network to allow supply and demand to meet.

Demand, and therefore energy price pacesetting, is shifting from the United States to the developing countries, particularly China and India. This will create new competitive rivalries and alliances. Many energy executives focus their public statements on the difficulty they are experiencing in gaining access to conventional oil resources in traditional places. The implication is that they wish to continue to pursue the same sort of opportunities in the same ways as in the past. It is our belief that this approach will encounter strong head winds and will not lead to success.

So the trade winds favor new value propositions from IOCs and INOCs for accessing conventional oil resources. The trade winds are also blowing towards the transformation of stranded gas and solids to liquid fuels, the incorporation of biofuels in blends and the adaptation of refining systems to accommodate them. And finally, there are signs for systemic energy integration by extending the range of feedstocks and product mix so that refineries become energy transformation complexes providing a full range of transportation and power products with rapid deployment of emerging technologies.

Producing country NOCs must determine whether this phase change will create a market response even more profound than the 1973 to 1985 up-cycle, dethroning oil as the strategic energy source and marginalizing longer-term oil prices. But the more interesting opportunities for them may well be the many industrial development niches in specific markets and technologies that could leverage their resource positions and create significant value for their government shareholders and society

1   T.S. Eliot. *Four Quartets: Little Gidding.* Quoted by Frank McEachran, 1992. Somerset: Greenbank Press, p. 268.
2   World Bank. Policy, Research Working Paper No. WPS 3341. Chen and Ravallion, 2004. *How have the world's poor fared since the early 1980s.*
3   Jamal Saghir, 2005. World Bank. Energy Working Note No 4. *Energy and Poverty: Myths, Links and Policy Issues*
4   United Nations, 2005. *The energy challenge for achieving the millennium development goals.*
5   World Business Council for Sustainable Development Sustainability Project, 2004. *Mobility 2030: meeting the challenges to sustainability.*
6   IEA World Energy Outlook 2004 (WEO, 2004).
7   World Energy Outlook 2005, International Energy Agency, OECD/EIA, 2006.
8   World Energy Outlook 2004, International Energy Agency, OECD/EIA, 2004.
9   International Energy Outlook 2006, June 2006. Energy Information Agency, U.S. Department of Energy.
10 ExxonMobil, The Outlook for Energy—A View to 2030, December 2005. Estimates from published charts.
11 Crude oil stated as 50 MBD plus estimated NGLs of 5 MBD; total estimated from published graphic.

12 Matthew Simmons, 2006. *Twilight in the Desert*. Hoboken: John Wiley and Sons.

13 John Mitchell, 2001. *The New Economy of Oil*. London: Royal Institute of International Affairs and Earthscan Publications, Ltd., p. 25.

14 http://www.eere.energy.gov/hydrogenandfuelcells/presidents_initiative.html, p. 1.

4 Society's Evolving Expectation

Setting the Stage

This chapter explores the evolving relationship between energy companies and society. It addresses the world of geopolitics, the impact of civil society, the industry's negative image, and a new mindset for energy leaders. We learned in chapter 1 that society is made up of individuals with distinct sets of needs and wants that evolve over time. Maslow's hierarchy of needs provided a useful framework to see the role that energy plays in this hierarchy, satisfying lower level needs around basic standards of living, such as the comforts at home, mobility, less toiling jobs, and so forth. American and European societies have advanced well into the ego and occasionally onto the self-actualization levels of Maslow's hierarchy. This ego level is illustrated by the term "me generation." Consumer expectations have risen on every dimension, including time, quality, and price; there is a growing expectation of personalization. Self-actualization needs are elevating the environmental debate from tangible issues of local air and water quality to more abstract issues such as protecting the planet. Large developing countries like China and India are climbing this hierarchy at a faster and more efficient pace than their predecessors. The plight of less-developed countries, mired at the lower level needs for survival and economic security, presents a difficult moral challenge to the global community.

We also saw in chapter 1 that energy is consumed indirectly by the customer; it is used through some form of machinery. It is the automobile, airplane, or train that actually transports a person. It is the stove, dishwasher, dryer, heater, cell phone, or air-conditioning unit that consumes energy to serve our domestic needs; the fuel source that powers these appliances is several stages removed from the actual consumer need in advanced economies. People can grow emotionally attached to the physical unit, such as their automobile, but the type of energy used to propel the car has far less significance. The availability of sufficient quantities of energy, such as gasoline, is taken for granted. But when energy or a fuel becomes unavailable or unaffordable, people quickly take notice, grow dissatisfied, and demand rectification. Energy consumers have little bargaining power as individuals; only by acting collectively through government can consumers hope to affect energy availability and prices. When prices rise or availability is curtailed, consumers make their displeasure known, and their representatives feel obliged to do something, even when they know that it may be best to let the market sort out the imbalances that have caused the dislocation.

In the larger societal context, people combine in groups with shared interests, form institutions and start bargaining with and confronting other institutions to further their interests. Examples of institutional confrontation are: government versus business, labor versus management, consumers versus producers, environmentalists versus industrialists, and so forth. These group alignments representing their constituencies are often referred to as stakeholders, and over time they often develop a public personality. In aggregate they have become known as *civil society*. As American and European countries prospered, the shift towards social needs gave rise to issues beyond the workplace into the communities where people worked. Consumer advocates emerged who wanted to hold companies accountable for the safety and performance of the products they sold. Environmental and safety concerns brought forth a new stakeholder to represent the individual's social needs on these issues. Human rights activists sought to extend the advances made in developed countries to newly emerging economies. In each case, a wide array of non-governmental organizations (NGOs) emerged to represent individuals' interests. These powerful new stakeholders argue for a triple bottom line to measure an energy company's overall performance: economic, environmental, and societal.

The largest aggregations of people with shared interests are nations. We have learned from chapter 1 the inseparability of energy from national and economic security. National governments inevitably are engaged in forming energy policy and in bargaining with and sometimes confronting other nations to assure secure supplies of affordable energy for their citizens, and to assure also that their own natural resources are developed in ways that further their national interests. In these roles, governments can be allies or adversaries to the energy companies that actually develop and deliver the energy products and services. International institutions have been formed to facilitate resolution of disputes among nations and to build consensus on how to address global commons issues such as sustainable development and climate change. The United Nations is the most comprehensive of these, and is supplemented by narrower groupings such as the World Trade Organization with a specific mission to promote trade among nations, and energy-focused institutions such as OPEC and its counter-balancing consumer organization, the International Energy Agency.

Energy's core societal role, coupled with the mix of these different stakeholder groups spread across different cultures and societies globally, gives rise to a complex set of conditions for energy leaders to navigate.

As society ascends the needs hierarchy, value takes on a much broader and more complex meaning than the basic notions in micro- and macroeconomic theory, which apply more narrowly to economic security. This evolution through the needs hierarchy creates a different stakeholder landscape that, properly understood, offers an opportunity to transform the negative image of the industry into a positive, forward-looking, public persona. We propose in the final section of this chapter that energy companies supplement the common theory of profits with an understanding of value creation that incorporates the perspectives of the various stakeholders who, after all, are the various individuals that make up the society. As Thierry Desmarest, Total CEO, says: "The principle purpose of the firm is to create value, particularly for shareholders. You're using resources, and if you're not adding value to the resources, you're not fulfilling any useful function. You just need to be sure that you're taking account of *all* the resources you're using, and not dumping pollution in the sea—you have to factor in the externalities, and also you have to add value in the countries you're working in. If you focus only on short-term metrics, you may maximize short-term profit but won't create long-term value."

In summary, geopolitics sets the parameters for whether and how energy companies will be allowed to gain access to international resources and markets; civil society sets expectations for how they should behave once they have gained access. We deal with each in turn.

Geopolitical Considerations

Societal expectations coalesce to form a country's national interest, and the global interaction of various nations' interests constitutes geopolitics. Energy companies are impacted by the interaction of nations and sometimes suffer collateral damage when national interests clash. In some cases, energy companies can influence the course of events, but in all cases they benefit from a deeper understanding of the drivers and motivations of the nations that are in any respect stakeholders in their operations.

Clearly, energy and geopolitics have always been joined at the hip. The energy industry has both national and global dimensions. It is buffeted by what British prime minister Harold MacMillan called "the wind of change" in a speech to the South African parliament in 1960: "The wind of change is blowing through this continent, and whether we like it or not, this growth of national consciousness is a political fact. We must all accept it as a fact, and our national policies must take account of it." At that time, he was referring to the end of colonialism and the rise of nationalism across Africa. It was this wind of change that inspired the resource rich countries to seize control of their petroleum from the "colonial" international oil companies. These were powerful head winds, and they inspired the major IOCs to look for oil in politically friendlier places, leading to discoveries in the North Sea and Alaska in the late 1960s. A graphic displayed recently by Tony Haywood of BP shows clearly the magnitude of the change for that company (fig. 4.1). As winds of change go, this was clearly a "Category 5" hurricane.

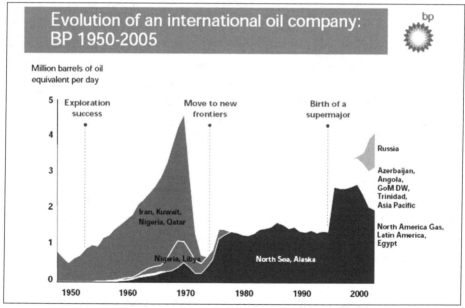

Figure 4.1: International oil companies needed to reinvent themselves in the 1970s.
Source: historic data from BP archives: 2005 BP estimates as presented in 2005 results

And politics goes both ways. ConocoPhillips worked hard to win the rights to develop acreage off Iran, only to find that the U.S. State Department would not or could not allow them to circumvent the Iran and Libya Sanctions Act. So the opportunity passed on to the French company Total. Similarly, Libya was out of bounds for U.S. oil companies until recently, when Muamar Gaddafi renounced terrorism and settled with the victims of PanAm Flight 101. It is clear that geopolitics impacts energy companies' business opportunities.

The "wind of change" is shifting again, but the forecast is obscure. The trade winds seem to be propelling globalization and opening up markets. However, there are numerous squalls that may signal coming doldrums, or another shift in direction towards resource nationalism, or may just be local problems to be steered around. Since we don't know what the future will hold, energy companies need to understand the underlying geopolitical forces as best they can in order to prepare for future course corrections.

The Yin and Yang of Globalization

The last 50 years have seen an exceptional growth in world trade. Merchandise exports grew on average by 6% annually. Total trade in 1997 was 14 times the level of 1950. GATT (General Agreement on Tariffs and Trade) and the WTO have helped to create a strong and prosperous trading system contributing to unprecedented growth. Most national economies have been growing robustly, and the World Bank expects them to continue to grow through 2015 (fig. 4.2).

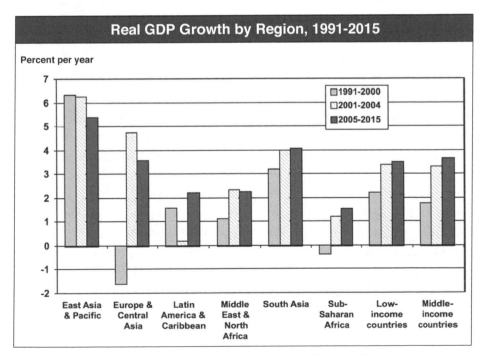

Figure 4.2: Economies are expected to continue growing. *Source: Global Monitoring Report 2005*

And a key ingredient for economic growth is the supply of affordable commercial energy products.

The fall of the Berlin Wall in November 1989 ended the Cold War that had dominated international relations for 50 years. It ushered in a new era with a new world order and a surge in the phenomenon known as globalization. The 1990s were a time of great hope and optimism, as countries dismantled the burdensome superstructure of state ownership and regulation that had been suppressing economic growth worldwide.

The collapse of communism provided an enormous boost to international trade. During the high oil prices in the late 1970s, Chinese premier Deng Xiaoping declared "to get rich is glorious" and ignited the explosive growth in the Chinese economy that continues a quarter of a century later. Kenichi Ohmae has analyzed the Chinese phenomenon: "In the 1980s, the Chinese government opened up a number of special economic zones aimed at attracting foreign direct investment. . . . The current boom in demand by the Chinese economy is mostly based around businesses in these successful regions." This potentially sets a new stage in which regions rather than national governments become the primary drivers of economic development. In some cases, regions may also wish to set their own geopolitical agendas, further complicating the task of energy companies wishing to build productive relations with both regional and national authorities.

The World Bank has documented that when economies grow, poverty levels gradually decrease, life expectancies rise, education levels improve, fewer children suffer from stunted growth due to malnutrition, and more children are immunized against diseases. As we have seen, hundreds of millions of people in China and India were lifted from extreme poverty in the past two decades. However, there are still 2.6 billion people living on less than $2 per day while the rich are getting richer. Even though the proportion of the world's poor population is falling, many people believe economic injustice is growing. Many people in the developed countries also believe that jobs are being exported through "off-shoring." So, although the statistics show improving average incomes and living standards, there are enough people who have not benefited or have been harmed by increasing trade to form a powerful and vocal opposition.

The process of globalization appears altogether too Darwinian for many people when the primary beneficiaries seem to be developed country corporations. In order to prosper, developing countries need markets for their products, which are often agricultural. However, the Doha round of trade liberalization has at the time of writing broken down over the issue of agricultural subsidies in developed countries. These subsidies prevent emerging economies from accessing the largest markets with their produce and even allow subsidized farm produce from developed countries to undermine the economics of small farms in developing countries. By the end of the 1990s, the bloom had faded from the rose of globalization and the thorns were becoming more prominent.

When the WTO met in late 1999 in Seattle, it was greeted by violent protests from an eclectic mix of opponents of globalization, including anarchists, labor activists, peace proponents, native rights organizers, and developing country representatives. The common theme was that globalization and free trade were destroying traditional cultures, endangering jobs in developed countries, and not leading to improvement in living conditions for the poor. Oil company projects were prominent targets for the demonstrators' wrath.

Why does it matter to energy companies that trade rounds are moving slowly? It matters because the energy industry is global, and energy companies need access to global resources and global markets in order to prosper and grow. Countries wishing to become WTO members have to commit to its rules, which require opening of markets. Open markets create opportunities for energy companies. In particular, India and China are huge and growing markets to which international energy companies historically had little access; but they may become more accessible as they integrate with the global economy. Further, the process of opening markets and reducing tariff and non-tariff barriers is consistent with opening resource rich areas to exploration by international companies. Conversely, increasing opposition to trade and opening is consistent with closing resources to international companies. As described in chapter 2, IOCs are facing strong competition from NOCs and INOCs, fueled by increasing resource nationalism. This loss of competitive position will likely continue at least until the prior trend of market opening and globalization returns.

Understanding Middle East Geopolitics

There are important hydrocarbon resources in all continents. Energy leaders are advised to immerse themselves in the history of resource policies in each region where they desire access. They will find that the history of resource development will form an important component of the national identity, and will be intertwined in many cases with the history of colonization and of relations with the developed world. Chapter 2 chronicles the development of national oil companies with attention to Latin America, where resource nationalism is flaring again particularly in Venezuela and Bolivia. But with such a significant share of oil reserves sitting in the Middle East, it is particularly important for energy leaders to have a deeper understanding of the geopolitics in that part of the world. Since the Middle East is a place where past and present are tightly intertwined, understanding the current situation requires some familiarity with recent history.

Rise of Arab nationalism

Colonel Gamal Abdul Nasser overthrew king Farouk of Egypt in 1952 and trumpeted a new pan-Arab nationalist message. Nasser attempted to annex Sudan, and turned to the Soviet Union for weapons. In July 1956, the Egyptian army seized control of the Suez Canal. Britain and France invaded Egypt in October 1956 in combination with Israel and the Suez Canal was closed. After Britain and France withdrew, the canal opened again in 1957 under Egyptian control.

Ten years later, the Suez Canal again closed after the "Six Day War" between Israel and Egypt, Jordan, and Syria. This time it would remain closed for eight years. As a result, a new class of very large crude oil carriers (VLCCs) was introduced, each capable of carrying more than 200,000 tons of cargo, about double the size of the previously largest tankers. In response to the Israeli attack, Arab oil ministers called for an oil embargo against countries friendly to Israel. But in this case the lost production was replaced by increases from Texas, Venezuela, and Iran, so the move was ineffective.

Throughout the post-World War II period, the major oil companies supplied most of the growth in global oil demand, primarily from their Middle East concessions (fig. 4.3), increasing their share to over half the world's production by the late 1960s.

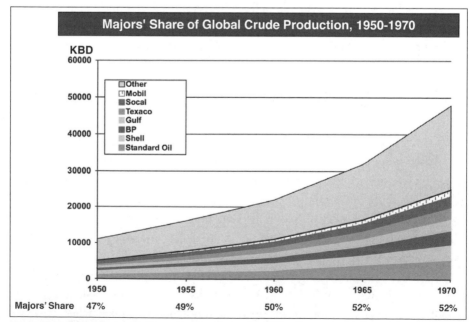

Figure 4.3: Seven Sisters' oil production through 1970. *Source: British Petroleum and Global Oil 1950–75 by James Bamberg*

But the wind of change was still blowing, this time in Libya. Another Arab nationalist, Colonel Muammar Gadaffi deposed king Idris of Libya in September 1969 and quickly demanded an increase in the posted price of Libyan crude oil as well as increased royalties and taxes. Iran continued a game of leap-frog and increased its share of profits to 55%. The Tripoli and Tehran agreements were the start of a slippery slope toward transferring the economic rent on oil production from the IOCs to the producing countries.

Arab oil embargo and aftermath

The Suez Canal was still closed at the time of the next Arab-Israeli conflict, the "Yom Kippur" war of 1973. This confrontation also resulted in use of the "oil weapon" as Arab producing countries embargoed the United States and certain European countries. The Arab producing countries agreed to reduce production by 15%, but for the most part they did not implement it. Although the embargo was again not very successful, it was followed by acceleration in the pace of nationalization. By the end of 1975, Arab oil production was largely nationalized. As shown in figure 4.1, this was a catastrophic turn of events for the big oil companies, but in particular for BP.

In his 1976 book entitled *The Seven Sisters*, Anthony Sampson discussed the widespread distrust of the motives of these "Anglo-Saxon" companies. Sampson's book followed in the footsteps of Ida Tarbell's criticism of the Standard Oil Trust, as outlined in chapter 1. Sampson alleged that the Seven Sisters were acting as a cartel and that their OPEC national oil company successors had every interest in continuing this practice. So some of the public distrust of oil companies shifted to OPEC, which was vilified in the press. However, the public even today continues to distrust the new generation of "super-majors," and this can escalate quickly to anger and hostility.

Political Islam

Political disturbances in the Middle East took a new form with the Iranian revolution of 1979—that of Islamic fundamentalism. This dislocation began with the departure of the shah in January, continued with the repatriation of the Ayatollah Khomeini in February, and culminated with the taking of hostages at the U.S. embassy in November. Islamic fundamentalism had been fermenting in the background for decades and had been repressed by Egypt and Syria; but now it had a representative at the highest level of influence in Khomeini, who had avowed "the purest joy in Islam is to kill and be killed for God."[2]

Where political Islam has either gained control or acquired strong influence, it has pushed for a theocratic state based on conservative Qur'anic principles and dominated by the clergy. In Afghanistan under Taliban rule and Iran, the Islamist-dominated governments have been authoritarian, intolerant, and hostile to America and Western influences. The situation is a bit different in Saudi Arabia, where the monarchy remains in control of the government while conservative clerics rule the social life of the nation. The Saudi government is complex with aspects of consultation within a dynastic, authoritarian framework, but it maintains its close relationships with the U.S. and Europe.

Islamists in several other nations, notably Egypt, form a political opposition to the nominally democratic, secular governments, which are in fact highly authoritarian. Here the Islamists argue for more representative democracy and truly open elections, seeking to use the ballot rather than bullets to gain political power.

In contrast to theocratic fundamentalism, and with pressure or sponsorship from the U.S. and Europe, there are movements towards democracy in Lebanon, Palestine, and perhaps in Iraq if it survives as a nation. Ruling families in Kuwait, Qatar, the United Arab Emirates, and Saudi Arabia are allowing increased participation in government. Saudi Arabia and other Arab countries are serious about addressing deficiencies in higher education and job creation. These are encouraging trends within the Middle-East. The authors believe Arab nations will continue to evolve driven by the forces of globalization. The transformation, however, will not be another Western society but one that fits within their Arab cultural heritage. But complications from the hostility of Islamic fundamentalism to all forms of modernism are slowing the process of political reform.

Youthful demographics

Middle Eastern nations have some of the highest population growth rates in the world (fig. 4.4), and consequently very youthful populations.

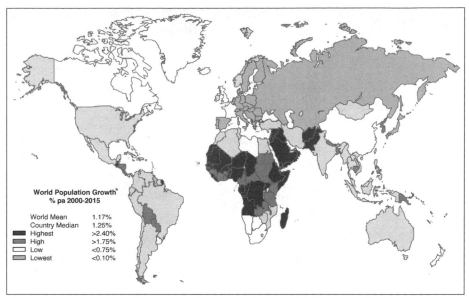

Figure 4.4: Highest population growth rates are in the Middle East and Africa. *Source: CRA International, based on United Nations Data*

With 35% of the region's people below the age of 15 (approximately double the proportion in Europe), it is all but certain that rapid population growth will continue. When this rapid growth is coupled with the region's limited acreage of arable land, acute water shortage, autocratic governments, and conservative institutions, it is difficult to see how these countries can create enough jobs for their young people. Without the prospect of gainful employment, young people become frustrated, angry, and alienated. They offer a fertile field for the seeds of ancient tribal conflicts or religious radicalism.

Growing ties to Asia

Meanwhile in the 1970s and 1980s, the Western world had been adjusting to higher oil prices by substituting nuclear energy and coal for residual fuel oil, improving the efficiency of transportation vehicles, and finding and developing oil outside the Middle East; and at the same time oil demand in Asia was growing much more rapidly than in the West. As a result, the supply of oil east of Suez now closely matches the demand for oil east of Suez (fig. 4.5).

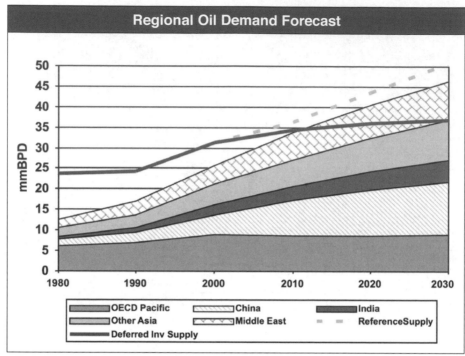

Figure 4.5: East of Suez oil demand could exceed supply by 2010. *Source: CRA International, based on IEA Data*

In the IEA deferred investment scenario, the Pacific Region could become a net importer from the Atlantic Region by 2010. Already, West African crude oils are being exported to Asian markets, although mainly for reasons of oil quality; but when the net trade shifts direction, there must surely be some repercussions. At the very least, Asian countries would appear to have an even greater interest in stable access to Middle East resources than do European countries and North America.

Geopolitical Outlook

There appear to be three main points of view on the world's future:
1. The western powers are doomed to decline, as previous empires have declined. This point of view is expressed by Paul Kennedy[4] and Jacques Barzun,[5] among others. The western powers would be replaced by a different set of powers that would dominate the world in a way similar to the Romans and the European colonial powers before them.

2. Technology has fundamentally changed the world order and, therefore, the future may not be a replay of the past. Kenichi Ohmae[6] has an intriguing version of this concept, based on his experience working with regional governments in China. There he sees regions becoming the principal economic entities, with national governments obliged by self-interest to support the regions. Tom Friedman[7] is certain that the world has changed but sees both promise and danger in the change. Technology can support rapid economic growth and global brands, but can also support al-Qaeda to grow globally.

3. The apocalyptic view proposed by Samuel Huntington predicts a destructive clash of civilizations.

Each of these points of view can claim basis in recent history. There are aspects of western civilization that do appear decadent, in the sense that most developed nations are now unwilling to allocate resources to defense. They are therefore potentially vulnerable, as the Romans were to barbarians. There is no doubt that computer and communications technology is shrinking the world and allowing corporations to assemble global supply chains, thereby becoming more global and less national in character. And there are obvious challenges to the current world order from Islamic extremists on the one side, and doctrinaire populist socialists on the other.

Capitalism and liberal democracy lead to economic growth and create surplus income that can be used to allow citizens to reach the self-actualization peak of the Maslow hierarchy. With the surplus income, they have the means to protect the environment and elevate more of their people from poverty. There is no evidence that Islamic fundamentalism or socialism can provide similar benefits; in fact, all the evidence points to the contrary. This does not preclude clashes of civilizations. Even if the odds in such contests will favor those with economic strength, the fall-out could be very damaging. Also, there is a risk of a new "Great Game" in which China and India compete destructively with the West for access to Middle East and Central Asian resources.

Steve Lowden, CEO of Suntera, sees the rise of China and India as a risk and an opportunity:

> The major new economic growth markets and the other parts of the world are becoming more concerned about energy security. And the flame has just been fanned everywhere you go. The U.S. is fanning the flames. The UK is fanning the flames. The net result is, is that those booming economies have been led to believe that if they don't have energy security, they will damage their economic growth possibilities.

So they're going to get it. The only way they can get it is by going out to collect it. So they are funding and motivating their national oil companies to go and collect more resources. Once again, you have to be able to play with those companies to be able to sustain growth—in fact the biggest, hottest, most aggressive competitors in the marketplace today are the Indians and the Chinese, by miles. I think it was clear to see it five years ago, but nobody did anything about it.

We believe that the critics of the current state of the world are not necessarily wrong in their criticisms. When candidate Hugo Chavez of Venezuela spoke out against the corruption of the incumbent political parties and contrasted widespread poverty in the nation with the wealth of the few, he was factually correct. The problem is not in the diagnosis, it is in the proposed cure, and most opponents of liberal democracy propose cures that create new elites that are intent on protecting their power. They can soon become indistinguishable from their predecessors in relative wealth and corruption. Even worse, their form of government, whether theocracy or socialist, will bring neither long-term prosperity nor an attractive standard of living to their nations, and it is dependent on the invention of enemies outside. Nevertheless, the criticisms need to be heard and addressed, because if they are not, there will be instability.

International energy companies cannot significantly influence the political context; they can only adapt to it and try to inform policy makers of the trade-offs that will be necessary. Most energy companies recognize in their public statements that improved efficiency and alternative energy forms have important roles to play. They rightly state that the goal is not to create energy independence, because interdependence with a variety of suppliers actually contributes to supply security. However, the goal should be to take some pressure off global supplies, so the system can function properly and not be buffeted by every minor supply setback. To the extent that large energy companies are seen as adding value to their customers and to the communities and national governments where they operate, they will be more welcome and accepted. To the extent that they find ways to work with global institutions and support transparency, property rights and strong political and legal institutions, they gain credibility. To the extent that they engage with civil society, they can expect to be better understood. We believe that new value propositions and new business models will be required by IOCs to navigate the changing currents of geo-political trends. It will not be easy, but the authors are persuaded that energy companies are not paying enough attention to the need for change, and that this is a risky policy.

Civil Society

While geopolitics set the stage upon which governments interact to further their national interests, civil society provides forums in which groups of individuals with shared interests seek to advance their causes by influencing public opinion and bring pressure to bear to cause other institutions—governments, international organizations and corporations—to change their behaviors.

Environmentalism

Concern for the environment is not a new phenomenon. Londoners grew angry about smoke from burning coal in the late 1600s. The seminal work of the current era was probably Rachel Carson's *Silent Spring*, published in 1962, in which she highlighted the potential risks of pesticides. During the 1960s, concerns over pollution of air and water rose and culminated in the first Earth Day on April 22, 1970, where concerned citizens demonstrated strong support in the United States for environmental improvements. Since that day, both support for protecting the environment and opposition to almost every specific proposal to mitigate damage have grown and matured.

Cleaner environment

Later in 1970, the Clean Air Act was passed in the United States, mandating a reduction in the sulfur content of fuels burned by industries and power plants. Significantly, it was the first major environmental law to include a provision for citizen lawsuits. The Environmental Protection Agency (EPA) started up in December 1970 with a mission to protect human health and the environment, as well as the provisions of the Clean Air Act as well as water, land, communities, and ecosystems. Also in the 1970s, Japan and several European countries reduced the maximum sulfur content of fuels that could be burned by industries and power plants.

Other environmental legislation followed during the 1970s. In 1972, the EPA launched a program to phase out the use of tetra-ethyl lead from motor gasoline on grounds that automobile emissions of toxic lead compounds could end up in citizens' blood streams. Further, lead poisons the catalysts in catalytic converters that were mandated for automobiles in the 1970s to remove carbon monoxide and unburned hydrocarbons from exhaust gases. Other countries followed the same path later. In 1975, the Energy Policy and Conservation Act mandated corporate average fuel economy standards for new passenger cars. In Europe, governments used the price mechanism to achieve efficiencies by adding sales taxes to pump prices of gasoline and diesel. Gasoline and diesel prices in Europe are approximately two to three times their United States' levels, and the automobile fleet is about 50% more efficient in Europe than in the United States.

These actions had a profound effect on air quality in the United States and around the world (fig. 4.6). In the United States, emissions of sulfur oxides and volatile organic compounds have dropped by 50% since the EPA was formed, and those of nitrogen oxides have declined by about 25%. This was during a period when primary energy consumption rose by 40%. Similar positive trends can be produced for all the developed nations and for water quality and forestation (fig. 4.7).[8]

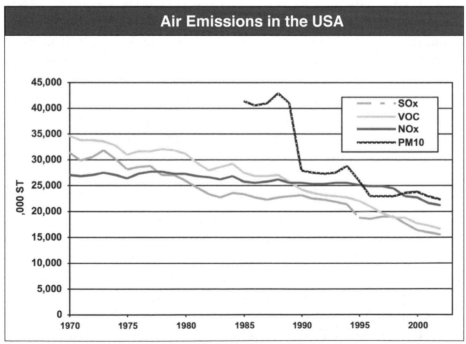

Figure 4.6: Air emissions of major pollutants have halved in absolute terms. *Source: CRA International, based on EPA Data*

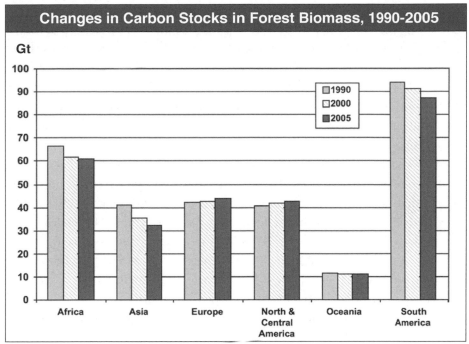

Figure 4.7: Deforestation has been reversed in developed economies. *Source: United Nations Food and Agriculture Organization*

Looking at these environmental results from a distance, one would conclude that environmental legislation has delivered real benefits. Initially, affected industries complained that their welfare and that of their shareholders would be damaged, but enough voters indicated their support to spur policy makers to action. After enactment, the affected industries applied their ingenuity to lower the cost of compliance, and often found new ways of operating that lowered cost below levels of prior practices. Inefficient players were forced to shut down, improving the outlook for the stronger companies. Jobs were redistributed causing dislocation for the affected workers, but the economy advanced and emissions declined.

Indeed, the most egregious examples of environmental degradation that we have come across have been in countries with monopoly or near-monopoly state-owned enterprises, where the government is both owner and regulator. In a society where the poacher and the gamekeeper share the same employer, the ecology will be the loser.

Conundrum of greater measurement precision

So, if the vital signs in the developed countries are all showing improvement, why do so many people think that things are getting worse? One reason is what has been called the "tyranny of small numbers": measurement technology allows us to detect impurities in parts per trillion, and to set specifications in parts per billion, so we do. The ability to measure very small concentrations of toxic materials allows us to see possible dangers of which we were previously unaware. Our higher living standards also allow us to fund increasing levels of research in medical and ecological areas. This research has enabled us to understand connections between the physical environment and human health that were not previously recognized, and to track endangered species.

However, it has been shown that we are poor at weighing risks and consequences, often overestimate the danger of small risks, and adopt policies with unknown, unintended side effects. An example is the decision to mandate the inclusion of oxygenates in gasoline, which was designed to reduce emissions of volatile organic compounds from older vehicles. This led to the widespread use as a gasoline blendstock of the chemical methyl tertiary butyl ether (MTBE) that eventually contaminated water resources. The costs of adding MTBE to the industry were large, the benefits were uncertain, and the consequences unfortunate.

But technological advance also allows us to reduce pollutants still further. So diesel fuel in the early 1970s was regulated in Europe and the United States at a maximum of 0.5% sulfur content (5000 parts per million). Now, the sulfur content will be limited to a maximum 15 parts per million in the United States, and in Europe to 50 parts per million, reducing to 10 parts per million in 2009. From 5000 parts to 10 parts per million is a reduction of more than two orders of magnitude, and will further improve air quality by reducing emissions of sulfur oxides and the small soot particles that are believed to be carcinogens. In this case, the cost to the industry is again large; however, the benefits are more certain and, based on prior experience, there is a low risk of unexpected consequences.

Challenges remain

But the second reason people think things are getting worse is that in some areas of the world, things actually are getting worse. For example, deforestation continues in Africa, parts of Asia, and South America (fig. 4.7), and even if things aren't getting worse, the current state is unacceptable in terms of the social conditions of the world's poor.

The key environmental issue is: "compared to what?" What is the proper baseline for measuring environmental degradation or progress? The fact that we are making progress does not satisfy the self-actualization needs of environmentalists. Bjørn Lomborg has been pilloried by his former environmental friends for his book

The Skeptical Environmentalist, but expresses the issue with admirable clarity: "By far the majority of indicators show that mankind's lot has *vastly improved*. This does not, however, mean that everything is *good enough*. The first statement refers to what the world looks like whereas the second refers to what it ought to look like."[10] Lomborg's thesis is that it is vitally important to establish the facts so we can determine what is working and what is not. Failure to be honest about the current reality that progress is being made in the developed world will blind us to the policy moves that have been successful. This will make us less likely to apply these lessons in places that desperately need further improvement.

Three different lenses complicate dialogue

The difficulty in agreeing on solutions is that people see the world through different lenses. There are various ways of describing these, but a simple annotation that we have found useful is *red*, *green*, and *blue* (fig. 4.8):

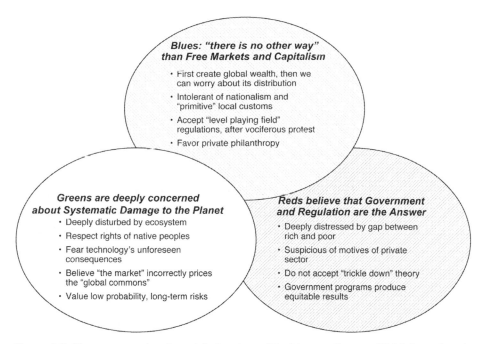

Figure 4.8: There are overlapping global socio-political lenses. *Source: CRA International*

- Those looking through a red lens tend to be deeply suspicious of corporations and strongly believe that governments should direct what happens in their territory with the goal of delivering an equitable result.

The extreme position of this group would be totalitarianism (communist, radical Islamic, or fascist).

- Those looking through a green lens may be very worried about the escalating human population and the potentially destabilizing effect that it may have on regional and global ecologies. The extreme view in this group would be those believing in Gaia, a reverence for the earth as it once was.

- The group looking through the blue lens believes strongly in the power of "the invisible hand" of the market. At the extreme, they may be libertarian and believe that government intervention inevitably does more harm than good.

There is a great deal of overlap among these positions in their more moderate articulations, which allow for compromise, legislation, and regulation that can appease if not satisfy a majority of citizens in a democracy. There are many *blues* who support greater government authority in the name of national security, and others that favor tightening environmental regulations so long as the mechanism includes emissions trading. There are *greens* who recognize that a strong economy is a prerequisite for environmental improvement. And many *reds* and *greens* merge on a platform of greater government intervention to force environmental improvement.

In prosperous nations there is an appetite for continuous improvement and a belief that technology should enable it. So *greens* continuously strive for more rigorous standards, *reds* embed them in legislation, and *blues* find ways to deliver the higher standards at prices that people can afford.

The process sounds quite simple, but as Count Otto von Bismarck said. "Laws are like sausages. It is better not to see them being made." And the process by which environmental improvements are made can be quite messy. The problem is that people seeing the world through different lenses tend to be mutually suspicious and carry with them a serious "left hand column"[11] that colors their understanding of the intentions of the other side. It is easy to imagine an unproductive conversation between the tri-polar positions on the topic of opening the Arctic National Wildlife Refuge (ANWR) to oil and gas exploration:

What we were thinking	What we were saying
I can't believe anyone could question the logic for reopening.	Blue: Oil prices are high, people are suffering, and we need to open up ANWR for drilling to increase supplies and dampen price increases.
He just wants to make an exorbitant return and doesn't care what mess he leaves behind.	Green: There are serious risks to a pristine environment, to the traditional ways of the indigenous people and to the breeding grounds of the caribou.
The government could really use the license fees, and the Trans-Alaska pipeline is losing volumes that need to be replenished.	Red: Surely there are regulations that can make this acceptable.
I can imagine a permitting nightmare, where we have to secure the approval of everyone with a remote interest before being allowed to do anything.	Blue: We are prepared to accept reasonable regulations, so long as they do not make the project uneconomical.
I don't trust the oil companies to deliver on their promises. They will find a way to obey the letter but ignore the spirit of any agreement.	Green: I'm not sure there are regulations that can assure that the interests of the indigenous people and the caribou are fully protected. This is an important issue for us.
He is just using this issue to fire up his supporters and increase donations to his organization.	Blue: The caribou herd increased after the Trans-Alaska pipeline was built, and many indigenous people welcome the influx of money into the region.
I am not sure where the voters are on this, so had better hedge my bets.	Red: I think we need more studies.

So long as the debate is framed in a way that results in clear winners and losers, and as long as the industry believes that the war is unwinnable so they are better off not participating, it will be difficult to make progress. As we describe in chapter 8, the key is to search out win-win solutions using the precepts of collaborative game theory elaborated by Professor Barry Nalebuff and others.

Sustainable Development and Global Warming

Civil society believes that people should have some purposes or goals that transcend our concerns for current economic well-being. Attainment of material comfort or affluence finally is not the measure of all things. We aspire to something more. This thinking reflects an implicit belief in the stability of prosperity and the availability of surplus wealth to enable attainment of the higher levels of the Maslow hierarchy.

This social orientation started to take shape in the 1980s, and was crisply summarized by the 1987 World Commission on Environment and Development, chaired by then Norwegian prime minister Gro Harlem Brundtland, which provided the definition that is now widely accepted: "Sustainable development is development that meets the needs of the present without compromising the ability of future generations to meet their own needs."[12] The importance of the commission's report was that it recognized there were regions of the world where poverty, disease, and environmental degradation were linked; there were serious issues of the "global commons" that needed to be addressed holistically. It also recognized that economic development was necessary for advancement and that government and inter-government action would be necessary for improvement. Thus, it came up with a definition of sustainable development that could be accepted by *reds*, *greens*, and *blues*.

The report cemented the shift of the environmental movement from local to global. It took a while for the business community to recognize the importance of this shift, but as preparations for the ambitious 1992 UN Conference on Environment and Development (the Rio Earth Summit) gathered momentum, its secretary general Maurice Strong, former chairman of Petro-Canada, sought to involve major corporations. In 1991, Stephan Schmidheiny, a Swiss industrialist with a vision of involving the private sector in sustainability issues, put together a group of 50 companies to form a fledgling Business Council for Sustainable Development. This was the precursor of the World Business Council on Sustainable Development, whose mission is "to provide business leadership as a catalyst for change toward sustainable development, and to promote the role of eco-efficiency, innovation, and corporate social responsibility." This organization sponsors studies such as the sustainable mobility project cited in chapter 3. It participates in policy development, highlights best practices in environmental and resource management, and "tries to contribute to a sustainable future for developing nations and nations in transition[13]."

Sustainable development as defined by the Bruntland Report clearly embraced economic development, the environment, and society. This was taken further in the "Triple Bottom Line" concept attributed to British environmentalist John Elkington.[14] This concept proposed that companies should report not only their financial results, but also their impact on or stewardship of the environment and their contribution to the communities with which they interact.

Local community needs

Energy companies, by virtue of their wealth and the nature of their business, have always had problems with local communities affected by their operations. For example, the Unocal oil spill off Santa Barbara in 1969 contributed significantly to the current moratorium on drilling offshore in California. There were also follow-on problems with the siting of pipelines. This is a good example of people in an economically, advanced region putting their self-actualization needs far ahead of local economic benefits. Beyond the local economy, the potential economic benefits of offshore drilling would have flowed to the oil companies, to the state in tax receipts, and to the federal government in tax receipts on profits. The nation would have had the advantage of the refined products. Too often though, the local communities bear all the environmental risk, while others take the economic benefits; this poses a conundrum for agencies charged with development, and increasingly also for the IOCs.

However, other areas, such as Alaska, support oil industry development, although environmental activists have stopped exploration on adjoining federal lands. In still other cases internationally, environmental activists have been encouraging protests by indigenous people (e.g., Ecuador) that may not be supported by their national governments. In other cases (e.g., Nigeria and Colombia) national governments have provided security assistance to oil companies that have then been portrayed as repression of local tribes. Or oil companies threatened by local tribes or bandits may have hired security guards that have treated local communities harshly. In Canada, attorneys have encouraged indigenous peoples to negotiate substantial economic rent from companies wishing to build pipelines crossing their tribal lands. In many developing countries there are overlapping and unclear jurisdictions, and skirmishes and even wars can break out as different interests jockey for position.

The largest companies have all taken up the challenge of society's increasing expectations and now produce annual "social responsibility" reports that are available for inspection on the company web sites:

- ExxonMobil has its Corporate Citizenship Report
- BP has its annual Sustainability Report

- Shell has the Shell Report—its progress in contributing to sustainable development
- Total has its Corporate Social Responsibility Report
- Chevron has its Corporate Responsibility Report
- ConocoPhillips has its Sustainable Development Report

While some environmentalists consider these reports self serving and "greenwash," they do represent a higher level of transparency than was previous practice, and in many cases mark real progress towards environmental improvement and social awareness.

National needs

One frequent complaint is that energy companies enter a country, extract the national resource, sub-contract to international companies, and pay taxes to a corrupt national government. At the end, the local communities have suffered some environmental degradation, few jobs have been created, and the tax moneys have ended up in numbered accounts of government officials in Zurich. More and more, countries are finding a justifiable demand for greater local content in the capital and operating budgets of the energy companies, and energy companies are among the advocates of greater transparency on governments sources and uses of funds.

Energy companies and national governments are increasingly including international institutions, governmental (United Nations and World Bank) and non-governmental (NGOs), in development scenarios. The objective of bringing international institutions into the picture from the energy company point of view is to create a stable foundation and reduce the risk for a major investment program. The objectives of the international institution are to stop immature governments running off with the money, encourage the provision of local employment, safeguard the local environment, and as far as possible leave indigenous ways of life undisturbed. The resource-owning government should in theory have the same objectives as the international institutions, but in some cases may be more preoccupied with cementing its power in the name of national sovereignty.

It is a fact that fiscal terms change when oil prices change dramatically. Obviously oil companies would prefer it to be otherwise, but they have accepted this for decades in countries such as the UK. As long as they receive a reasonable risk-adjusted return on their investments, they will continue to invest. Most large companies are prepared to enter into a dialogue to help shape new terms so they are not confiscatory and resist the cardinal sin of retroactivity. With international institutions as part of the deal, there is no longer a dialogue, but a multi-party international negotiation, and the increased complexity will undoubtedly be challenging.

Global Warming, Hot Issue

Greater transparency in social and environmental stewardship has been an important advance. Even more recently, a growing number of people, largely from developed countries, have become increasingly concerned about global warming. This issue has the potential to have a profound impact on the energy industry and substantially change the relative values of fossil fuels to the benefit of low-carbon fuels (e.g., natural gas) and detriment of high-carbon fuels (e.g., coal). There is general agreement that global warming is occurring (fig. 4.9), a strong suspicion that man-made emissions of greenhouse gases is contributing to the change, and a fear that the consequences will be problematic.

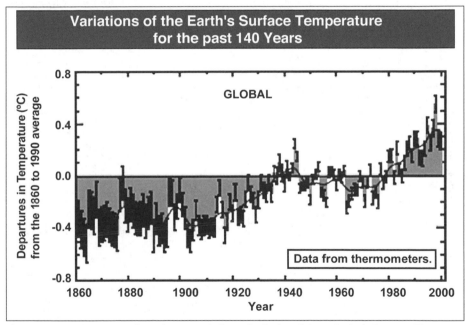

Figure 4.9: Temperatures have increased since the industrial revolution. *Source: The Third Assessment Report of Working Group I of the Intergovernmental Panel on Climate Change (IPCC), Summary for Policymakers*

The scientific opinions on climate change, as expressed by the UN Intergovernmental Panel on Climate Change (IPCC) and endorsed by the national science academies of the G8 nations, is that the average global temperature has risen 0.6 ± 0.2 °C since the late 19th century, and that "most of the warming observed over the last 50 years is attributable to human activities." The concern is that some of the earth's most fragile ecosystems are at risk of being most severely affected. The fearsome phenomena associated with global warming include melting of the polar ice caps and flooding due to higher ocean water levels, droughts, stronger storms and floods, and migration of tropical diseases to higher latitudes. Insurance companies are concerned that climate change may be causing an increase in weather-related natural disasters.

The 1992 Rio Earth Summit adopted the UN Framework Convention on Climate Change. In December 1997, a follow up convention in Kyoto, Japan negotiated specific targets for reductions in greenhouse gas (GHG) emissions for each country or region. Vice President Gore signed the treaty on behalf of the United States, but President George W. Bush has indicated that he does not intend to submit it to Congress for ratification. Australia has also abstained from the treaty. Russia has signed the treaty. China and India are not signatories. The abstainers create a "global commons" issue, in that it is thought that countries that do not accept Kyoto will have a global competitive advantage in energy costs over those that adopt Kyoto targets. The GHG that has been rising most steeply is carbon dioxide, mainly from burning fossil fuels (fig. 4.10).

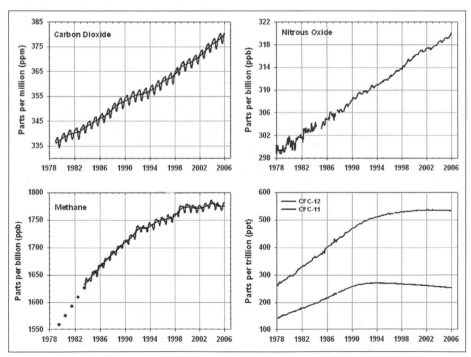

Figure 4.10: Greenhouse gas concentrations have been rising. *Source: National Oceanic and Atmospheric Administration*

Europe has adopted the Kyoto targets and has instituted a carbon trading system to encourage companies to choose the most economic methods to reduce carbon dioxide emissions. Of the energy companies, European BP and Royal Dutch Shell have made the strongest affirmations that global warming is an issue that must be dealt with, and that companies have a responsibility to be proactive in reducing their GHG emissions.

Technologies to prevent CO_2 emissions from entering the atmosphere, referred to as sequestration, are new in the sense that there have not been long-term tests of their efficacy; but they are old in the sense that they all require well-established compression, drilling, and pipeline technologies. The main proposed sinks are in existing oil and gas fields, new salt domes, or deep ocean areas where CO_2 would be trapped by the cold, high-pressure environment.[15]

There are some concerns that the CO_2 injected into underground formations may leak back into the atmosphere. Also, power plant operators are very concerned about the potential cost of sequestration from coal-fired power plants. The world's three largest coal users, the United States, China, and India, are not bound by Kyoto provisions. So those that are so bound may be at a disadvantage to these powerful and growing economies. Power generation companies in the United States are actively engaged in understanding the risks of future restrictions on GHG emissions and are studying the various possible forms of regulation to sort through potential unexpected consequences. Northeastern and Pacific states are moving ahead with GHG controls, but since they do not generate much power from coal, this will have little impact. As popular opinion in favor of GHG controls mounts, the likelihood of the United States adopting these controls increases.

Effective controls require collaboration between the public and private sectors. The critical factor is that everyone understands and agrees on the likely outcomes and potential risks of any mitigation measures. It is not just in the United States that such collaboration is proving difficult. Helge Lund, CEO of Statoil, described a situation in parts of Norway where a power deficit is looming. In this region, Statoil and Shell have launched a feasibility study of a project for using the CO_2 from a gas-fired power plant for pressure support and increased oil recovery offshore. He has a pragmatic view of the mounting public pressure for action on climate change:

> I think the different players in the industry are taking different views on this. Statoil's view is that we have seen enough evidence that the use of hydrocarbons is a challenge to the global climate. Therefore, we would like to attack the issue proactively.
>
> The project we have launched with Shell is challenging, both technologically and commercially. The aim is to be able to inject

CO_2 into the reservoirs to increase oil recovery. We also aim to electrify some of the oil and gas installations in the Norwegian Sea and thereby reducing CO_2 emission from the platforms. In addition we can improve the long term power balance in the area and provide electricity not only to industry but also households in the region. This is one example where I think Statoil is in the lead to come up with a possible solution.

Like several of my colleagues, I believe we can do this from a technical point of view, but it is not yet commercially feasible. If the government believes this is important, it has to contribute to the solution. In this special case, the most likely alternative is to build new power lines into this particular geographic area. However, there is no power balance in Norway and we have to import electricity from Europe made by nuclear and/or coal that is emitting more CO_2. This makes little sense to me from an environmental or climate perspective. Getting a gas-fired power plant with CO_2 capture and storage would be much better for the climate and for society at large.

CO_2 and climate change are likely to continue moving higher up on the political agenda regardless of which position you take on the substance and the scientific evidence. Moving forward, I can only see a development where the ability to handle CO_2 issues will be more and more important for oil and gas companies. Over the time, this could develop into a competitive advantage.

I believe that technology, and the systematic use of appropriate market mechanisms to bring forward the best solutions, are the best answers to this challenge. That is also exactly the intention in the Kyoto Protocol. Statoil is putting capital and human resources on this, because it is part of our environmental responsibility, but also because we think this is an area where we can turn an obligation into a competitive asset.

So the solutions are feasible, but the institutional frameworks that will allow those solutions to be realized are not yet in place. As an indicator of growing public concern, a *Houston Chronicle* editorial, from the center of the global oil business, and the home state of former president George H.W. Bush, said about the UN Climate Change Conference in Montreal: "At Montreal, developing nations, including Brazil, proposed to limit the ongoing destruction of rain forests that absorb carbon dioxide and generate oxygen. China indicated a willingness to work under future Kyoto limitations to curb its expanding industrial emissions, so long as the targets were voluntary. In one of the few positive developments involving the U.S., the American delegation agreed to continue nonbinding talks with other nations on ways to slow down and eventually reverse global warming. In doing so, the Bush administration at least kept the door open to future international collaboration by the world's largest polluter. U.S. emissions constitute a quarter of the world's total output. Unfortunately, that action was only a baby step on an issue that demands bold strides. The longer the United States waits to join the rest of the world and act decisively to curb greenhouse gas emissions, the more likely our children will face a future where deluge, droughts, and rising oceans will be the norm rather than the exception." It is a sign of the times when such a viewpoint comes from "the Energy Capital of the World."

Industry Image

When the public gaze falls on the energy industry, it is usually for something negative: an abrupt price increase, an oil spill, an electric power failure or blackout, an accident with fatalities, or environmental damage. Even occurrences that are generally favorable for other industries, such as high profitability, typically receive negative news coverage when the energy industry is involved. The energy industry, and particularly the oil industry, has always suffered from a negative public image that it seems powerless to change.

A number of factors probably contribute to the negative image. There is first the industry's immense size and perceived economic and political power. Second, the public uses a lot of energy, and it is basic to the way that we live. We are thus both sensitive and vulnerable to any availability and price issues. There is also the tainted industry history, beginning with the infamous antitrust case against Rockefeller' Standard Oil Trust and later oil titans, as well as the caricaturing of the Texas oilman J.R. Ewing in the T.V. show *Dallas*.

Perhaps the most basic issue, however, revolves around the mindset of energy leaders, who typically have engineering educations and technical backgrounds. They think methodically and systematically, focusing on facts and reasoned explanations while trying to reduce the emotional element. The public, on the other hand, is driven by needs that have high emotional content; they are easily swayed by snapshots and sound bytes in the media and become bored with long technical discussions which don't get to the key points. Energy leaders on the whole have not been trained to engage in a continuous dialogue with the public. They often make public appearances only when the situation absolutely demands it, and then approach the task reluctantly.

For example, one of the authors led a study for the National Petroleum Council in 1995 based on interviews and workshops with environmentalists, human rights activists, government officials, and energy industry representatives. The study found that "the industry bears a particular responsibility to take a lead in improving relations with its many stakeholders. It can do so by communicating clearly (what it is doing, why it is doing it, and how its efforts will benefit other stakeholders), balancing other constituents' needs, and building bridges whenever possible."[16]

The same study concluded that "The perception of a negative public image of the oil and gas industry was reflected in the interviews and workshops conducted for the study. While it is tempting to ignore evidence of this perception, the importance of public policy to the industry in the future requires that the issue be taken seriously." This rather lukewarm endorsement of a mission-critical issue is regrettably still part of the mindset of industry leadership. At the time, one interviewee cited in the NPC report said, "I almost despair of changing public perceptions of the industry." Ten years later, Nick Butler, group vice president for strategy for BP, again expressed an industry perception that the battle may not be winnable: "You haven't asked why we're so unpopular. I think it's partly size, partly arrogance, partly apparent success and large profit numbers, people feeling that we're taking something away from them, using our market power against them. And then what we haven't got across properly. Petrol is still seen as a grudge purchase. Does this constrain our business? It does to some degree. It's not a totally comfortable position to be in, but maybe it's inevitable." It is also true, however, that well orchestrated image programs, such as the long running Plastics Council campaign, have made significant changes in public perceptions.

Establishing a Dialogue on Sensitive Public Issues

One of the best ways to learn about the need for greater public dialogue about sensitive issues is to examine a particular case in more detail. The events surrounding the disposal of Shell's Brent Spar facility in the North Sea offers good insight into the dynamics of such issues. Shell provides the following timeline:

June 1976—Brent Spar installed in the Brent Field in the UK sector of the North Sea, a unique design for oil storage and tanker loading.

Two of six tanks were later damaged in operation. The structure was also later found to have been stressed during installation, creating major difficulties in reversing the procedure to raise it from water.

Sept 1991—Brent Spar ceases operating.

1991–1993—Detailed decommissioning studies are carried out by Shell and independent external organizations and contractors to assess options. Two compared in detail—horizontal onshore dismantling and deep-sea disposal. Deep-sea disposal emerged with six-fold lower safety risks, four-fold lower cost, and minimal environmental impact.

Feb 1994—Independent Aberdeen University study (AURIS) endorses choice of deep-sea disposal. There were formal consultations with conservation bodies and fishing interests. Draft Abandonment Plan submitted.

Dec 1994—Shell submits final Abandonment Plan to UK Government Department of Trade and Industry and receives approval.

Feb 1995—UK Government announces approval for deep-sea disposal and notifies 13 other contracting parties (12 nations and EC), signatories to the Oslo Convention covering protection of the marine environment. No objections were raised within normal time limit. Shell announces deep-sea disposal plan.

April 30, 1995—Greenpeace activists occupy Spar, wrongly alleging Spar is "a toxic time bomb"; "14,500 tons of toxic rubbish," or contains "over 100 tons of toxic sludge." Over the next few months they say Spar will be "dumped in the North Sea" rather than disposed of at a carefully selected site in the deep Atlantic and suggest "more than 400 oil rigs in the North Sea" might also be "dumped." They say Spar contains 5,550 tons of oil.

May 5, 1995—The UK Government grants disposal license to Shell UK.

May 9, 1995—The German Ministry of the Environment protests against the disposal plan.

May 13, 1995—Independent UK scientists begin stating support for deep-sea disposal for Brent Spar.

May 23, 1995—Activists are removed from Spar. Greenpeace calls for Shell boycott in continental Europe.

June 8–9, 1995—Fourth North Sea Conference at Esbjerg, Denmark. Several European countries now call for onshore disposal for all oil installations. The UK and Norway, the countries with the largest, heaviest, and most difficult deepwater structures, argue for "case-by-case" decisions.

June 11, 1995—Shell UK begins to tow Spar to deep Atlantic disposal site.

June 15–17, 1995—Public opinion in continental northern Europe is strongly opposed. Chancellor Kohl protests to UK prime minister John Major at the G7 summit.

June 14–20, 1995—Protesters in Germany threaten to damage 200 Shell service stations. Fifty are subsequently damaged, two fire-bombed, and one raked with bullets.

June 20, 1995—Several continental northern European governments now indicate opposition. Shell UK decides to halt disposal plan in view of untenable position caused by European political shifts, increased safety risks from violence and need for more reasoned discussion.

Late June 1995—UK scientific debate intensifies, with growing support for Shell's approach to environmental decision-making based on reason and sound science.

June 26–30, 1995—Eleven states call for a moratorium on sea disposal of decommissioned offshore installations at meeting of Oslo and Paris Commissions. Opposed by Britain and Norway.

July 7, 1995—Norway grants permission to moor Spar in Erfjord while Shell reconsiders options.

July 12, 1995—Shell UK commissions independent Norwegian marine classification society Det Norske Veritas (DNV) to conduct another audit of Spar's contents and investigate Greenpeace allegations.

July 12–18, 1995— UK Government makes clear that any new plan for which Shell UK seeks approval must be at least as good or better than deep-sea disposal on the Best Practicable Environmental Option criteria.

Aug 26, 1995—UK television executives admit to lack of objectivity and balance in coverage of the Spar story, and to using dramatic film footage from Greenpeace that eclipsed the facts.

Sept 5, 1995—Greenpeace admits inaccurate claims that Spar contains 5,550 tons of oil and apologizes to Shell.

The importance of this incident is that the technical, engineering approach adopted by Shell, and their dialogue with the UK government to gain approval, proved insufficient to forestall an international crisis. Civil society intervened, and essentially told Shell and the British government that their (secretive) cabal was unacceptable. Civil society demanded a higher level of transparency and public debate on a matter that Shell believed was a technical problem requiring a technical solution to be implemented with government permission. Equally important, civil society informed other governments of the issue and raised questions of the "global commons," illustrating that national governments may no longer be ceded the authority to act as they deem right in matters that have implications beyond their domains.

Greenpeace was wrong to spread disinformation on the magnitude of the problem, and this dual standard of requiring companies to be exactly accurate while accepting material inaccuracies from NGOs should be ended. We will only find solutions to issues of the global commons through honest dialogue and sound science. However, the fact remains that decision-making has become more complicated for companies like Shell that need broad public franchise and approval of their operations.

A far more serious case of an energy company mishandling public relations arose from the massive *Exxon Valdez* oil spill. Exxon's initial response to the spill has been widely criticized, first for the apparent indifference of then-CEO Larry Rawls, who made no public comment for a week, and second for the "scorched earth" approach the company adopted toward litigation that has endlessly delayed settlement of the various claims against the company. This incident has been the poster child example of how not to handle public relations following a crisis or negative incident. The public memory of the Valdez spill is surprisingly strong and entirely negative because of management's lack of responsiveness.

Other companies have learned lessons from prior disasters. BP CEO John Browne, Lord Maddingly, immediately flew to Texas City following the explosion in the company's large refinery there that killed 15 workers, and pledged a deep investigation followed by remedial actions. But BP later released an interim report that placed the primary blame on workers not following company rules. That alienated unions, workers, and local citizens to the point that *Houston Chronicle* correspondent Loren Steffy wrote:

"Before it overhauls its equipment, BP must overhaul its attitude. It must make safety paramount. It must care about more than money.

"For the past eight months, BP has been paying the price of its revenue-first philosophy. It's shelled out hundreds of millions of dollars in legal settlements, and it's been smacked with a $21 million government fine. Its reputation has been sullied, and its safety is now under scrutiny by an independent panel. That panel is led by former secretary of state James A. Baker III, who has staked his sizable public reputation on the fact that he will hold BP accountable.

"Through it all, though, BP's statements have lacked a degree of sincerity. It issued a report that tried to blame workers for its problems. Internal memos showed even after the March explosion, managers decided to keep units running under unsafe conditions. And, of course, there were two more explosions this summer."

Poor Job of Public Education

Civil society and the press have high expectations of energy companies but little understanding of how the industry operates. By and large, the public is poorly informed about all the activities and processes needed to get gasoline to a retail station where they can pump it into their cars or to get electricity to a switch they can flip to create light. The 1995 NPC study did conclude that "Industry has a strong history of supporting educational programs, yet more could be done in support of science, economics, and energy education." It also proposed that "Industry should improve and expand communication with stakeholders outside the industry." Implementation of these proposals has been disappointing, but the tide seems to be changing.

In the United States, the public became incensed with the high gasoline prices after devastating hurricanes in 2005 disrupted refining and upstream operations in the Gulf Coast. The financial reports of record profits in that same year by the super-majors fanned the flames. Having previously endured onerous legislation, labor shortages, and protracted litigation, the oil and gas industry developed a greater urgency to improve its public profile. A plethora of programs began to develop at national, regional, and local levels, but they did not coordinate with one another to gain critical mass. Energy leaders need to be sure that there is an ongoing public education strategy that is effective and leverages the industry's resources.

Perception of Poor
Environmental Performance

One of the toughest image challenges lies in the perception of a poor environmental track record in the oil industry. Consider the following partial list of major spills (table 4.1):[17]

Table 4.1: Major Oil Spills

Date	Location	Vessel	Oil Spilled (Tons)
March 18, 1967	Cornwall, UK	Torrey Canyon	119,000
January 28, 1969	Santa Barbara, CA	Unocal Platform	10,000
August 9, 1974	Magellan Str. Chile	Metula	50,000
December 15, 1976	Nantucket, MA	Argo Merchant	28,000
March 16, 1978	Brittany, France	Amoco Cadiz	223,000
March 7, 1980	Brittany, France	Tanio	13,000
August 6, 1983	Cape Town, S. Africa	Castillo de Bellver	50,000
March 24, 1989	Prince William Sound, AK	Exxon Valdez	37,000
January 5, 1993	Shetland, UK	Braer	85,000
February 15, 1996	Milford Haven, UK	Sea Empress	72,000
December 11, 1999	Brittany, France	Erika	20,000
November 13, 2002	Northern Spain	Prestige	63,000

Thirteen major spills over nearly 40 years, and three since 1990, is not a bad record, particularly if we compare it to the frequency of train derailments or auto accidents. In fact, in many respects it is amazing with the incredible volume of oil transported daily to have about one major spill every three years. Most of these involved tankers on the sea as opposed to the public view of oil spills on the homeland. However, these were accidents with horrible consequences for wildlife and for fisheries, and there were also a number of smaller incidents. In the final analysis, a spill every three years is still too many. The goal clearly has to be zero tolerance, but in practical terms civil society is looking for performance improvement in the absolute number of catastrophic incidents. From an image viewpoint, once every three years is enough to maintain the public's jaundiced skepticism about the oil industry's environmental record.

Looking Forward

Societal demands are not going away. If anything, they will become more stringent. Companies are going to have to learn to deal with them. The fact that many advocates do not share the political views of many energy executives does not change the fact that their criticisms resonate with many people. The fact that environmental advocates' concerns for the sustainability of economic growth and on climate change cannot be mathematically proven does not alter the fact that many people share these concerns. An Algerian client once said to us: "We need to stop kicking the players and start kicking the football," and that advice is good for both sides.

We have good evidence that certain market-based policies work well in reducing emissions, and that property rights and democratic institutions provide a good framework for improving living conditions. We know that corruption is corrosive and will destroy any society. We know that low tariff and non-tariff barriers allow greater economic growth than is possible with high barriers. The global economy is not a "fixed cake," and too much attention to how it is sliced limits the potential for expanding everyone's economic condition.

Exploring Future Scenarios

"Without measureless and perpetual uncertainty, the drama of human life would be destroyed," according to Winston Churchill.[18] There is considerable uncertainty about how the discontinuity in supply and demand trends described in chapter 3 will be resolved, and about the technologies that will emerge to reduce demand and enhance supply. There is also uncertainty in the geopolitical outlook, and on what form the passions of civil society will take in the future to influence environmental and societal policies. If we were in a period of relative stability, then we might be able to forecast the future with reasonable confidence, and develop strategies and plans that create competitive advantage in this expected environment. However, we would argue that the energy industry has seldom been predictable since its inception, and strategies based on historical trends would have failed to deliver value. The best way to make strategic decisions under uncertainty is to enrich the leadership's insights through scenarios.

There have been many substantial discontinuities in the business environment over the past 40 years that have demanded changes in strategy to capture opportunities and respond to threats:

Table 4.2: Major Energy Industry Discontinuities (1970-2005)

Timing	Change	Consequence
1970	U.S. Clean Air Act	Rise in demand for low sulfur fuel oil empowers North African crude oil producers
1973–76	Nationalization of oil fields	End of integrated business model; power companies shifted from residual fuel to coal and nuclear energy
1978	Natural Gas Policy Act	Prior natural gas supply contracts made obsolete; destruction of embryonic U.S. LNG business
1980–87	Loss of residual fuel oil demand	Need for residual conversion refinery process equipment
1980s	Increased computer power enables 3-D seismic processing	Exploration success rates improve; new business models emerge in the Gulf of Mexico; natural gas bubble depresses gas prices
1986	Oil price crash	Need to massively reduce costs throughout the oil industry
1980–95	Privatization, opening and deregulation	Opening of new business opportunities in countries previously closed to international investment; utilities expand beyond borders
1985–2000	Rapid decline in cost of LNG supply chain	LNG costs become competitive with other sources of natural gas and with alternative fuels; LNG becomes a growth segment.
1990s	Combined cycle gas turbines double the efficiency of new gas fired power plants	Massive investment program in GCC plants increases gas demand and strengthens prices
1992	United Nations Rio Earth Summit	Global endorsement of sustainable development concept and subsequent ratification of Kyoto Protocol; European carbon trading
1998	Second oil price crash	Industry consolidation and emergence of the super-majors
2000–2005	Accelerating Chinese and Indian oil demand	Rising oil prices; emergence of new Chinese and Indian international competitors

Royal Dutch/Shell recognized this problem in the early 1970s, and its chief planner, Pierre Wack, and his team built on prior work by Herman Kahn of the Hudson Institute to apply the concept of scenario creation to the oil industry. The value of the scenarios in Wack's view was as much in the process of developing them as in the final products. Scenarios are stories, and "a story is a letter the author writes to himself, to tell himself things that he would be unable to discover otherwise."[19] It is in the writing of the story that the discovery takes place. Wack's challenge in the early 1970s was to help Royal Dutch Shell executives understand that the future would be different from the past and that business models that had been successful in the prior period would not necessarily be successful in the future. This effort was extremely valuable for Royal Dutch Shell, that at the time considered itself disadvantaged relative to its peers with large Middle East production, in its inability to set up a fully integrated and optimized supply system from the production well to the gasoline pump. Through the scenario development process, Wack and the Royal Dutch Shell leadership came to the realization that there were powerful geopolitical forces under way that made the integrated model unsustainable. The same forces of nationalism that were undermining the majors' integrated models would favor companies that identified strongly with the host countries in which they did business.

Rather than create a new forecast, which would most likely have been rejected by executives with fixed mental models of how the world worked, Wack described various possible futures, and through a series of internal conversations, the executives gradually came to realize that a continuation of the past trends and structures was less likely than the new scenario that Wack had developed. It was helpful that Royal Dutch Shell was already distinctive in its capabilities to "go native" and identify with the countries in which it did business and had traditionally had a less centralized business model than its peers. Royal Dutch Shell was, therefore, in a strong position to further develop its decentralized business model by granting even greater authority to its "country chairmen" and by introducing powerful regional Coordinators to manage interactions within regions. This decentralized business model served Royal Dutch Shell very well until the 1990s, when the forces of globalization and the need for cost reduction across the entire system made the decentralized Royal Dutch Shell model less competitive than its peers.

Wack's work was taken up by his successor, Peter Schwartz, who later formed the Global Business Network with the goal of connecting scenario thinkers across industries and governments. Schwartz said: "Why do scenarios work? Because people recognize the truth in a description of future events. The story resonates in some ways with what they already know, and then leads them from that resonance to re-perceive the world. Observations from the real world must be built into the story. The only way they can emerge there is for the storyteller to sample evidence from the world before spinning the tale."[20] Scenarios are at the same time literary and analytical exercises. In an energy company populated by engineers and scientists, scenarios must be deeply rooted in quantitative analysis in order to muster respect. But they also need to be presented with flair, so they become real and clearly imaginable.

An example of scenarios that have stood the test of time is the suite developed by the World Business Council on Sustainable Development in the late 1990s under the leadership of Ged Davis, who was then responsible for scenario development at Royal Dutch Shell. It depicted three geopolitical futures, with "Jazz" clearly providing the most favorable framework for technology-led solutions to supply and environmental issues.

FROG !

"First Raise Our Growth!" depicts a strongly nationalistic future in which the underdeveloped countries move forward rapidly so that the total impact on the planet's eco-structure becomes unsustainable. Neither technological advances nor global institutions are strong enough to respond effectively. "People react like the proverbial frog: when placed in boiling water, the frog leaped out of danger; but placed in cold water that was heated to the boiling point, the complacent frog was boiled to death."

GEOpolity

Recognition emerges that "the market has no inherent incentives to protect the commons, social welfare, or any other non-economic values." New forms of global governance emerge to "design and enforce global standards and measures to protect the environment and preserve society."

Jazz

As in the music form, orchestration and improvisation coexist. "This is a world of social and technological innovations, experimentation, rapid adaptation, much voluntary interconnectedness, and a powerful and ever-changing global market." Quick learning based on transparent information within free markets, sound legal systems, and respect for property rights deliver rapid solutions to emerging environmental and social issues.

Ged Davis was responsible for Shell's 2002 two "People and Connections" scenarios, which also provide a durable framework. The "Business Class" scenario "explores what happens when the *connected freedom* of the globally interconnected elite and the only remaining superpower, the United States, lead the world towards greater economic integration and a dream of economic prosperity for all." The "Prism" scenario questions the monochromatic world of global integration and explores, instead, the persisting power of culture and history—*the connections that matter*—and the pursuit of multiple modernities as they emerge in a *new regionalism.*" "Business Class" has natural gas as "the great game" in energy. "Prism" relies more on oil.

Scenarios are useful if they enable companies to adjust their mental models and shift their strategies to take advantage of changing trade winds. They are not useful if they present polar opposite futures that can simply paralyze organizations into inaction. They are also not useful as an academic exercise conducted by outsiders on behalf of the organization. They have to be internalized by executives for them to be useful, and therefore the executives must be part of the scenario development process. The current social and geo-political environment is clearly in flux, and it is a major challenge for leaders to determine whether to plan for a specific scenario or to put in place options that provide flexibility to adjust as the future unfolds.

Focus on Value Creation

Energy companies need to have confidence in the expanding global economy and seek win-win opportunities out of self-interest as well as out of altruism for the greater societal good. Companies benefit when the institutions of government are honest, stable and predictable. They benefit when the communities where they do business believe that they are adding value. They are better off when Greenpeace picks up the phone to talk before it tries to occupy a production platform. Their stock prices will do better if all segments of the population consider them as adding value to society, rather than having some segments put them in the same category as tobacco companies, organizations unworthy of investment.

Energy companies are beginning to understand this. BP has been proactive in both its position on climate change and its willingness to engage NGOs. As Nick Butler noted:

"We engage with advocacy groups all the time. We listen. They're part of society and we can't ignore them. What we would like to do and are beginning to do is to work with them. We have very good relations with Greenpeace. They don't agree with everything we do, and we don't agree with everything they do. But John Browne has spoken at their conferences. He was the first oil industry executive to occupy a Greenpeace platform. So we have a good, healthy dialogue with lots of groups."

First and foremost, energy companies need to perform operationally and financially. Operational reliability is essential for financial success as well as for the safety and health of workers and communities. Financial strength is necessary to provide the funds needed for investment and proper maintenance. Nick Butler of BP again:

"You can't just focus strategy on the future; you've got to pay a lot of attention to operational excellence and the efficiency of what you're doing: supply chain, costs, and the capacity to do it. The forward-thinking pieces are really just an add-on to that. If you lose sight of the current business, you risk losing everything."

The triple bottom line is not an either/or proposition; it requires a both/and frame of mind. Energy companies have to be leaders not only in financial results, but also by adding value to communities and governments where they operate through the quality of their participation in addressing environmental and sustainability concerns. Further, there is no clear destination. Civil society will always want more, as long as there are communities that are affected by operations and concerns about the environment. The point is to be more effective in your use of resources in this as in other areas than your competitors. The leaders, though, should be far enough ahead that they can find ways to collaborate with each other and with their stakeholders to shape rules and regulations so that they achieve the desired results without threatening economic growth.

Societal pressures are strengthening. As we look forward, geo-politics will continue to create challenges for energy companies. The apparent 1990s trend toward opening of closed markets and resources appears to have at least been suspended. Historians are talking more of the "clash of civilizations" than the "end of history." The expansion of the international national oil companies (INOCs) has a mercantilist flavor, which hopefully will not lead to world wars as in the past. At the same time, societal pressures continue to mount in developed countries that have moved far beyond the subsistence levels of Maslow's hierarchy of needs and have high expectations not only of what energy companies do but also how they do it.

A New Mindset

What is the advice to energy leaders going forward? Too often we become victims of past solutions that worked well. When given a new problem, we use the old paradigm, not realizing a new model is required to solve this new problem. Having their roots in the industrial revolution, energy companies have been mesmerized with the economic rational man theory. That model worked well as developing countries climbed Maslow's hierarchy. It fit the very technical mindset of the energy leaders of that time and era. Economics and standard of living were the primary value drivers. Energy was the backbone, and relationships were clear and linear. Value could be easily calculated by traditional accounting measures of profit and loss. Revenues were recognized when they were realized. Costs were recorded based on their cash outlay. Things were rational and economic. At the time, there was no need to measure societal and environmental performance, or to look at the human resource as an asset on the balance sheet. That rational man theory worked because at the time society did not expect or need more.

Somewhere along the way profits acquired some unsavory associations. Too much profit became "obscene." That the public was expressing such views suggested that there was a new problem not being addressed by the old model. These views were a barometer of need dissatisfaction. Diehards argue that the old economics is still valid today. That may be true: it is necessary, but it is not sufficient. Many of us are content with a Seiko watch. But why are some people willing to spend thousands of dollars on a Rolex? Others are willing to take exotic vacations or buy Ferraris primarily for their garage. These are not good economic choices. At the ego level, value is much more complex and much more diversified. And those at this ego level are not rational in an economic-man sense, because assignment of value is based on more than just actual cost. At the ego level, needs go beyond physical goods and basic services. For example, the issues around sustainability represent ramifications of people thinking beyond their own comforts to the needs of future generations. They are driven more by beliefs of moral responsibility.

So what is the new model? In our opinion, the emerging paradigm focuses on building a *value bank* with customers, publics, and nations that recognizes the fact that there are concerns at the higher levels of Maslow's hierarchy. The industry's approach of focusing totally on its ability to deliver at levels two or three is not sufficient to build positive capital in its value bank. Rolex has banked a lot of emotional content that brings people to pay their higher price. A startup watch company could not hope to bring this value because there is nothing in the bank.

Today, stakeholders still demand energy supplies at affordable prices, but with no environmental impact; they call for companies to report a triple bottom line and not just financial success; they require sensitive treatment of local communities and assurances that government oil revenues are used wisely, as well as technical excellence; they will not tolerate environmental or safety lapses; and they will demand greenhouse gas caps. But more than that, the civil society of more advanced countries deplores wild pricing fluctuations, insecurity in supply availability, and lack of sustainability for future generations. If you ask the public, do they support sustainability? The answer is an easy yes. If you ask them will they pay for it, the answer is more mixed. But in many respects, that is not the public's problem; it is the industry's problem, and the industry must figure out innovative ways to satisfy this need. The more the public sees the industry tackling the sustainability problem, the more value that will be banked in the relationship. Similarly, as we have argued, developing countries are expecting more than simply an oil lease deal. They need help in growing and developing their country. Those who can help achieve that create value in their banked relationship.

In the end, though, at the ego and self-actualization levels of the Maslow hierarchy the public expects to be treated as equals. The rather arrogant approach often taken by energy leaders in solving problems that have come into the public limelight drains huge amounts of value from the bank. Without building public relationships on an adult-to-adult basis over sustained periods of time, the industry will continue to find itself under attack from the public's discontent. Windfall profits taxes do not make economic sense, but they resonate with a bruised public ego.

In chapter 8, Pat Yarrington of Chevron describes how her company is encouraging informal debate among stakeholders in response to this challenge. Several of the leaders we interviewed live this new model of banking public value. And they are finding their bottom line profits growing and their stock prices rising!

1 Kenichi Ohmae, 2005. *The Next Global Stage.* Upper Saddle River, New Jersey: Pearson Education, Inc. Publishing as Wharton School Publishing, 98.

2 Daniel Yergin, 1991. *The Prize.* New York: Simon & Schuster, 711.

3 As this is being written, Lebanon is consumed by war between Hezbollah and Israel, the outcome of which is unpredictable.

4 Paul Kennedy, 1987. *The Rise and Fall of the Great Powers.* New York: Random House.

5 Jacques Barzun, 2000. *From Dawn to Decadence.* New York: HarperCollins.

6 Kenichi.

7 Thomas L. Friedman, 2005. *The World is Flat: a Brief History of the Twenty-first Century*

8 World Bank, 2005. Global Monitoring Report.

9 Howard Margolis. 1996. *Dealing with risk: why the public and the experts disagree on environmental issues* Chicago: University of Chicago Press.

10 Bjorn Lomborg, 1998. *The Skeptical Environmentalist.* Cambridge, UK: Cambridge University Press.

11 Peter M. Senge, Art Kleiner, Charlotte Roberts, Richard B. Ross and Bryan J. Smith, 1994. *The Fifth Discipline Feldbook.* New York: Doubleday, p. 246.

12 Report of the World Commission on Environment and Development to the United Nations General Assembly, August, 1987, 54.

13 http//:www.wbcsd.org.

14 John Elkington, 1998. *Cannibals With Forks: The Triple Bottom Line of 21st Century Business.* Gabriola Island, BC: New Society Publishers.

15 See, for example, "Carbon Capture and Storage from Fossil Fuel Use" by Howard Herzog and Dan Golomb, Contribution to Encyclopedia of Energy. http://sequestration.mit.edu/pdf/ Massachusetts Institute of Technology Laboratory for Energy and the Environment.

16 Arthur D. Little report to the National Petroleum Council Committee on "Future issues for the U.S. oil and gas industry."

17 International Tanker Owners Pollution Federation Limited, http://www.itopf.com.

18 Wintson S. Churchill, 1959. *The World Crisis,* quoted in Longhurst, Henry *Adventure in Oil: The Story of British Petroleum.* London: Sidgwick and Jackson Ltd.

19 Carlos Ruiz Zafón, 2004. *The Shadow of the Wind.* Luca Graves, ed. New York, The Penguin Press, p. 363.

20 Peter Schwartz, 1991. *The Art of the Long View.* New York, Doubleday, p. 64.

5 Persuasive Investor Value Proposition

We have shown in previous chapters that the global energy mix goes through transitions and that current high prices may be signaling a change ahead. Internationalizing NOCs (INOCs) are competing aggressively for available resources, and resource rich countries are favoring their domestic companies, whether NOCs or private. Recent supply and demand trends for oil appear unsustainable, and the future energy mix must become more diverse. Concern over a business cycle downturn should be subordinated to the recognition that the energy industry is in a secular growth trend bearing greater resemblance to the 1950s and 1960s than to the 1970s and 1980s. Geopolitics has always been central to energy developments and societal demands on energy companies for clean energy, ecological protection and exemplary behaviors are increasing in intensity.

In this chapter we discuss how the energy companies may need to adjust their value propositions to investors as they seek sustainable growth and returns in an energy market in transition. The super-majors are riding a wave of success that they caught in the 1990s, but the sustainability of their current business model is questionable. We suggest that they need to change their shareholder value proposition from capital discipline expressed as screening out projects to capital effectiveness in managing the risks of a diverse and growing portfolio of projects. We relate seven growth stories and a turnaround to illustrate several new, focused business models that appear to have staying power. We also review an example of a company that chose to monetize its successes through a thoughtful exit. We are convinced that companies will rise and fall in the transition ahead: leaders of change will create extraordinary value, while those trapped by obsolete shareholder value propositions will miss the boat.

Purpose of the Energy Firm

A firm's existence depends on its ability to satisfy needs within society and by so doing create more value than it expends. An energy firm generates revenues by fulfilling the demand for products at prices that reflect their market values. It incurs direct costs to find, transform and deliver the energy. A firm's investor value proposition has traditionally focused on the financial, operational, and technical means it employs to produce a greater surplus of revenues over costs than its competitors. Now, however, a sustainable investor value proposition must also instill confidence that a firm can productively deal with its myriad stakeholders, including customers, employees, suppliers, partners, resource owners, local communities, and the broader public, as well as local and national governments.

In the previous chapter, we discussed how these evolving societal expectations impact not only direct costs to be incurred, such as a potential windfall profits tax or the additional costs associated with environmental or social policies, but also indirect lost opportunity value caused by restrictive laws and policies resulting from negative public and government sentiments. In many respects, the net value that is created by the firm represents the degree to which the firm has satisfied societal needs. The final stakeholder, the shareholder, captures the residual interest in the value of the firm, after all the other stakeholder interests have been accounted for. Shareholders take the risk that the firm may not succeed in creating net value. So an energy leader must continuously assess whether the resources committed or value expended in furthering the firm's relationships with other stakeholders are likely to add to or subtract from the net value of the firm.

As Michael Porter observed, "In this corporate competitive context, the company's social initiatives—or its philanthropy—can have great impact. Not only for the company, but also for local society. . . . I used to see this area of corporate social performance as the last thing on my agenda 10 years ago, but now I agree that social and economic issues are intertwined."[1]

The energy leaders we interviewed all recognized that they must satisfy this wide community of stakeholders. At the same time, the shareholders enjoy a special position in this community for very good reasons: boards of directors are established with a primary fiduciary duty of ensuring that management is protecting shareholder interests. Originally, that duty was focused mainly on assuring that the accounts accurately reflected the company's financial activities and current position. That is still necessary, as witnessed by the financial scandals of the merchant energy companies and the reserves bookings controversies in the oil sector, but it is no longer sufficient. There are huge risks that shareholders' value may be destroyed by management failures to satisfy other stakeholders. This can occur through accidents (Exxon Valdez and BP Texas City refinery), ill conceived decisions (Shell Brent Spar), disgruntled local communities (Nigeria), problems with security forces (Colombia), and sanctions (Myanmar, Libya, and Iran). Making money and accounting for it honestly is only part of the CEO's job. The super-majors are particularly exposed because of the scale of their operations and their high political profiles. We have quoted in the previous chapter Total's Thierry Desmarest, who sees the purpose of his firm in the context of a broad concept of value creation; two other definitions from CEOs of super-majors of the purpose of their firms also reflect the need to satisfy a broad set of stakeholders.

John Browne, CEO of BP, articulates the purpose of BP quite expansively: "The purpose of the firm is to provide energy products to people who want to buy them at a cost they can afford. To do that, you need to work with a large number of people: primary among whom are the shareholders, whose trust you need so they will provide you with the money you can then use. But they're not the only ones you have to work with. You have to work with governments, the staff, local communities."

Dave O'Reilly, CEO of Chevron, looks back from the customer: "The purpose of the firm at the very beginning is to provide energy in its broadest sense, and to do that in a manner that meets the needs of customers—here we're not talking about micro customers, we're talking about macro customers including societal needs—in a way that is sustainable, and make a return while we're doing it, because we have to satisfy the shareholder at the other end."

Investor Perspective

So there is widespread agreement in the super-majors that the shareholders are the crucial "barometer" of a company's success, but also recognition that the shareholder's interest can only be fulfilled by satisfying the other stakeholders. Shareholders are looking for financial returns, which in turn come from a company's ability to increase the value of the firm and to distribute any economic surplus to shareholders effectively. Shareholders in many cases need energy stocks to play specific roles in their portfolios; energy company leaders are well advised to understand what investors expect from them in terms of risks, returns and growth and to communicate carefully any planned changes in direction.

Shareholder Objectives

Ultimately, shareholders want a strong, sustainable investment performance from the stocks that they own. They want some combination of appreciation in the value of the stock and income from dividends. Some investors focus mainly on stock appreciation, while for others it is dividends that matter. National tax policy can influence the preferences of investors and shift investors from one set of preferences to another.

Management has four main choices for allocating the free cash generated by the business after paying all costs and taxes: it can pay back debt, it can reinvest in the business, it can pay dividends, and it can buy back stock.

If a company is over-leveraged with debt, then paying back creditors can be the best use of free cash, since it can improve investors' perceptions of the sustainability of the firm. By lowering its perceived financial risk, the company's future cash flow prospects will be discounted at a lower rate, which clearly can enhance value. Beyond a certain point, generally considered about 30% of total capitalization, further debt reduction does not increase value.

Companies with a shareholder value proposition centered on growth generally reinvest their free cash flow and raise further capital through new debt and equity offerings. Companies focused on high returns generally pay down debt and return surplus cash to the shareholders through dividends or share buybacks. In each case, it is important that the value proposition can be sustained—that there is a pipeline of growth opportunities that a growth company can capture, or that the return company can replace value lost as existing assets deplete or depreciate with new opportunities of equivalent value potential.

Alfred Rappaport insists that companies should return cash to shareholders when there are no credible value-creating opportunities to invest in the business, either by buying back stock, or "when a company's shares are expensive and there's no good long term value to be had from investing in the business, paying dividends is probably the best option."[2]

Steve Lowden, CEO of Suntera and Sun Energy Resources, a private capital company, believes there are structural problems with public companies' excessive scale. He claims that many investors want to be closer to management than is possible in a public company:

> Companies are generally getting bigger. And they are getting to the point where the investor and the investment decisions are too far apart. So management is no longer in touch with the investors about what they are doing. And they are working through a public

market screen which has become nearly impossible to use unless you talk to every shareholder at the same time about everything, which means you basically say very little, as a result.

That's why the private equity market is booming, because it reflects the need for the investor to be closer to the management. It's the desire to have a private dialogue, and not through this useless public screen that gets so filtered that there is nothing on the other side of it that differentiates it.

So it forces people into the private equity sector. And private equity wants to get into energy. And so what that will lead to is a number of energy companies that will have their stakeholders very close to their management. And I think that's a big value add.

The Value of the Firm

Theoretically, the value of an enterprise (enterprise value) should be the present value of future cash flows (after investments, interest payments and taxes), discounted at the firm's weighted average cost of capital (WACC). The risk profile of the firm should be a determinant of its WACC. The value of shareholders' equity should be the enterprise value less the value of the firm's debt. The stock price should be the value of shareholders' equity divided by the number of shares. Again, you find this rather mathematical view in any introductory finance textbook. And for the aficionados, the calculation becomes more complex, because there may be different classes of shares, convertible shares and stock options that may affect the proportion of the value that is due to common stock holders and there may also be long-term obligations that are described in footnotes that need to be considered as if they were debt obligations.

This short treatise is quite easy to state in theory. The problem is that we don't know what the future cash flow stream will be because we don't know what sales volumes, prices or costs will be. These are all just guesstimates. Moreover, the calculated WACC is a flimsy reed to cling to in judging the risk-adjusted rate of return that we should use to discount future cash flows to calculate the value of the enterprise. The few analysts who dissented at the time that Enron's stock was soaring did so on the grounds that they could not understand what the future cash flow would be, and were uneasy that the full risk of the enterprise was properly expressed by calculating the weighted average of its cost of equity and cost of debt. It later transpired that Enron's future cash flow was compromised and its marginal cost of capital was very high; however, these conditions were illegally hidden from the marketplace. Most analysts were still touting the stock until shortly before the crash, illustrating that it is not easy from the outside to fully understand future cash flow prospects or risks.

Even a firm's management is uncertain of future cash flows. At any given time, prices are uncertain and costs can change; demand for energy products can surge or ebb; events such as hurricanes can intercede, wars can break out, and tax rates can be changed. Consequently, SEC regulations are justifiably strict on making "forward looking statements," and investors are left with the necessity of making their value judgments based on the past track record of the firm and the apparent logic of its plans. These are the tea leaves that analysts sift through in their attempts to make "buy, hold, sell" calls to investors.

Stock Analysts' Metrics[3]

Stock market analysts focus on past and expected future stock price appreciation. They compare stocks on some simple ratios to try to uncover potential anomalies that could suggest a stock is undervalued and may catch up with its peers in the future. They include dividend yield as a variable in looking for potential anomalies. Then they relate their observations on value anomalies to what they call fundamentals, which are a series of indicators from the financial results posted by companies. The job of corporate management in a public company is to build the highest possible value of the firm; the investor relations job is to try to persuade markets to fully reflect this value in the stock price; the analyst's job is to advise clients to buy the stock before the value is fully reflected in the stock price or sell the stock if it appears significantly overvalued.

Valuation

Price/Earnings Ratio (P/E): This ratio allows analysts to compare how the market value of the firm compares to its earnings. A firm with a high P/E ratio is worth more than a company with the same earnings and a lower P/E ratio because investors believe that earnings will be higher in the future or are more predictable at the former than at the latter firm.

Dividend Yield: This is the cash dividend per share divided by the market price of the stock. In general, high growth stocks pay low or no dividends, since they reinvest their cash flows in growth.

PEG ratio: is the P/E ratio divided by the future expected earnings growth rate. This is a cross-check on the P/E ratio. If the PEG ratio is low, then the stock price may not fully reflect the value of future growth. Alternatively, the market may have determined that the consensus view of future earnings growth should be discounted because of high risk.

Fundamentals

Profit Margin: Net profit after taxes divided by total sales revenues. Companies with high profit margins are generally preferred, since high profit margins can be an expression of strong competitive position or attractive industry dynamics.

Return on Equity: Net profit divided by book value of equity expressed as a percentage. In principle, this measures the effectiveness with which capital is invested.

Debt to Equity: Long term debt divided by common shareholders' equity measures the firm's leverage. High debt to equity can raise return on equity, but increases risk.

EPS Growth: Annualized earnings per share growth rate, generally for the preceding five years. A high growth rate in earnings, if it can be sustained, is valuable.

Stock analysts also look carefully at the consensus view of a stock's investment potential. If a number of analysts have all recently concluded that a stock is a "buy," then a surge of investment can produce stellar appreciation. However, if the surge takes the stock too far, the price may start to fall back before the consensus records its overvaluation.

Stock analysts also look at technical data to assess whether a stock has momentum or not. Based on extensive and sophisticated mathematical analysis of cycles in stock price movements over time, technical analysis is generally used for short-term trading; but technical signals can provide an early warning that a growing body of investors is becoming concerned with the fundamental value of a stock.

Investor Portfolio Considerations

Investors are generally interested in building a portfolio of investments. Energy stocks must fit into portfolio segments representing various categorization variables, for example:

- Industry classification
- Performance profile (cyclical or steady)
- Technology (high or low)
- Growth (high or low)
- Size (large, mid or small cap)
- Geography (US, international, emerging markets)
- Social and environmental performance

Who are the investors that companies need to persuade to buy their stock? There are mutual funds that combine different segments in a bewildering number of combinations, seeking to appeal to specific segments of the investing public. An extreme form of mutual funds are hedge funds, which often use very sophisticated mathematical models to take positions on international commodities and currencies, and can from time to time take positions on individual stocks that can add to price volatility. There are also institutional investors, who are stewards of capital on behalf of a variety of institutions, including pension funds, insurance companies, college endowments, and charitable organizations. And then there are private investors who like to pick stocks rather than buy mutual fund products. In practice they are often represented by stockbroker firms and their analysts, who must make recommendations to their clients.

These investment managers and advisors try to provide their customers with a portfolio of investments that satisfies their aspirations for risk, returns and investment time horizon. They undertake fundamental analyses of the economic and business outlooks for specific sectors with the objective of advising customers on the best mix of stocks and bonds, and within the stock portfolio which sectors seem most promising. They often start from a picture of the allocation by industry type

Table 5.1: Sector Allocation of U.S. Market

(Composition of Wilshire 5000 Index at 6/30/06)	Percent Market Capitalization of Companies
Consumer Non-Durables	25%
Finance	22%
Technology	17%
Materials & Services	10%
Energy	10%
Utilities	6%
Capital Goods	6%
Transportation	2%
Consumer Durables	2%

in the overall market (table 5.1). A total market index fund will include representatives of sectors close to the total market proportions.

As described below, energy decreased as a proportion of the overall stock market as the economy grew faster in other sectors in the 1980s and 1990s. Also, the share of public companies in the overall energy market has declined as government-owned companies have increased their share. Nevertheless, the energy sector remains a very large component of the economy and of the stock market. As noted in the previous chapter, the National Petroleum Council study on future issues facing the petroleum industry demonstrated that the petroleum industry was still a vital part of the U.S. economy, but that the industry was at a critical juncture in establishing public trust. Since then, the industry has increased in importance; however, public trust has continued to prove elusive. This lack of trust affects the value of energy stocks in general by causing a segment of the population to avoid them.

From time to time, investment professionals will recommend that their customers "overweight" one sector over another on grounds that the sector either appears currently undervalued, or has more exciting future prospects than most investors perceive. In recent years, several analysts have been recommending "overweighting" the energy sector, presumably influenced by their reading of the forces described in chapters 2 through 4.

The Role of Energy Stocks

No one has a perfect crystal ball, however, and many portfolios retain some exposure to a variety of sectors. Energy stocks in the 1980s and 1990s have provided lower returns but also lower variability than most other sectors. As we learned in chapter 1, this historical period was a down cycle for the energy industry. The utilities portion of the energy chain has likewise had low returns, but coupled with high variability (risk). This is a signal the utilities industry is due for some substantial restructuring. The question for investors is not the past but the future. Will the past performance continue or are there upsides? What is the right balance of energy in a portfolio?

Energy companies can rely on investment professionals to include some energy stocks in their portfolios. However, it is up to the individual companies to make the case that their stock rather than another energy company's should be included. In times gone by, this used to entail quiet conversations between the CFO or investor relations executive and investment professionals. An analyst would float a trial balloon about what he thought the next quarter's results would be and the assumptions behind the forecast. The company representative would then give a subtle nudge by asking, "Have you thought about this?" never giving any real information but indirectly encouraging the analyst to go back to the drawing board. Since the analysts earned respect (and a bonus) for accurate predictions, this process worked reliably. However, this practice is now illegal and, as Steve Lowden observed, all investors must be provided with the same information. This has elevated the importance of road show investor presentations in conveying information about a company's prospects to the investment community; and it has also diminished the value of the insights of individual analysts.

Risks and Returns

There are numerous segments within the energy sector. John S. Herold, Inc., probably the most knowledgeable analyst of energy company performance, divides the 371 public companies that it tracks into 17 segments depending on size and area of geographic and functional focus.

As with the different sectors of the economy, the different energy segments fulfill different roles in investors' portfolios. The year 2005 was kind to energy stocks, and segment averages showed high appreciation. Nevertheless, within some segments the gap between the highest and lowest performance was very large (fig. 5.1).

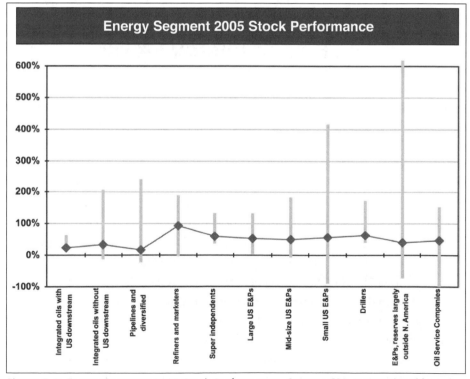

Figure 5.1: Energy Segment 2005 Stock Performance. *Source: CRA International based on data from J.S. Herold*

The highest variability in performance within a segment was among companies with reserves largely outside North America. If you were lucky enough to have picked Det Norske Oljeselskap, your investment would have septupled in nine months. By contrast, if you had picked First Calgary Petroleum or Sydney Gas, your investment would have declined by two-thirds. In general the segments covering smaller companies show the greatest variation in performance, indicating high risk in picking a single stock from this segment, but also the potential for high rewards for picking right.

The role of energy stocks as providing decent, predictable returns with modest growth is under threat from new competition and increasing difficulty in gaining access to conventional resources, as well as from public distrust and suspicion. The success strategies of the 1990s were based on increasing market share by acquisition and improving returns through intense cost reduction, at a time when supply growth from OPEC and countries of the former Soviet Union was depressing oil prices. These strategies will be difficult to sustain in the next decade as asset values are bid up and shortages of skilled personnel make cost reduction a challenge.

Majors at a Turning Point

Rising oil and gas prices created opportunities for the entire oil industry to earn record profits in 2005 and 2006. The largest oil companies, the majors or super-majors, delivered truly jaw-dropping results, spectacular enough to trigger public backlash and political watch-dogging. It is far from clear, however, that the majors can sustain this kind of performance. Indeed, the majors even more so than smaller energy companies would seem to be threatened by the profound changes now underway in the global energy industry. In our view, the majors are at a turning point, and the decisions their leaders make in the next few years will in many cases involve their companies' very survival.

Return to Favor

Energy stocks were not favored in the 1980s and 1990s because they were seen as cyclical and low-growth. They were often characterized as part of a dying "smokestack industry" segment of the economy. Indeed, the number of energy companies in the New York Stock Exchange (NYSE) largest 25 companies by market capitalization fell dramatically in the 1990s (fig. 5.2).

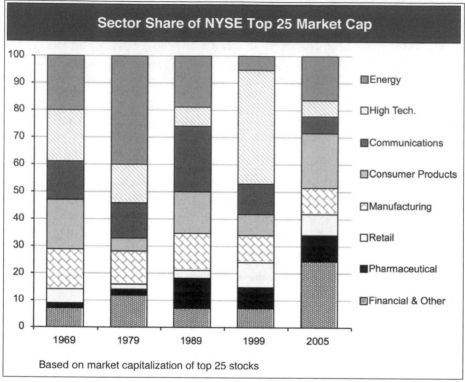

Figure 5.2: Sector Share of NYSE Top 25 Market Cap. *Source: CRA International, Inc.*

By 1999 there was only a single energy company (Exxon) in the top 25, down from 11 of the top 25 in 1979. But 1999 was the height of the dot-com bubble, and since then consolidations in the energy sector and decline of the "irrational exuberance" have helped energy stocks regain positions among the largest companies in the NYSE. The other sector that increased its share of the largest companies substantially in recent years was the financial sector, following its deregulation, while "high tech" stocks and communications companies declined.

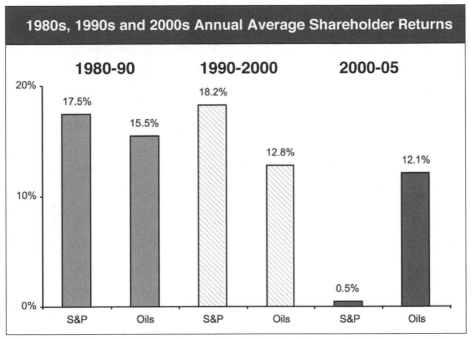

Figure 5.3: Relative oil TSRs have improved in the 2000s. *Source: CRA International, Inc.*

As petroleum companies became a smaller part of the total market, the total shareholder return[4] (TSR) of energy stocks lagged key market indices (fig. 5.3). In the 1980s, Exxon was the best performing super-major stock. In the 1990s, BP took over the sector leadership. In both periods, the leaders provided higher TSRs than the S&P 500, but a broad oil portfolio of oil stocks lagged the S&P 500 for the full 20 years. This reflected the growing maturity of the oil sector in this period, a time when there was strong growth in other sectors of the global economy and a boom in high technology stocks

Since the year 2000, energy stocks have substantially outperformed most market indicators as oil prices have once again risen, while the tech stock bubble burst. ENI and Total have been the leaders of the super-major sector so far in this decade.

Some Warning Signs

How does a company's market value relate to the value of its assets? The relationship can change over time. J.S. Herold regularly calculates the appraised net worth (ANW) of each company it follows by estimating the liquidation value of the assets it owns (that is, the value they should command if they were sold at prevailing asset prices). Their Web site details their methodology, which inevitably includes a significant element of analyst's judgment, since the value depends on uncertain future events. Herold places most weight on its assessment of company future cash flow prospects. However, they also take account of P/E ratios, EBITDA multiples, comparative transaction values, replacement values, balance sheet worth, public, and private market values.

Herold's Comparative Appraisal Reports allow its clients to compare the calculated appraised value of assets (AVA) with each company's enterprise value (EV), which is the sum of the market capitalization and long-term debt. Herold's calculation provides an idea of what each company's assets are worth in terms of their future cash flow potential, but also taking account of recent transactions for comparable assets. The stock market also is trying to set a value on future cash flow potential. But the two sets of values rarely coincide. This is because investors may believe that the assets are worth more (or less) in the hands of the company that owns them than in the hands of a third party.

At the end of 2004, despite strong stock performance relative to the general market, each of the super-majors showed an AVA that was above the enterprise value of the firm. That is to say, investors valued those assets less highly in the super-majors' hands than they were worth on the open market (fig. 5.4).

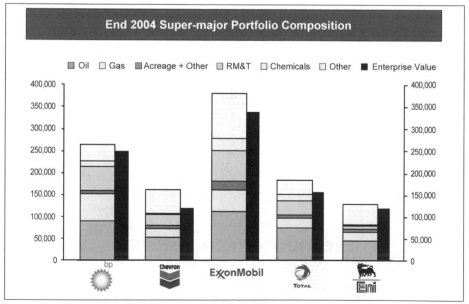

Figure 5.4: By end 2004, all the super-majors showed parenting disadvantages. *Source: CRA International based on data from J.S. Herold*

The difference between enterprise value and asset market value is known as "The Parenting Advantage," a term coined by Michael Goold and Andrew Campbell and developed in a series of books from 1984 to 2002. Their proposition is "that multi-business corporations only make sense if the corporate parent has a clear strategy for adding value," and that "the objective of corporate strategy should be to create 'parenting advantage,' that is to say, to add more value to the company's business than any other owner could."

In the early 2000s, the super-majors did deliver a parenting advantage in that their enterprise value was higher than the value of their assets at market prices. But there was an abrupt change in 2004 as escalating asset values overtook increases in super-major stock prices (fig. 5.5).

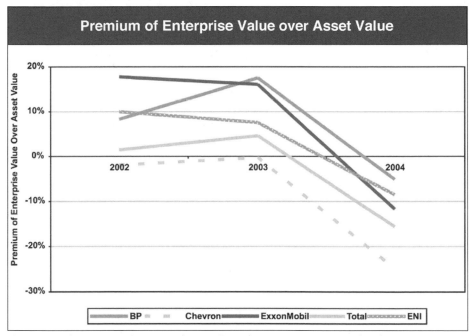

Figure 5.5: The super-majors turned parenting advantage to parenting disadvantage in 2004. *Source: CRA International based on data from J.S. Herold*

This raises the question of whether the underlying methodology used by J.S. Herold has broken down in the recent increasingly heated market, or whether something has changed and the value proposition that served the super-majors so well prior to 2003 is no longer appreciated by stock markets. It could be that in today's market, in contrast to less frantic times, the excess demand for oil and gas properties over the naturally available supply creates a larger difference between the value of the marginal transaction and the value of a large pool of assets, such as those owned by the majors. In that case the Herold methodology may lose validity. Or it could just be a question of leads and lags, as suggested by the strong move of stocks in 2005 that had the highest discount against AVA at the end of 2004. Or it could be that growth-oriented companies or INOCs (see chapter 2) supported by their home governments can justify paying higher multiples of future earnings potential for assets than those with lower growth targets, and the super-major business model is no longer advantageous.

Figure 5.6: Super-majors TSRs. *Source: CRA International, Inc.*

Super-majors have provided good returns to their shareholders in this century (fig. 5.6). What has the super-major value proposition been? For many years their value proposition has centered on returns on capital employed and sustainability in the form of the predictability of results. They argued to investors that their size provided economies of scope and scale that allowed them to generate higher returns than other energy companies. These higher returns resulted in very strong balance sheets and generated excess cash that could be used to make acquisitions, pay generous dividends, and return money to shareholders by buying back stock. The value proposition to shareholders has been much less concerned with growth. Indeed, the super-majors have pointed to the fact that their capital expenditures after the mega-mergers were lower than the combined total of the predecessor companies as evidence of their capital discipline and selectivity.

The jury is still out on whether shareholders are refocusing on growth and sustainability rather than returns and sustainability. However, there are signs that growth has become more important.

The Case for Change

In the 1970s, the Seven Sisters' assets in OPEC countries were nationalized. BP lost the greatest portion of its assets, some four million barrels per day of production (see chapter 4, fig. 4.2). The company quickly invested in the North Sea and Alaska and grew production through much of the 1980s. Then in the low price environment of the late 1990s, BP led the consolidation charge with its acquisitions of Amoco and Arco. It has since made a pioneering step in its Russian joint venture TNK/BP. Other super-majors took similar steps in the 1970s to find new sources of production to replace their lost Middle East, African, and Venezuelan resources.

There is some evidence that the stock market has shifted its emphasis from capital discipline to production growth in valuing super-major stock prices. Indeed, the difference in shareholder returns among the super-majors during 1999–2005 can be explained almost entirely by changes in annual production (fig. 5.7), while returns on capital employed have ceased to be a differentiator of stock market performance.

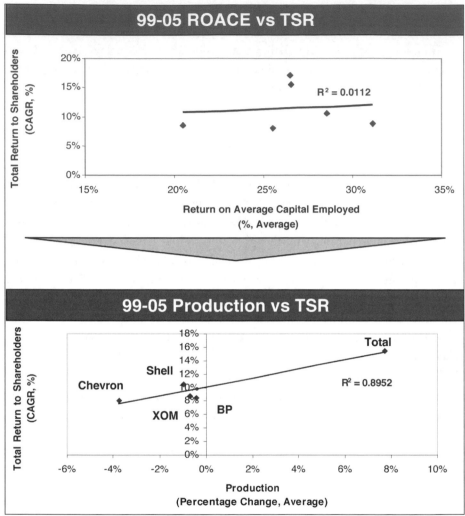

Figure 5.7: Production growth has replaced returns as a predictor of total shareholder returns. *Source: CRA International, Inc.*

That is not to say that ROACE is unimportant, but that the market is satisfied that all the super-majors have proper processes in place to assure capital discipline. Further, it seems that investors are no longer impressed with stock buybacks. Art Smith, CEO of J.S. Herold confirmed this: "Investors in the super-giant oils are no longer content with the share repurchase option as a relief valve for excess cash flow; increasingly they are demanding upstream growth in reserves, production—and strong financial performance to boot." As Smith says, investors' concern has shifted to whether the companies can sustain production. And the reality is that they have been quite unsuccessful in recent years. Over the period 1999–2005, only ENI and Total showed positive production growth, BP was flat, and ExxonMobil, Chevron, and Shell showed production declines.

The super-majors today face a challenge similar in magnitude to that of the late 1970s. They have established projects to secure growth through 2010, but the weight of their declining mature oil fields will make it difficult to maintain, let alone grow, their conventional production beyond that date. Enormous as the super-majors are, they pale in comparison with Saudi Aramco and Gazprom (see chapter 2), and have competitive disadvantages against several INOCs. As Clarence Cazalot, CEO of Marathon Oil, told us: "It is one thing to replace the volume of declining fields, it is another to replace their value." This is because the value of positions accumulated during the low price 1990s is just not replaceable in the high price 2000s.

There are new exploration opportunities in the Arctic and possibly off East Africa. However, should they yield major petroleum discoveries, it will take a decade or more to develop them, and even then they are unlikely to compensate for declining mature properties. Super-majors and large integrated companies will need to pursue the alternative organic growth avenues outlined in chapter 3, and consider the challenge laid down by John Wilder, CEO of TXU (see below), to enter the power generation business in order to renew their shareholder value propositions and make them more persuasive and sustainable.

This is not to say that the super-majors should abandon their previous business model and seek suddenly to become growth companies. They have a large group of stockholders who depend on them for reliable performance and steady dividend checks. It would be very dangerous for them to alter course abruptly since they would lose existing shareholders when the change became apparent, and would only gain new shareholders when the new course was demonstrably successful. In the meantime, the stock price would fall as more investors sold the stock than wanted to buy it.

Growth brings with it new risks, and the companies will need to demonstrate to investors that they are up to the task of managing greater risk by diversification, hedging, or laying it off to third parties such as governments. The story needs to change from capital discipline expressed as screening projects out to capital effectiveness expressed as managing the risk of a diverse and growing portfolio of major projects.

These are enormous companies and, like the very large tankers they use, there is a long time lag between turning the rudder and changing direction. Nevertheless, we believe there is value in moving the rudder firmly in favor of growth and change, and in putting even more serious capital to work in alternatives to conventional oil.

Seven Growth Stories, an Exit, and a Transformation

Some companies have been growing vigorously by acquisition and by building new businesses in unconventional resources. We profile seven of those companies whose CEOs and senior leaders granted us interviews. We also describe Burlington Resources's very successful exit. Companies that are concerned that their shareholder value proposition may not be sustainable have very attractive exit opportunities in this strong market environment. Finally, we note the emergence of a new type of power company in the transformation of TXU.

Several key themes emerge from examination of these cases. First, there remains considerable room for industry consolidation. The upstream segment, especially, is fragmented, and there are benefits of "scale in basin" that large E&P companies can capture. Second, there are huge opportunities in unconventional resources—stranded gas, tight gas, and oil sands. Stranded gas and heavy oil opportunities require an integrated value chain approach, but these resources seem to present substantial running room. Third, there is an emerging opportunity in power generation. Companies that have focused their value propositions on profitable growth in these areas have amply rewarded their shareholders.

ConocoPhillips

Just below the super-majors, the next tier of companies comprises the large integrated group of U.S. and international oil and gas producers. Both tiers promise growth primarily through international production and propose that their downstream assets provide them with important capabilities to monetize difficult resources. The division between U.S. and international companies in this segment is becoming increasingly irrelevant as their value propositions and the types of investor they try to appeal to converge.

The largest of the U.S. companies in this segment is ConocoPhillips, which is knocking on the door of the super-major group. One dubious reward for that status was that the CEO, Jim Mulva, received a summons along with the CEOs of ExxonMobil, Chevron, BP America, and Shell Oil to appear for a grilling by U.S. senators at the November 2005 hearings on rising oil prices. John Lowe, executive vice president of ConocoPhillips told us: "I think we're now clearly the smallest of the super-majors rather than the largest mid-cap. I think we're in a different class than the mid-sized companies, in our ability to take the steps to remain a long-term competitor. The companies in that smaller group are limited in their ability to take the steps they need to remain a long-term competitor, because access to reserves requires financial capacity, global scale, scope and knowledge, and they don't have those." Nevertheless, we have included ConocoPhillips within the integrated segment for historical reasons and because of the growth focus of its value proposition. Nearly all the companies in this segment had at the end of 2005 doubled the value of an investment made at the end of 2000, and the highest performer, Murphy, had quadrupled its value (fig. 5.8).

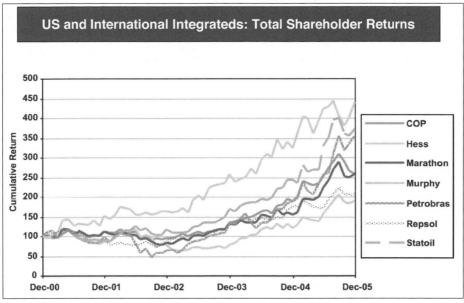

Figure 5.8: U.S. and international integrateds. *Source: CRA International, Inc.*

ConocoPhillips argues that it has all the capabilities of a super-major in size and scope, but that it has greater potential for growth by virtue of its smaller base. ConocoPhillips is also stressing unconventional resources and is investing aggressively in LNG and oil sands. Most recently, ConocoPhillips acquired Burlington Resources at what appeared to many analysts to be a high price. Yet ConocoPhillips' growth and transformation towards the super-major category has been based on aggressive acquisitions, each one building on the last and opening up new options for the next move.

As John Lowe said:

> What we thought might happen with that size and scale is that there are efficiencies that you didn't even know could exist. We created the commercial organization to look and try to optimize this whole system instead of optimizing each asset, I think that's worth

several hundred million dollars a year. Take the Wilhelmshaven refinery that we just announced the acquisition of. If you look at what Louis Dreyfus purchased their crude for and what they sold their products for, we receive better value for those products because of our size and scale. That allowed us to acquire that refinery: the value that we see there that the seller couldn't see.

John Lowe summarizes the pathway as follows:

First, you have to get competitive or give up and sell out; once we achieved scale, we saw that we could add value by getting even larger. Then if you look at how the majors trade, consistency of earnings, lack of volatility of earnings, financial capacity, the ability to fund projects through the troughs, the ability to have sustained dividend increases, all of those things are attributes that people who trade better than we do have that we did not have. So we felt that if we acquired those attributes, then we would trade better as well.

An investor in ConocoPhillips has been rewarded for sticking with an aggressive management team committed to growth and adding value by "getting even larger."

Statoil

The international integrated companies have value propositions similar to those of the large U.S. integrated companies. In several cases they also offer a strong position in their home markets. However, this can be a two-edged sword. On the one side, their strong home position provides them with higher margins and lower costs through the benefits of scale than their competitors. Also, because of their prominence they can benefit from diplomatic interventions by their home countries to gain access to opportunities. On the other side, they are potentially subject to political interference because of that same prominence in the economies of their home countries. The issue is whether they can hold political interference to the minimum, while benefiting from their advantages. One of these international integrated companies, Statoil, is leveraging its experience as a former national oil company to deepen relations with host governments and their NOCs.

Helge Lund affirms that Statoil's value proposition is centered on growth and profitability:

Recently I had discussions with investors in Oslo, London, Chicago, Boston, and New York. Since the IPO in 2001, the investors have achieved strong and competitive returns. Originally many were skeptical as to whether the Norwegian government with their high ownership share would act as a professional and transparent owner. Well, they have and I think the investors recognize the government for taking care of its interests like any other owner through the appropriate mechanisms and corporate bodies.

Further, I think we formulated quite aggressive targets that were communicated to the market in 2001 as a basis for the investment proposition. These ambitions were very high, but in 2004 Statoil delivered on the targets in a much disciplined manner. We have formulated new business targets for 2007, and we are on track to delivering on these as well.

A key part of our value proposition is growth. Today there are few international oil and gas companies that are delivering the kind of growth figures as Statoil. This is one key value proposition for our shareholders, combined with the fact that the company has achieved increased confidence in terms of strong capital discipline.

We also have a strong position on the NCS, a stable and predictable OECD region. The comparative value of this region has increased over the last few years as operational risks and political volatility are perceived to have increased in many other regions.

Portfolio choices

I believe these factors, in addition to the strong technology position that Statoil has, are key reasons why investors are looking at Statoil with interest.

Lund is sympathetic to the situation of the majors, but also emphasizes the vital rejuvenating role of growth:

I have a strong sympathy for what you are saying; but the issue the majors may be facing is that they are enormously big, and at the same time there are relatively fewer new big opportunities to pursue. Some of the areas which still hold massive resources remain more or less closed. I believe this is a key dilemma. Organizationally I feel a business—like in nature—that is not growing has started a process of decline.

Growth

So then the question is what kind of growth are you talking about? I am not sure that you talk about growth in production every year or growth in profitability. But it has to grow in terms of challenges and opportunities. Too many years of consolidation may be very dangerous for a company. Developing an "anorexic organization"—trying to save yourself to success—will never work.

It is an expression of the march of globalization that the top three performing stocks in this segment were Murphy (U.S.), Statoil (Norway) and Petrobras (Brazil) from 2000-2005.

Chesapeake

Large independent E&Ps have value propositions that fall in a broad spectrum of growth and returns, though the sector as a whole generally is expected to offer higher growth than the super-majors. At one end of the spectrum, there are companies that promise rapid growth, such as Chesapeake.

Chesapeake has an approach centered on strong organic and inorganic growth. Early on in our interview, Aubrey McClendon, Chesapeake's CEO, made it clear what he thought was unique about the company: "There is no playbook that says, 'Here's how to start with 10 people and $50,000, and 16 years later here is 2,700 people and $20 billion value, and oh by the way, you almost went under five times along the way.'" Certainly his investors have had a wild ride, but the stock performance since 2000 has been stellar (fig. 5.9).

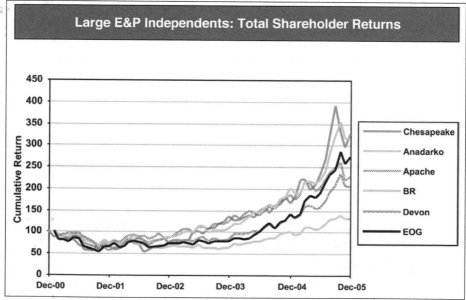

Figure 5.9: Large E&P independents. *Source: CRA International, Inc.*

He continued:

> No playbook, and if there were, I would like to have skipped a few pages! So when you talk about what's distinctive about the

company or how it works, I'd have to start with what's unique. The uniqueness is:

1. Leadership structure;
2. Continuity of senior leadership;
3. The distinctive history of the company.

We've always had to be scrappers, never had enough money until recently, not much respect in the industry, and that has made us be hungry and ambitious and determined, willing to take chances, have thick skin, criticism doesn't bother us much.

So he offered investors a higher risk, higher growth package that obviously was aimed at a different segment of investors, or a different role in an investor's portfolio, than the propositions of the super-majors (fig. 5.10).

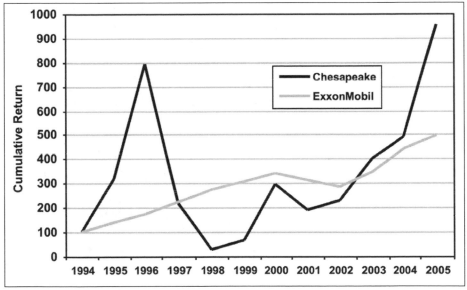

Figure 5.10: Chesapeake's value proposition appeals to a different segment of investors than ExxonMobil's

Chesapeake's initial play was in the Austin Chalk in the 1990s, and the company was very successful for a while. It established two themes that are still key components of Chesapeake's value proposition: geographical focus and leveraging technology.

We made a bet on technology enabling us to drill gas wells to 10,000 or 14,000 feet and to drill horizontally in an environment that had previously been too hostile for horizontal tooling. Down dip Austin Chalk was the original gas resource play. Everybody knew that hydrocarbons were there; you just didn't know how to get them out or you didn't know if you could get them out. We proved we could in Texas but couldn't in Louisiana.

This upset made them rethink the strategy, but they retained the growth-based investor value proposition.

Now, we drove it in the ditch and we were going 80 miles an hour when we did! But still, being in the ditch made us confront some truths that were only becoming barely visible then and are now obviously self-evident to anyone.

We restarted in Oklahoma, as this is what we know best, and it's a region that Houston companies seem to ignore. We looked at the top 20 list of producers, and no one was drilling very much, and we said, you know, it's unconsolidated, it's produced lots of gas, everybody thinks it's dead; but maybe it's not, and if we hit it with the new generation 3-D and if we hit it with a lot of people, not two geologists working in Oklahoma but 20 geologists working in Oklahoma . . . That's where we got to this kind of unified vision of the geology in Oklahoma, because we had so many people dedicated to figuring it out and great new 3-D seismic that didn't exist ten years before."

So with classic entrepreneurial spirit, McClendon and his long time partner Tom Ward rapidly accelerated to 80 mph again and built competitive advantage in their focus area.

I think a lot of it came through achieving scale. Every industry that I know of around the world (I'm sure there are exceptions but most that I think of today) consolidates for scale, generally to gain advantages on the cost side. What we found through scale is that there are other benefits besides just cost. We quickly found out that you could matter to service companies, get better pricing, better people, become a beta-tester for new products. And we found we were able to attract talent in that we had apparently a lot more on the go than companies that were all spread out, because we were so highly visible in what we were doing putting all these pieces together for the Oklahoma jigsaw puzzle.

There is a question on whether the growth is sustainable, and whether the business model that has been so successful in the mid-continent in a period of low prices can be as successful in an expanded geography at higher prices. McClendon thinks so. He notes that, while acquisition prices have risen, the margins on acquisitions are higher than ever for those who can lock in future commodity prices. Further, he aims to build functional expertise in "black shales" that he can leverage more broadly. The ride is not for the faint-hearted, but the Chesapeake value proposition has provided excellent returns to the shareholders so far.

Suncor

In the upstream oil business, a key to success has always been to capture a "legacy asset" that will provide steady income, cash flow and growth opportunities for many years. In the 1970s, the majors lost their Middle East legacy assets but were able to replace them with new ones, primarily in the North Sea and Alaska. As we have described in chapter 3, discovery of new conventional reserves has become more difficult, and companies are looking for new resources in unconventional and stranded gas, and in heavy oil and oil sands. Three companies, Calgary-based Suncor (oil sands) and EnCana (tight gas), and UK based BG Group (LNG), have become substantial and successful by focusing intently on a specific resource type and have, in our opinion, formed a new industry segment. Each of these companies has more than doubled its shareholders' investment in the past five years (fig. 5.11), and the two Canadian specialists have quadrupled their shareholders' investments.

Figure 5.11: Focused Resource Players. *Source: CRA International, Inc.*

Rick George has been CEO of Suncor for 15 years and has led the development of a new value proposition centered on growth in oil sands. It has not been an easy road. "I remember my first day on the job—I'd just come out of the North Sea, where just one platform we operated was producing as much as this oil sands complex and was done by 120 people. And the oil sands plant at that time was about employing about 2,500 people." He had serious doubts about whether the business model could ever work. "So, what started off was a sense of emergency and necessary change, which is the mother of invention. Probably the toughest capital appropriation we had was back in late 1991 or early '92, when we went to the board and asked for $120 million to move from an old bucket-wheel mining system to a truck and shovel."

George continued:

Now, you would look at that today and say that was a pretty easy decision to make. It wasn't when you had a 4% or 5% return business with not a lot of hope, and not a lot of faith by people that had been around the business for 20-25 years of really changing that much. And so I can tell you it's easier to get a $2 billion dollar project approved right now than it was $120 million in 1991-'92. But that was kind of the kick-off of the whole thing, in the sense that you had to get your costs down so you could get your return on capital up, so you could attract capital; and then you could work on technology and growth and scale and size and a whole bunch of other things.

We were still wholly owned by Sun at the time that appropriation was approved. By the time we had it implemented, we were on our way to becoming a public company. We became a public company in March of '92 and then we were a totally public company by '95.

George explained that this took away the option to say: "Oh, we can't get money to make change."

George also believes that it is the integrated approach that differentiates Suncor from other oil sands companies:

So if you're looking across Suncor, the leadership team knows the strategy. And so, if you're working in Denver, or Sarnia, or the market group, or in oil sands production, part of our job is: how do we get oil sand streams through to the customer; how do we produce the best products for the marketplace? And so that's where we differ.

The other differentiation George cites is the huge resource they are working with and how that forces a "long view" on him and the management team.

> I think this comes from strength around this vision of Suncor— we're going to be here producing oil sands for a hundred years. We have reserves in the ground for a hundred years. And you just put that in your brain and kind of think about it. So where a typical oil company will be in a certain country, they know they've got about a 20-year run on their leases. Or you're in East Texas and you drill those fields, and you're Exxon, and it gets down to a certain point, and you can't be an effective operator, so you sell it to some small player. But in Suncor we're going to be producing oil sands for a hundred years.

We asked why Suncor's assets may be worth more to Suncor than to another company:

> You know, that's a really good question. I think the track record. So let me take you to our share price for our shareholders versus any of the big five, I don't care who. Okay, they're up 400% in a 10-year period, and good for them; that's a great return. We're up somewhere between 1200% and 1500%. And so what I would say is, if I were a shareholder of Suncor, why would I want to move into a low return operator?

EnCana

Gwyn Morgan has recently retired from the CEO position at EnCana and is now vice chairman. However, he was at the helm as EnCana was formed through the merger of Alberta Energy (AEC) and Pan Canadian. Morgan reflected on his career: "I feel very fulfilled. When we started Alberta Energy Company in 1975, and I was responsible for getting the oil and gas business going from zero, organized the drilling of the first well, which we got done in '76, who would ever have known what happened from there?"

EnCana recently divested major portfolio segments in the Gulf of Mexico, Ecuador and the North Sea to focus intently on resource plays in North American tight gas and oil sands. Morgan explains:

> What appeared to be a dramatic portfolio shift was actually returning to our roots. Because when we started AEC back in '75, the very first rights we obtained were to a very large, tight gas field. At the time, we weren't thinking conventional, unconventional, like our annual report says now.

We were thinking that it was a tight gas field, and we had to do special fracing techniques to get the gas out of it. What we know now is that we were working in a classical resource play. These are plays where the hydrocarbon in place is so large that it is almost irrelevant as opposed to a conventional field, where it is very relevant. And so what we discovered over time, of course, was that the initial component of the hydrocarbon we actually contacted with our drilling and fracing techniques was a very small part of the ultimate recovery.

Altogether we've drilled about 6,000 wells in that field, I think, and will ultimately recover three or four tcf, which is a lot bigger number than we originally booked. But I guess the learning there is that we were dealing with a lot more gas in place than we ever dreamed of. We couldn't see it. We didn't know exactly where it was. And the other classic thing about the resource plays, because you work it so long, and so intensely, you learn more and more and more about it. And you keep applying those learnings. One of our key strategies in the company, one of our key philosophies and cultures is called 'Look Back and Learning.'

The classical resource play, every year you are going to drill more wells. And you are going to drill them for decades and decades and decades. And every year you can find out more. You can adapt your equipment. For example, the equipment we use now after 30 years, the nominal dollar cost to drill a completed well is roughly the same. And you know what has happened with the currency and service sector costs.

And over the years in Canada, we always gravitated towards fields that had low decline rates, long life. Well, low decline rates, long life basically meant the unconventional resource play model. Over the years we expanded that to other resource plays in Western Canada, more recently into coal bed methane, where we are in a leading position in Canada.

And then we entered the U.S. Rockies, which is all tight gas, at a time when the Rockies was fairly dormant. We moved into Denver, and it was a ghost town for the oil and gas industry. And we have had spectacular success there. I mean, in five years we have gone from zero to 1.5 bcfd; and we are the fastest-growing company and the biggest landholder in the U.S. Rockies, applying the core competencies and philosophies that we learned from the beginning.

And we built our culture and our knowledge base along with the company. We now have the strongest resource position for natural gas in North America, largest amount of undeveloped resources, unbooked resources. And a tremendous core competency in the best resource plays in North America.

Then we said, 'Okay, what about our international stuff?' We looked at it, and we looked at what was available as the price for what we could get for it. And we realized that—you know, the way I always look at, the way all of us in EnCana always look at the value of assets—we don't look at what we've got in it. We don't even look at what the full-cycle economics would be of developing a Buzzard or a deep Gulf of Mexico discovery like Tahiti. We say, "Well, the investment is what we could get for it."

For example, if our North Sea position were worth $2 billion U.S. today, before we actually put any more money into the development of it—there's a lot more money to go in—so, starting with $2 billion, how does this investment look? And when you do that analysis, you conclude that you are better off not to put more money into it; you are better off to sell it. Because your incremental rate of return is dramatically lower than the incremental rate of return of our North America asset base. So it's a question of allocating capital to your best returns.

What played into our strategy was the fact that good quality oil resources worldwide were getting in extremely short supply. And so we had tremendous competition for those assets. I mean, when we sold our totally undeveloped discoveries and land in the Gulf of Mexico for $2.1 billion U.S., it was going to be years before a barrel of oil comes out, and a lot of money to go in. And we had $300 million in the whole thing.

I've found that if you can sell something to a strategic buyer, you'll almost always do very well. And we had the luxury of being able to do that because we don't need the reserves, we don't need the growth, and we can now focus on being the best that there is at two things only, North American gas, especially North American unconventional gas, and North American in-situ oil sands. And we have the best resource position in each of those two things. And we have a culture, and a technology base, and a way of thinking, from

the top management down through our operating people, that's
lived and breathed that for a long, long time.

BG Group

British Gas was privatized in 1986 but continued to operate as a monopoly
until natural gas markets were gradually opened. Customers were allowed to bypass
British Gas starting in 1986 with large users, and spreading to smaller users in
1992. The Gas Act of 1995 introduced competition into the residential gas market.
Following passage of the Act, the UK initiated a free market experiment in natural gas
distribution by allowing a half-million residential and small business consumers in
three southwestern counties to choose their natural gas suppliers. This experiment
was then extended nationwide.

In 1997, responding to the new competitive environment, British Gas
shareholders approved a plan to split the company into two parts. The retail natural
gas sales business was assigned to a new company called Centrica, which added
retail electricity and home services to its product offering and has expanded beyond
the UK into U.S. and Continental European markets. The remaining business of gas
transmission and gas supply was split further in 2000 into National Grid and the BG
Group. National Grid now owns and operates gas and electricity transmission and gas
distribution networks in the UK and U.S. and electricity distribution networks in the
U.S. The BG Group is now a global E&P business that finds and develops natural gas
and connects the discoveries to markets.

Frank Chapman joined British Gas from Shell in 1996, as managing director of
exploration and production. He then became president of BG International when
Centrica demerged in 1997 and was responsible for E&P worldwide and non-
UK downstream businesses. He became CEO for what is now BG Group when
Transco demerged in October 2000. The company's main offices are in Reading,
about 40 miles west of London. We met him in BG Group's London office to discuss
the company's value proposition. Mr. Chapman smiles at the thought of heading a
company with a history that stretches right back to the Gas Light & Coke Company
of 1812:

> When I say we've been around for almost 200 years, it always
> causes a smile. But it is a fact that we dig up pipes today in Argentina
> and Brazil that our forebears laid 100 years ago—cast iron mains
> and Victorian-style installations. So ours is a background steeped in
> the history of supplying gas markets.

> Of course, more recently British Gas was the monopoly
> gas supplier and gas purchaser for the UK market before the

privatization and liberalization processes began during the 1980s and ran on into the 1990s.

Mr. Chapman went on to describe how he developed BG's value proposition from the company's legacy, its skills, and the opportunity set that he saw:

So, I started with that company and with a set of assets, which were largely gas-orientated. You may recall also that Enterprise Oil, which held the oil reserves within British Gas, was spun out as a separate, private company.

My perspective is always to see our business as being about opportunities and skills. There is a wide range of opportunities. Whether the reservoir fluid turns out to be oil, or gas condensate, or gas—well, that's a matter of history. We have a view as to whether it is oil-prone or gas-prone but we don't know with any certainty until we drill it.

In that sense, the idea of your being a gas company or an oil company is, to some extent, outside your control. You can have predominantly gas plays that you're drilling, but you might find oil. What do you do then? At that point, you ask the question: "Do we have the skills to monetize this opportunity?" If you say, that you won't drill it because we're a gas-only player, then that seems to me to be painting yourself into a corner.

So BG Group is predominantly a gas player but we do have oil and associated liquids in the portfolio. Fundamentally, the real starting point is, if you've got skills and you've got opportunities, you can make money. But our legacy is gas and we remain predominantly gas.

Of course, gas invariably takes much longer to commercialize than oil—particularly long-range gas—because of the complexity of the commercial arrangements and the amount of capital required to bring these things to fruition. The upside of oil is it is internationally tradable. It can move more quickly and it can provide a means of filling troughs between other programs.

Growth in this business has never been linear. There are periods of pause and consolidation. There are periods of taking off. And that itself is not something that is 100% under the control of a company. So, having oily projects that can be phased and plays of work that can be moved around is actually not a bad thing.

Out of this came the overall value proposition:

> If you were to go and talk to some of our employees, you'd come back to me in a week, and you'd say, as so many have, 'There's a buzz about this place.' There's a sort of energy and intensity. People are really engaged. I mean, we've grown operating profit from 36 million pounds in '96 to 2.4 billion last year. That's a 40% compound average annual growth rate in operating profit since we started out in 1996. It's quite an achievement.
>
> People know we are unique because we are this gassy company that looks at markets and works backwards. We are also great explorers and implementers. Our business proposition is about creating value in the gas-chain. The investment proposition is one which says: "We offer an investment vehicle that is growing at exceptional rates relative to other energy players."

BG found its "sweet spot," then balanced agility with discipline and focus to become a substantial player in the Atlantic natural gas industry segment, and continues to expand from that base.

Valero

The concept of an independent refining company is a U.S. creation. Indeed the most successful early example was Standard Oil, which controlled 90% of U.S. refining capacity in 1879 before Rockefeller decided to buy production assets in the late 1880s. Following that precedent, and the break-up of the Standard Oil Trust by Teddy Roosevelt, most oil companies had both refining and producing activities.

The global increase in crude oil prices in the 1970s and 1980s resulted in a sharp reduction in residual fuel oil demand, with two consequences. First, there was now an excess in refinery capacity globally. Second, there was a need to invest in new processing units to convert residual fuel oil components into transportation fuels. In addition, public opinion forced a series of product quality improvements, from lead phase-out, to sulfur reduction, to refinery emissions restrictions, to proliferation of single-state specifications for motor gasoline. These mandates required refiners to invest large sums of money (fig. 5.12), with no means to recover the cost from the market.

Figure 5.12: The refining industry has been burdened with massive, low return investments to keep up with increasing environmental regulations. *Source: National Petroleum Council*

However, the ill winds that caused integrated incumbents distress breathed life into the independent refining sector. The disenchantment of the majors with the financial performance of refining caused them to divest plants that did not fit with their forward strategy. Financial troubles in mid-size integrated companies led in several instances to a decision to split the company into two "pure play" parts, one for upstream and one for downstream. Finally, the aftermath of the mega-mergers among the majors led to Federal Trade Commission demands that certain refining and marketing properties be divested for competition reasons. These factors resulted in a massive rearrangement of the ownership of refineries in the U.S. in favor of independent refiners, paving the way for substantial increases in stock value (fig. 5.13).

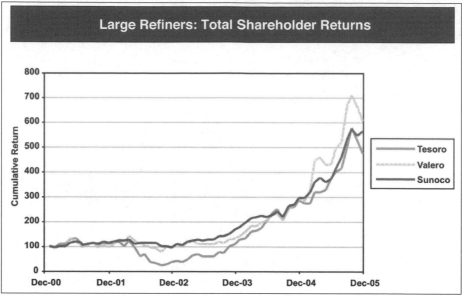

Figure 5.13: Large Refiners. *Source: CRA International, Inc.*

Valero was born in 1980 out of a forced spin off of Coastal States' Lo Vaca Texas gas gathering system. Its first, and until 2006 only, CEO, Bill Greehey, then took an enormous risk in the early 1980s by investing a billion dollars to build a heavy oil upgrading project at a small refinery in Corpus Christi, Texas. In 1997, Valero spun-off and merged its natural gas-related services with PG&E Corporation. In conjunction with this merger, the company placed its refining and marketing operations in a new independent company, retaining the Valero name. In the same year, the company acquired Basis Petroleum, Inc. (formerly Phibro Petroleum), making Valero the largest independent refining and marketing company on the Gulf Coast. At that time, Bill Greehey formulated his value proposition of growth through acquisition, focusing mainly on coastal refineries capable of processing heavy, sour crude oils. The company is now the largest U.S. refiner, with capacity of about 3.3 million bpd in the U.S., Canada and Aruba. Most of this capacity was acquired at a fraction of replacement cost, because of the factors described previously. However, Valero's most recent acquisition of Premcor was reported to have been at 90% of replacement cost, reflecting the improving outlook for worldwide refining described in chapter 3.

Bill Greehey reflected: "I've always been criticized for being too aggressive. Not buying back stock. Not increasing dividends." He invited us to compare the Valero and Sunoco value propositions:

Their strategy is, you buy back stock; you increase the dividend; you process primarily sweet crude; you like retail more than you do refining. And then they got into chemicals. You look at Valero. We don't particularly like retail. We like wholesale retail, but we don't like company retail. So that is in contrast to them. We're sour. They're sweet. We don't like chemicals. They like chemicals. We love acquisitions. We love to spend money. We don't care if we leverage up. We got up to 55% with the UDS acquisition. But every time we've gotten up, then we pay it off and we go down. I mean the rating agencies are always on my case. They are probably so happy that I'm retired as CEO because they were always wondering, 'What's his next deal?' because they knew I was going to leverage the company. I'm criticized all the time by the rating agencies and a lot of the analysts. They like boring companies, and they'll say, you know, Sunoco had a meeting. It was kind of a boring type meeting. They are doing their same thing. But, having said all that, look at what shareholder return has been with these two companies.

And the reality is that the more conservative Sunoco value proposition provided its shareholders with better returns from 2000-03, but Valero's more aggressive value proposition has resulted in surging shareholder returns since then; and over the full period, Valero's approach has provided its shareholders with the highest returns in this sector.

Going forward, the company's value proposition continues to promise growth in refining, but more from organic investments than acquisitions. Greehey told us:

"And we do have a lot of organic growth. We've got $4 billion that we can spend in the next five years increasing capacity 400,000 barrels per day. I mean, that's two world-scale refineries. I really feel that the next couple of years are going to be really great for refining. But then when you start looking at all of this expansion and upgrading, the discounts aren't going to stay as great, and margins probably will not be as good. And so what you need to do is ask, 'Where are we going to be two or three years from now?' and figuring out how to come up with additional revenue to increase shareholder value. What we really need to do is, we need to focus on a new business line. So what we are looking at is, how do you get into a new business line where there is synergy with your existing operations, there is synergy with your existing customers; and we are looking at pet coke gasification. If you look at where there is a shortage of electricity, pet coke may be one of the things you look at doing.

So Valero is looking at extending its value proposition into related opportunities, but with a continued strong discipline to its thinking. Greehey continued:

We don't know anything about chemicals. Now, gasification, we're getting a lot of experience at Delaware. We've hired some people. That's more aligned toward a processing operation. So that's not going astray. Now if we would decide to go upstream, I think it would be an absolute disaster.

Exit—Burlington Resources

At the other end from the growth companies are organizations that emphasize returns and sustainability, such as Burlington Resources (BR). There is an important role for both types of companies in serving the needs of different investor segments. The key question for both types is whether their value proposition is sustainable. Can the growth companies sustain their growth, or will they "hit a wall" after they reach a certain size? Can the companies offering superior returns sustain returns through the cycle as costs inflate? To inform this discussion, we looked closely at BR and Chesapeake, and while we were doing that, BR agreed to be taken over by ConocoPhillips.

The value proposition in the E&P segment has always included growth. Ellen DeSanctis, head of investor relations and strategy for BR, told us:

The reason most people historically invested in the E&P sector is because it offered the growth part of the energy spectrum. Even as the sector matured, I don't believe the E&P thesis ever got terribly far away from growth. If returns were important in your portfolio, typically investors looked at the larger integrated companies. At BR, we made a conscious decision in the late 1990's to strive for modest growth, not high growth, in order to provide higher returns. At the time this differentiated us from both the E&P and the integrated sectors. We had to convince the marketplace that they were getting the best of both worlds.

Bobby Shackouls, BR's CEO, set out the new value proposition of moderate growth and sector leading returns at a meeting with financial institutions in 1998:

> There were about 100 analysts there. I told them that we were going to stop investing in the Gulf of Mexico, and there was an outcry! At least two of the analysts left the meeting half-way through the presentation and downgraded the stock! As I explained what we were doing, I told the analysts that the majors, especially Exxon, focused on returns more than growth, and that was what we were going to do at BR. They hated it, since the whole emphasis in the 1990s was on growth. But I told them there was no reason why we couldn't run our business the same way. One analyst said 'You mean you're going to try to run your podunk company like Exxon?' and another analyst, a friend who has retired now said, 'What's wrong with that?' It was the best compliment I was ever paid! And you know, I did the calculation the other day. Investors that bought our stock immediately after that meeting would have received over a 30% annual return up to now.

Interestingly, we interviewed Bobby Shackouls of BR and John Lowe of CoP the week before they announced their deal. Later both Shackouls and BR CFO Steve Shapiro confirmed that the deal was not done out of weakness in BR. In Shapiro's case, it was a product of the difficulty he was having visualizing BR's position in the future structure of the industry. For Shackouls, it was quite simply a reflection of his responsibility to the shareholders. After the second meeting with CoP, "We went to our board, told them what had been proposed, and got permission to sign a confidentiality agreement and do bilateral due diligence. And that was happening that week, and we had our first board meeting on December 10. And the second one was on December 12. The news actually leaked on Sunday night and we announced it on Monday night."

Shapiro has a rule of thumb for the current times: "You know I'm a rule of 10% guy. Somewhere between your volumetric growth and your dividend and return of capital, you need to show the marketplace something approaching a 10% value growth every year. Initially E&Ps could grow at 10%, but that became more difficult. And we felt amongst the E&Ps, although there is room for niche players, it would be hard for the big guys to show that kind of growth." Shapiro felt that market conditions would inevitably force the E&P segment away from growth and toward yield:

But the integrated model, I think, by definition has more certainty around that. And hence a better risk return to a shareholder should the market go that way.

And if I'm an integrated player, and I've got the chemical engineering technology and the major project capabilities, I can go into LNG and oil sands. So you can actually put growth into your integration. And so you have that horizontal diversification option, and you have the skills and the ability of doing that. If you don't, you need to get the money back to your shareholders versus competing down returns in your core business to a level that's value destroying. Because one of the risks in our business, and one where the E&P companies have traditionally destroyed value, is by overcapitalizing the business, reinvesting too heavily in the business, in a cyclical business.

Shackouls explained the structure of the acquisition. He and CoP CEO Jim Mulva made a tour to explain the deal to investors, and Shackouls was asked by one investor why he had agreed to be taken over. He explained:

I kind of looked at it in a pretty simplistic way, and that is that we've provided our shareholders returns of 15%, 31%, and 59% annual returns over the last three years, respectively. We entered this year (2005) with the stock price at $43.50. ConocoPhillips offered us $46.50 in cash and an equally-valued piece of paper that keeps our shareholders exposed to the exact same business they've already been in and to the commodity prices that they've had exposure to. Plus, it gives them exposure to a much more expansive international portfolio that we on a stand-alone basis can't provide them. And finally, it gives them exposure to a world class refining and marketing operation that we don't have any way of providing them. So, to me, it's pretty simple. You know, they get to take their returns. They get a piece of paper that keeps them in the same business if they want to be in it. And that's kind of how I look at it. And I think that's how the employees look at it, with a degree of sadness that they hate to see the franchise go away, and they know things will be different.

In the eyes of the BR executives, this was a win-win transaction. They speak with great respect of CoP's prospects, yet are pleased with the terms of BR's exit. They also provide food for thought on the future role of the independents and the sustainability of their value propositions going forward.

Turnaround—TXU

TXU was the Dallas-based regulated electric power utility serving North Texas. Texas has undergone a successful progression toward the current deregulated market. In 1981, ERCOT (Electric Reliability Council of Texas) became the central operating coordinator for Texas. Its website describes the progression:

> In 1995, the Texas Legislature Voted to Deregulate Wholesale Generation. In 1995, the Texas Legislature amended the Public Utility Regulatory Act to deregulate the wholesale generation market. The Public Utility Commission of Texas (PUCT) began the process of expanding ERCOT's responsibilities to enable wholesale competition and facilitate efficient use of the power grid by all market participants. On August 21, 1996, PUCT endorsed an electric utility joint task force recommendation that ERCOT become an independent system operator (ISO) to ensure an impartial, third-party organization was overseeing equitable access to the power grid among the competitive market participants.

> On May 21, 1999, the Texas Legislature passed Senate Bill 7 (SB 7) which required the creation of a competitive retail electricity market to give customers the ability to choose their retail electric providers, starting January 1, 2002. From 1999 to 2000, ERCOT sponsored a stakeholder process to address how ERCOT's organization would administer its responsibilities to support the competitive retail and wholesale electricity markets while maintaining the reliability of electric services.

TXU followed the pathway of many regulated utilities at the end of the 1990s and plunged into deregulated markets. John Wilder has been CEO of TXU since early 2004 and describes the situation he inherited: "I mean, we were taking 300 seconds to answer the phone for a customer. We had $14 billion in debt. We generated a billion and a half of EBITDA. We had a $4 billion business go bankrupt in Europe. We had billions of dollars of potential litigation against us. We had $2 billion of underwater commodity contracts. And I tried to just create a sense of urgency, a burning platform."

Wilder executed a huge transformation of TXU from a bureaucratic utility to a dynamic new energy company, and the stock market was impressed (fig. 5.14). From being one of the poorest power company stocks prior to his appointment, TXU has recovered to be a sector leader. In fact, he prefers to consider TXU as an industrial company, to distance it even further from its past: "I started coining this term the first day I joined, of an industrial company. You know, we are not a utility. We are an industrial company. We just can't see it yet. We have heavy manufacturing. We have a consumer business that provides industrial company kind of products; we are capital intensive. And what we want to try to become is a high-performing industrial company."

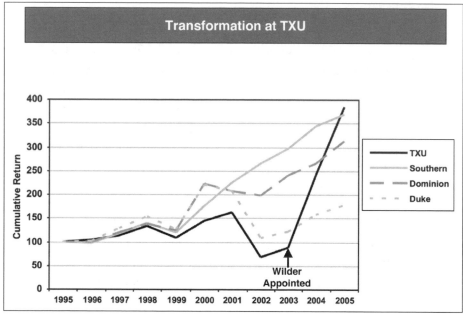

Figure 5.14: Transformation at TXU. *Source: http://finance.yahoo.com*

He worked on a three-phase turnaround. "The first phase was portfolio rationalization, which is, get the 'where and how to compete' right. What business do you want to stay in? The next phase is the performance improvement phase. And that phase will be zeroed in on driving each of these businesses. And when each business gets to top quartile—our ultimate goal is top-decile performance – then that business has earned a right to grow. Until then, no capital. Performance improvement, period."

Wilder is now in the final growth phase of his turnaround:

> So our goal was, we drive high performance here, and then—
> we call it the 'second-owner strategy'—we go buy underperforming
> generating facilities. Now what we blindly missed in the process is,
> we allowed ourselves to be suckered by the old expensive customs
> of new-build costs. We thought, 'Gee, should we build it or buy

it?' Let's calculate: $1,500 per KW to build, takes about this long, it is better to buy an old one and fix it up than it is to try to build a new one.' Well, our critical-thinking skills weren't good. In 2005, we announced publicly that now we were going to do an in-depth strategy review of each business. And it's through that review that the new-build idea unfolded. Because we just said, 'Okay, let's not take our word for it. Let's bring in Bechtel. Let's bring in Fluor. Let's put a hundred engineers on the floor. Let's bring in these lean construction guys from Stanford that were quite good. And let's really see what it costs to build, and how quickly you can build.' So then as we did that, then 'buy it and fix it' shifted to 'build it.' Strategic shift took place.

TXU has now announced an enormous capital investment program, with a focus on building new generating plants at a much lower cost than the utility norms. Simultaneously, he is working on customer-facing strategies to allow customers to choose between different price and service packages.

The irony is that in a deregulated market, Wilder (who had a successful prior career in the oil sector with Shell) believes that the oil companies may be better placed to execute such a capital program: "That next breed of oil company management knows that our coal plants are synthetic 80 million MMBTU, 40-year gas fields, and don't have a depleting resource, a heavy capex requirement to continue to generate that commodity exposure, or a heavy tax burden." However, our interviews suggest that the major oil companies do not see it that way, so for the moment, Wilder will be able to sustain his growth strategy without having to look over his shoulder at what the oil companies might be doing.

Conclusions

We have seen that energy stocks underperformed market indicators in the 1990s, but have surged in the 2000s as prices and margins have increased. Value propositions always try to find a balance between growth, returns, and sustainability. Value propositions centered on growth by asset acquisition may be difficult to sustain as asset prices have risen above the enterprise values of the companies that own them. If that situation persists, some companies will come under pressure from corporate raiders to sell off assets into the strong asset market and return the proceeds to shareholders.

The super-majors are at a turning point, in our view. It is unlikely they can sustain their historical value propositions without expanding the scope of their business and embracing the new resources and new technologies identified in chapter 3. The biggest strategic mistakes are always found in failures to define the industry properly, and we firmly believe that the industry is in the process of redefinition that will require new business models and modified value propositions to shareholders.

At the smaller scale (though these are substantial companies with market capitalizations north of $20 billion each), our growth stories demonstrate that determined leadership can find growth by consolidating the industry (ConocoPhillips, Chesapeake, Valero); by developing massive unconventional resources in gas (EnCana, Chesapeake and now ConocoPhillips) or a legacy asset of oil sands (Suncor); by leveraging a distinctive set of competencies (EnCana, Statoil); or by capturing a disproportionate share of a growth segment (BG Group). In the current environment, there is no obvious limit to these growth propositions except for rising costs. But Aubrey McClendon believes that acquisition costs have been lagging commodity prices, such that it is still possible to create value for shareholders. Lee Raymond's gruff reaction to the Chevron acquisition of Unocal—"I can never remember an industry consolidating at high prices"—is symptomatic of a sentiment that will deter many buyers. Nevertheless, strategic acquisitions that enhance the value of existing businesses and open up a new set of opportunities will still be rewarded.

Those companies that persuade themselves and their shareholders that the valuation gap will be transitory, or that this is the wrong part of the cycle to make major acquisitions, will be looking to enhance organic growth. And that is what higher prices are signaling—there is a greater need to increase production than to further enhance returns. The problem is that the supply increases will have to come from unconventional resources and non-traditional places, and must be accompanied by careful attention to society's demands for environmental and ecological protection and exemplary ethical behaviors.

There will still be room for a spectrum of value propositions, from those that offer a thrilling ride with high growth potential but the risk of "driving into a ditch," to those that promise "steady as she goes" with good returns and efficient ways of returning excess cash to their shareholders. Whichever proposition they choose, we believe that companies will need to express it with clarity as an authentic story and offer "proof points" to assure investors of consistency and sustainability. That being said, we also believe that even the most conservative companies will do well to tilt their value proposition in favor of organic growth in the decade to come. Any company that cannot see its way to sustaining its shareholder value proposition, whether growth or returns, should bite the bullet and choose the right time and right partner for a dignified exit.

So there is the challenge for energy company leaders. Tight energy supplies provide tremendous opportunities for growth as well as for profits, but intense competition is bidding up the cost of all the inputs of land, materials and services, especially for conventional resources in traditional places. Companies have the opportunity to rebalance their value propositions to investors in favor of growth by investing in expanding the global resource pool, as their predecessors did in the 1970s, when opening the new provinces of the North Sea and Alaska allowed the majors to recover from nationalization by resource owners. This is not a time to be timorous, or wring hands over difficult access, or learn the wrong lessons from the downturn of the 1980s. The real lesson from prior cycles is that energy company stocks performed better than the market in the 1970s and 1980s, because the industry leaders of that time adjusted to the cycle, delivered both growth and returns and created lasting value by deploying new technology to find new resources and convert them into saleable products. Further, there is a strong probability that the anticipated next down cycle in the 2010s will be overridden by the rise to materiality of China and India, and by the desire of resource rich countries to conserve their resources. Yes, it is a cyclical industry; but this is the exciting growth stage, the next cycle will be different from its predecessor and fortune will favor the brave.

---

1 Michael E. Porter, 2003. European Business Forum, Issue 15, Autumn 2003.

2 Alfred Rappaport, 2006. "Ways to Create Shareholder Value." Harvard Business Review, September 2006.

3 We have adopted the categories used by Charles Schwab as indicative of typical analysts' metrics.

4 TSR is defined as the total return to shareholders over a given period from appreciation in the stock price and from dividends paid.

5 John S. Herold provides a 43 page explanation of its valuation methodology on its web site at http://www.herold.com.

6 Market capitalization of equity plus debt.

7 Michael Goold and Andrew Campbell, 2002. *Designing Effective Organizations.* San Francisco: Jossey-Bass, p. xi.

6 Evaluating Strategic Choices

Formulating strategy is one of the most far-reaching activities that any leader undertakes. Corporate strategy is about choosing the right businesses to be in and leveraging a "parenting advantage." Business unit strategy focuses on competing successfully in the chosen marketplaces by creating a competitive advantage through core competencies. Both are essential for long-term success, which depends on being in the right businesses and competing successfully in each of them.

A continuous iterative process takes place between deciding on a persuasive shareholder value proposition and evaluating strategic choices. There must be viable strategic choices in support of the value proposition that can allow the leader to set a clear direction. Otherwise the value proposition will not be persuasive and a new shareholder value proposition may be required. Conversely, recognition of new strategic options can lead to a more attractive shareholder value proposition. In this chapter, we will continue our transition from analyzing the challenges facing energy companies to addressing how leaders are responding to those challenges. We will first look at some models for making strategic choices consistent with the overall framework outlined in chapter 1 (fig. 6.1).

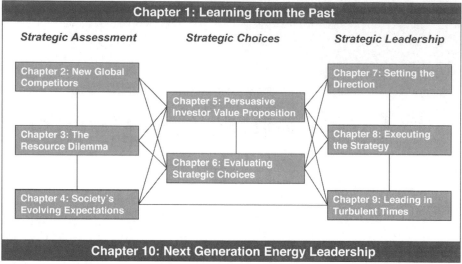

Figure 6.1: The architecture of energy leadership

Understanding Fundamental Strategic Choices

There are two fundamental strategic choices companies can pursue to offer value to their customers: cost leadership and differentiation. These are found in every strategy textbook for MBA students. Since energy products are largely commodities, cost leadership plays a pervasive role in this industry. Within this cost orientation, we will see some selected differentiation strategies emerge as ways to create competitive advantage. This drive for cost leadership also substantially impacts the service companies trying to support these operating companies, but opportunities exist there too for differentiation based on superior technology or service provision to support the customer's cost-leadership position. Beyond the fundamental value proposition, strategic choices involve selecting the parts of the value chain to participate in and deciding whether to pursue a broad scope or go after niches in the selected businesses.

Successful strategies provide real focus and direction while being robust as the competitive environment evolves. Simple cost leadership and/or differentiation rarely provide enough strategic sophistication in today's world to outperform competitors. Treacy and Wiersema[1] expand the two-pronged model by defining three market disciplines that companies can focus on: operational excellence, product leadership, and customer intimacy. Operational excellence is really the next stage of cost leadership in providing best value, at least cost, and without hassles. Wal-Mart over-ran its competitors with this strategic model. Product leadership represents a form of differentiation that persuades customers to buy the product because it is the best available, period. Intel offered a good example of this differentiation in the computer chip business, driven by fast cycle markets. Its leadership was continuously challenged by competitors and its ongoing success reflects founder Andy Grove's mantra "only the paranoid survive." That is, the moment a company becomes complacent, its position is likely to be threatened. There is a lot to learn from evolutionary biology that is relevant to the competitive world of business.

Energy products are commodities, and energy strategies start and often end with a drive towards cost leadership through operational excellence. Only in regulated energy businesses where companies have local monopolies and are permitted by regulators to pass on costs to customers can high cost producers survive and prosper. Oil companies have made attempts at product differentiation for gasoline, but none has found lasting acceptance in the marketplace. In the exploration and production segment, a "product" is the ability to monetize a host country's resource. To the extent that a company can create a technologically-differentiated approach, this has the potential to create substantial value for the host country "customer" and for the company. Finally, customer intimacy involves providing solutions that satisfy customer problems. There is no doubt that some companies are better at partnering with national oil companies than others, and this is a form of customer intimacy that can bear fruit.

Companies cannot be all things to all people, hence the need for focus; yet many companies run adrift pursuing various ancillary activities without knowing what their core competencies really are. Jim Collins in *Good to Great*[2] proposes the "Hedgehog Concept," in which companies figure out what they can be the best in the world at; they understand deeply what drives their economic engine, and they focus on the activities they are passionate about.

There are inflection points when strategy must change. In fast cycle markets this happens more frequently than in slow cycle industries like energy. However, the pace of change, even in mature industries, is continuing to accelerate and outcomes are becoming less predictable. Many companies drive forward while looking back through their rear-view mirror. The more successful a company has been, the harder it becomes to make a course correction from past business models. But, only by innovating ahead of the pack can any advantage, whether of cost or differentiation, be sustained.

This has been a continuous theme in the writings of Clayton Christensen stressing the importance and risks of disruptive technology advances, which became the major theme of management books in the late 1990s during the dot.com boom. Though that boom cycle turned to a bust, some of the observations on the pace of change, the importance of new business models and the need for experimentation remain valid. For example, *Blown to Bits*[3] highlighted the importance of advances in information technology, which in fact have had major productivity impact on the way energy workers do their jobs. Perhaps the next big impact will come from nanotechnology.

Creating a focused strategy that is also robust remains an elusive goal. Alfred Marcus profiled 18 best sellers that "purport to provide managers with the secrets of sustained competitive advantage"[4] and concluded that they tend to be contradictory and one-sided. He believes they stress too strongly the virtues of agility, on the one hand, or discipline and focus, on the other. Marcus maintains that you need both hands to succeed, as well as the favorable winds that we have mentioned earlier, which he calls finding "the sweet spot." Persistent winners find a sweet spot, and manage the tension between agility, discipline, and focus. The problem is that external conditions change. There are times of stability when discipline and focus are vital, and there are times of rapid change when agility and the ability to be innovative and move quickly are essential. There are times when acquisitions create enormous value, and there are times when it is better to "grind it out" with continuous improvement programs and modest organic growth. Energy leaders need to be looking forward enough to sense when the winds are changing direction to find the "next sweet spot" and avoid moving too soon or too late.

In the end, there is no simple "magic bullet" that works for all companies all the time, especially not in the energy industry. There are great successes built on taking cost leadership to the next level, such as ExxonMobil and Chevron. The first page of Chevron's presentation at its December 2004 Security Analyst Meeting is entitled "Focus and Discipline" and states its objective of delivering industry leading performance through its "4 + 1" business model: profitable growth, operational excellence, cost reduction, and capital stewardship, all underpinned by organizational capability. ExxonMobil describes its business model in its March 2005 analyst meeting as being disciplined investment and operational excellence, leading to industry-leading returns and superior cash flow, the whole system resulting in growth in shareholder value.

BP and ConocoPhillips differentiated themselves by their agility in deal making, as exemplified in their entry into Russia, as well as some specific competencies in exploration (BP) and heavy oil processing (ConocoPhillips). Both companies emphasize the value their commercial skills add to their asset portfolios. Total and ENI have distinctive asset portfolios and an ability to work in difficult countries. Shell traditionally has had a similar distinctive competence in blending in with the different cultures of the countries where it operates, and both Shell and BP have positioned themselves as more sensitive to environmental issues than the industry norm. We can map the different super-majors in terms of their positioning with respect to cost leadership and differentiation, focus, and agility.

By contrast, among the independents, Chesapeake is clearly adopting a niche strategy with its strong focus on the U.S. mid-continent. EnCana (resource plays), Suncor (Canadian oil sands) and BG Group (Atlantic LNG) have also specialized, which enables them to be more dominant in these plays. These companies have positioned themselves downwind of some strong trends, and have shown discipline in sticking to their areas of focus and agility, as well as innovation in expanding their opportunity sets within their areas of focus.

Choosing Strategic Navigation Aids

It is important to distinguish between corporate strategy and business unit strategy. Corporate strategy is about selecting the businesses that the corporation wants to be in, and figuring out the basis for a "parenting advantage" by which these businesses will be worth more as part of the corporation than they would be alone, or in someone else's portfolio. Business unit strategy is about building competitive advantage in specific industry segments through an aligned set of functional initiatives.

It is useful to be able to access a set of analytical frameworks and definitions to help structure our thinking about strategy. Good strategy is always founded on an honest appraisal of the current reality and benefits enormously from careful segmentation. In times when change is happening slowly, this reality constrains the strategic choices that organizations can make. In times of rapid change, incumbents may find it more difficult to protect their positions, and organizations necessarily must consider a wider array of choices. Current reality is always grounded in analysis of the dynamics of the specific industry segment in which the company competes.

Michael Porter developed some very powerful ideas in the early 1980s around the concept of aligning strategy with the industry value chain and structure, emphasizing a much more thorough industry analysis than was common at the time. The basis of competitive advantage according to Porter lay essentially in barriers to entry and mobility in a given industry. We characterize Porter's approach to strategy as "condition-driven" in which companies need to face up to the realities of their competitive position since these realities define the range of "natural strategies" open to them.

Porter's ideas and models were admirably suited to industries where complexity was low and change took place at a "normal" pace—by evolution rather than revolution. There were recognizable rules in the game of competing in a given industry, and winners were those who played hardest by the rules. As the end of the 1980s approached, however, the pace of change picked up and innovation became a vital factor in business success.

Recognizing this shift, Prahalad and Hamel forged a new model that set aside the notion of playing the game in the same way as the rest of your industry and instead focused on leveraging your core competencies to change the rules of competition. This required a new kind of flexibility and adaptability that a strategic plan based mainly on industry and competitor analysis would not permit. Prahalad and Hamel proposed the primary driver of strategy should not be a detailed plan, but a strategic intent that everyone in the company (in theory) could use to shape their actions and decisions. They further recommend a decentralized, more autonomous organization capable of adapting quickly to the changing business environment, guided by the strategic intent. Strategic intent is similar in concept to vision and mission as described later in this chapter.

Going still further, Shona Brown and Kathleen Eisenhardt[5] studied high technology companies in the late 1990s and concluded that these companies practiced strategy as continuous experimentation. They set a "semi-coherent strategic direction," then made progress through working "at the edge of chaos," between structure and chaos; "at the edge of time," between the past and the future; and with an acute awareness of "time pacing." By doing that, companies produced "a continuous flow of competitive advantages." This approach worked for companies such as Intel and Cisco Systems that were inherently disciplined, but was a recipe for disaster in companies that were less disciplined as the dot.com boom turned to bust. The new competition (chapter 2), the resource dilemma (chapter 3), growing societal expectations and geo-political changes (chapter 4), and questions on the historical shareholder value proposition (chapter 5) will combine to challenge energy companies to find a new balance between discipline and innovation in the future.

Porter's concepts are applicable for business unit strategy in established segments of the energy business, where thorough analysis of industry value drivers and of competitive behavior are essential. In the 1990s, companies in the downstream refining and marketing business, for example, were well advised to look carefully at their cost structures relative to the competition, and to develop strategies to become cost leaders. Similarly, oil companies benefited from looking hard at their competitive strengths in the convenience store segment, where they discovered that they lacked important skills, were burdened with high costs, and had the wrong governance processes to compete with new entrants that were specialists in this area.

However, as we find in the next chapter, leaders of some of the companies we spoke to that had growth-oriented value propositions spoke more in the terms used by Prahalad and Hamel of strategic intent and competencies. For example, Valero would never have been successful in becoming the largest refiner in North America if CEO Bill Greehey had been locked into analysis of his competitive position. EnCana might have seen itself as a small player in the international upstream segment rather than spotting the opportunity, as CEO Gwyn Morgan did, for rapid growth in resource plays. In both cases there were a clear strategic intent and deliberate and purposeful strategies to strengthen the competencies required for success.

As leaders reflect on the technologies of the future, the unsustainability of recent trends and the rapidly changing energy complex described in chapter 3, they will be wise to consider the concepts of Brown and Eisenhardt of strategy by experimentation. This is because the winning resources, technologies, and companies are not yet known. Several of the super-majors are in the process of developing experiments in renewable energy forms. For example, BP has a substantial business in photovoltaics, Suncor is making a significant investment in corn-based ethanol, Shell has experiments with partners in cellulosic ethanol production, and Chevron has formed a biofuels business unit with an initial investment in a biodiesel plant. In each case, the companies will need to move from experimentation to placing big bets if they wish to build material positions in the new energy forms.

As Peter Drucker has said: "What is crucial in 'strategic thinking' is first, that systematic and purposeful work on attaining objectives be done; second, that 'strategic thinking' start out with sloughing off yesterday, and that abandonment be planned as part of the systematic attempt to attain tomorrow; third, that we look for new and different ways to attain objectives rather than believe that doing more of the same will suffice; and finally, that we think through time dimensions and ask, 'When do we have to start to get results when we need them.'"[6]

Defining a Strategy Lexicon

Unfortunately, there is a plethora of vague terminology around strategic concepts that confuse communications because people are not really talking the same language. To remedy this situation, leaders must define a logical hierarchy of concepts and terms that the organization understands and uses to communicate. The strategic logic must address the attributes that enable the organization to create a sustainable competitive advantage. These attributes, to be useful, need to be articulated into a structure that provides a sense of direction, some targets and milestones to measure progress, and some specific actions that will result in success. In that vein, the authors will use the lexicon set out in figure 6.2.

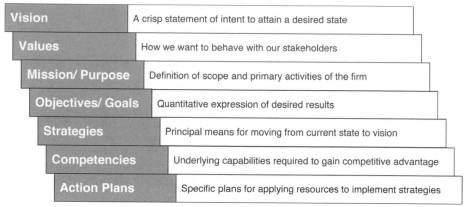

Vision	A crisp statement of intent to attain a desired state
Values	How we want to behave with our stakeholders
Mission/ Purpose	Definition of scope and primary activities of the firm
Objectives/ Goals	Quantitative expression of desired results
Strategies	Principal means for moving from current state to vision
Competencies	Underlying capabilities required to gain competitive advantage
Action Plans	Specific plans for applying resources to implement strategies

Figure 6.2: The lexicon of strategy. *Source: CRA International, Inc.*

What the company wants to look like in the future forms the basis for shared aspirations among its leaders (its hedgehog concept) and forms the *vision*. Key *values*, such as agility and innovation or focus and discipline as well as ethical principles, must be articulated to manifest the *vision* in a corporate culture. Values also must establish acceptable organizational behavior. Sometimes embedded in the vision and sometimes separately articulated is an intrinsic value proposition, such as Treacy and Wiersema's market disciplines of operational excellence, product leadership, or customer intimacy.

Mission or *purpose* describes the particular field on which we choose to play the game, and *goals* make specific what we hope to accomplish. *Strategies* are how we aim to reach those goals and attain the *vision*, through the *purpose*, in accordance with our *values*. *Competencies* describe the skill set we will need to pull the whole thing off. Core competencies are where we truly have a competitive advantage. *Action plans* then detail the steps that we want to take to deploy the *strategy* by leveraging our core competencies.

We like the descriptions of the terms in *The Fifth Discipline Fieldbook*[7] for the first four elements of the hierarchy:

"*Vision* is an image of our desired future; a picture of the future you seek to create, described in the present tense. Because of its tangible and immediate quality, a vision gives shape and direction to the organization's future." CEO Dave O'Reilly articulated a simple but powerful vision for Chevron: "To be the global energy company most admired for people, partnership and performance." This simple message defines a broad space in which the company will compete, and the key attributes that will bring competitive success.

"*Values* are how we expect to travel to where we want to go; a set of governing values might include: how we want to behave with each other; how we expect to regard our customers, community, and vendors; and the lines we will and will not cross. Values are best expressed in terms of behavior: if we act as we should, what would an observer see us doing? How would we be thinking?" Chevron CEO Dave O'Reilly notes that values are the product of history and have to include the acquisitions made over time:

> The thing that has become the most apparent to me has been the importance of a company's value system of the company, and how critical that is in an organization that is big and diverse with multiple cultures, multiple geographies, and multiple lines of business. You can't possibly micromanage everything. So it is critically important to communicate the value system in a manner that is fairly simple, and then check on it periodically to make sure that the other leaders of the organization live up to that value system. As CEO, it becomes very clear that this is one of the major roles—to be the keeper of the value system.

EnCana's Constitution articulates its values in shared principles of strong character, ethical behavior, high performance and great expectations (from each other, leading to the "thrill and fulfillment of being part of a successful team").

"*Purpose* or *mission* is what the organization is here to do." CEO John Browne defines it succinctly for BP: "The purpose of the firm is to provide energy products to people who want to buy them at a cost they can afford." So, whatever the vision is in terms of aspirations of a future state, it should be achieved by providing energy products at affordable costs. Within this purpose, BP has been able to conclude the TNK/BP joint venture in Russia, has launched a new renewable power business, and is working to improve its retail gasoline stations. But it could not provide medical products, nor could it provide energy products at unaffordable prices. ExxonMobil's purpose is wrapped into their guiding principle of being a petroleum and petrochemical company. Alternative energy does not really fit into this "petroleum" purpose.

"*Goals* are milestones we expect to reach before too long; goals represent what people commit themselves to do, often within a few months." But they can also be what Jim Collins calls "Big Hairy Audacious Goals" (BHAGs), such as those set out by president John F. Kennedy in his 1961 State of the Union address: "I therefore ask the Congress, above and beyond the increases I have earlier requested for space activities, to provide the funds which are needed to meet the following national goals: First, I believe that this nation should commit itself to achieving the goal, before this decade is out, of landing a man on the moon and returning him safely to the earth . . ." This is not a hedgehog-type vision, says little about values (other than the commitment to return the astronaut safely to the earth), and is certainly not the purpose of his organization. The rest of his speech contains comments that help us understand how he views the vision, values, and purpose of the U.S. government, but the excerpted passage is a fine example of a BHAG for NASA.

Energy companies like setting goals to focus the organization's attention on key elements of the shareholder value proposition. Burlington Resources, for example, focused on achieving sector-leading returns. Other companies have publicized volume growth targets. However, these are dangerous when they require timely completion of major projects. If the projects are delayed and the company misses its targets, then there will inevitably be market retribution.

Strategy derives from the Greek word *strategia*, or generalship. Webster's dictionary tells us that it is "the science or art of military command as applied to the general planning and conduct of full-scale combat operations." In a civilian sense, it is a broad plan of action, especially for attaining the BHAG. Navigationally, it charts the game plan for the voyage. In the business sense, the strategy outlines the overall business model and prescribes how all the pieces fit together within the context of the *vision* and *mission*. The business model covers front-line activities such as access to raw materials, marketing, supply, and manufacturing, as well as support activities such as human resources, information technology, finance, and external relations.

Companies are less inclined to publicize strategies, since competitors may use the information to frustrate the company's progress or to gain advantage by moving faster. Only when they believe that their strategy is so strong, and underpinned by competencies that others cannot replicate, will companies make them public. BG Group, for example, is quite open on its strategic intent to grow its LNG business, but does not signal in advance its strategies for accessing Nigerian resources.

Competency or *capability* means having the necessary skill or knowledge to do something successfully. A competency combines the ability to deploy technologies and the mastery of business processes and organization through well-trained and motivated people. A company may have a competency in drilling, or in trading and risk management, or in refinery operations, or in major project management, for example. Valero's competency in running refineries makes it attractive to super-majors as an acquirer of assets they wish to sell, since it treats its employees well. EnCana has a series of competencies around tight gas resource development that underpins its acquisition of acreage. It is important to compare the company's competencies against the competition to identify those that are superior or *core*. Competencies are required to execute strategies and action plans. However, competencies also enable new strategies, so there is a necessary iterative process in which companies reflect on whether they have the right competencies to execute their desired strategies, and whether their strategies fully leverage their core competencies where capabilities exceed those of competitors.

Action plans implement strategies. They specify the "who, what, when, and where" of the strategy. We find that the best practice is to rough out the action plans at the same sessions that design the strategy. In this way, individual executives are enrolled in delivering the strategy and are fully committed to specific targets. Goals that are set in headquarters and assigned to executives without their participation are seldom realized.

Formulating a Strategic Positioning Analysis

In order to "slough off yesterday" we need first to understand it through a thoughtful strategic positioning analysis. In our view, there are four preparatory steps to creating a situation analysis that will ground further thinking on strategic options and choices:

1. Define the industry and identify the segmentation of the businesses in which the company participates.
2. Develop an understanding of the maturity of the industry segments in which these business segments compete.
3. Evaluate the competitive position of the company's business segments relative to businesses of other firms with which they compete.
4. Combine the industry and competitive analysis with an assessment of competitive intensity in each segment into a strategic positioning diagram.

These steps provide the information and insight necessary to understand the company's situation relative to competitors, as well as its competitive strengths and weaknesses. From this sound foundation, we can build a viable set of strategic options, make choices, and set the corporate and business directions (fig. 6.3).

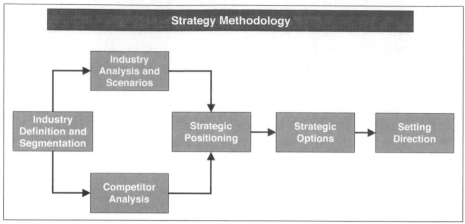

Figure 6.3: A strategy methodology. *Source: CRA International, Inc.*

Setting the Industry Space to Compete

Segmentation requires a deep understanding of the value drivers of an industry. Some of the most egregious mistakes arise when companies fail to correctly define the industry in which they compete. That is why the coal industry failed to participate in oil, and why the railroad industry failed to participate in air transport. If the coal industry had focused on its role in fuelling power plants, or the railroad industry had recognized its role in moving people across continents, there might be some different industry dynamics today. Service companies tend to be more flexible. General Electric uses similar competencies in jet engines and in combined cycle power generation turbines, successfully addressing two different industries. The authors believe there is a high risk that oil and power companies will make strategic mistakes in the near future by failing to understand the significance of the changes outlined in chapter 3.

At the business level, good strategy starts with good strategic segmentation. Segments should form the largest grouping of assets and businesses that is coherent in the sense that they satisfy at least some of the following criteria:

- There is a common set of customers
- Customer purchasing criteria are similar
- There is a common set of suppliers
- There are shared technologies and capabilities
- There is a common set of competitors
- The basis of competition is similar

Segmentation is not always obvious; what is important is to have a productive conversation on the topic in order to raise the level of insight and potentially open up a competitive advantage. We provide three examples of segmentation challenges below.

Gulf of Mexico exploration might appear to be a natural business segment, until we observe that the capabilities required for deep water exploration and production are quite different from those needed on the Shelf. Looking more closely, we find that some of the suppliers are different and that the competitor set is different. In fact, the deepwater competitors and suppliers more closely resemble those found in deepwater plays in West Africa. Should the segment be redefined as global deepwater exploration? But West Africa requires a whole new set of skills in dealing with foreign governments, and the economics are different due to different tax regimes, so perhaps that's a bad idea. But does that mean that we should combine Angola and Nigeria, or keep them apart for strategic evaluation purposes? We are not asking how we should organize the company, but how we should think strategically about the business segments. Would broadening the segment provide greater insight because we would have a better picture of what the super-majors are doing, or would it distract our attention from the strategic games that are being played in the Gulf of Mexico? There is not an easy answer.

European power and gas might form a business segment. The large, formerly national champions have extended their businesses as markets have been deregulated. Companies that previously defined themselves as providers of electricity to a discrete regional set of buyers all speaking the same language are now also providing natural gas to the same customers, but also providing gas and power in new countries beyond the language delimiter. Does it provide more insight to look at European gas and power as a single business segment and plot strategies to be a winner in the EU as a whole? Or would this be getting ahead of the game such that we would leave ourselves vulnerable in our home territory by diverting resources into international adventures, when deregulation has slowed to a snail's pace? John Wilder of TXU believes that the European companies are ahead of U.S. utilities in scaling up their operations: "So we're trying to scale up what we know, what we can do here. The Germans are doing a pretty good job of doing that. The French aren't bad. And Europe is ahead of America by a good margin across all these things. They started earlier. And they have stuck with it better than the U.S. has."

The global downstream business may be considered a logical segment. The super-majors mostly organize with global downstream (refining and marketing) as a discrete business. But are there really global strategies, or is downstream strategy fundamentally local? After all, the competitors and customers are quite different in Thailand from those in Italy. They're even different in Italy from those in Spain. On the other hand, refineries in Italy can supply Spain and vice versa. But wait, Valero has sliced the cake even finer. Valero has focused entirely on refineries capable of processing sour, heavy crude oil. Is that a business segment, or is it a strategy? The customers are the same as for other refineries, and the technologies are fundamentally the same; a lot of the competitors' refineries process both sweet and sour; but the suppliers of sweet and sour crude oils tend to be different. Also, the West Coast is clearly different from the rest of the U.S. Product specifications are different, and Arco in the 1980s perfected a distinctive business model based on integration back to Alaskan crude oil, which changed the basis of competition. And Wal-Mart and other "big boxes" are gaining gasoline market share in the South, but are not a feature in the Northeast. Perhaps we should have regional refining segments and a set of local marketing segments in the U.S. east of the Rockies; but in other areas where there is not a liquid wholesale market we may still need to bundle refining and marketing into a single segment. Sorting these questions out shapes the operating business model.

If the segments are too narrowly defined, we can miss important threats from competitors entering from similar segments in different regions or from natural extension of their current businesses in the same regions. If the segments are too broad, it may be difficult to come up with winning strategies in focused areas. The conversation around segmentation is one of the most important in formulating strategy, and we always find that careful review of the financial results of the different segments uncovers some surprises.

Having set the business space in which to operate, the next task is to understand the industry outlook and stage of maturity of each segment within that space. An important consideration is the competitive intensity of the segment, based on Porter's "Five Force" model, which works well for long-cycle industries. Finally, we stack up our capabilities and performance in these chosen businesses against our competitors, based on the key success factors for each segment.

Assessing the Business Life-Cycle Position

Strategies need to adapt as industry segments mature (fig. 6.4). This is easily seen in upstream oil. In the early, embryonic stages of a potential exploration play, strategies focus on acquiring geophysical data, developing an understanding of the petroleum system, and acquiring acreage in advance of the competition.

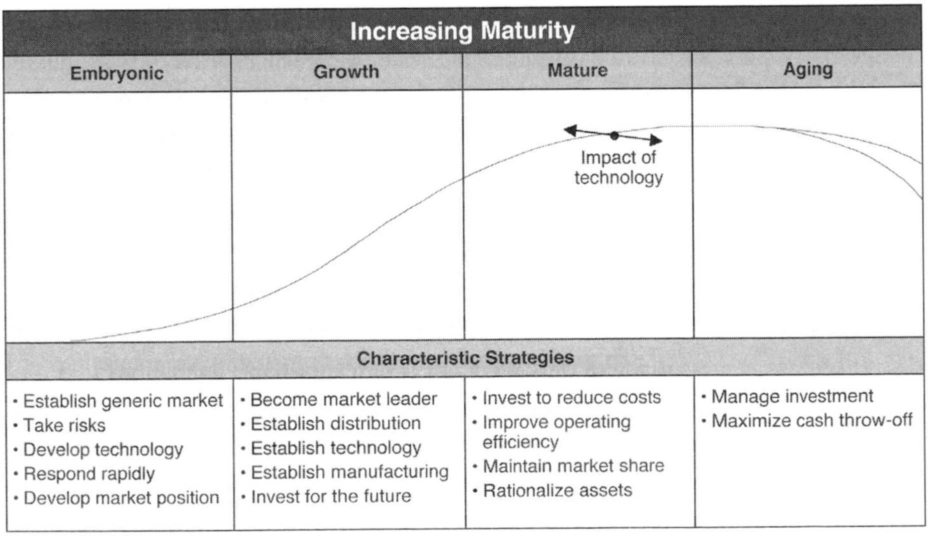

Increasing Maturity			
Embryonic	Growth	Mature	Aging
		Impact of technology	
Characteristic Strategies			
• Establish generic market • Take risks • Develop technology • Respond rapidly • Develop market position	• Become market leader • Establish distribution • Establish technology • Establish manufacturing • Invest for the future	• Invest to reduce costs • Improve operating efficiency • Maintain market share • Rationalize assets	• Manage investment • Maximize cash throw-off

Figure 6.4: Industry maturity measures the stability of the economic environment. *Source: CRA International, Inc.*

If hydrocarbons are discovered and development appears economic, the strategies shift toward acquiring access to the best acreage, drilling and infrastructure development. In the mature phase, when production and infrastructure are in place, the focus shifts again to cost-efficiency and consolidation. Then, as production peaks and starts to decline, some companies choose to exit while others consolidate to continue the cost-efficiency and consolidation strategies of the mature phase and maximize cash flow.

One strategic approach focusing on the maturity dimension explains BP's strategic actions of the 1990s. BP had "atomized" its organization to heighten accountability. Then it created "peer groups" of business units facing similar issues. Thus, mature producing areas were grouped together, regardless of geography, as were growth areas, recognizing the need for distinctive strategies for businesses of different maturities.

We have seen that the overall energy industry is growing slightly more slowly than the global economy. Therefore, in aggregate it can be classified as a mature to aging industry. Indeed, many segments of the industry are showing signs of maturity. However, there are also an increasing number of growth segments, and it is important for companies to recognize that different strategies are called for in the different segments.

It is also important to understand that technology, regulations and market changes can rejuvenate a sector. For example, technology advances in the liquefied natural gas (LNG) supply chain have transformed this segment from a ponderous, slow moving, mature business controlled by regulated utilities into a dynamic, rapidly changing segment with a number of new entrants. Similarly, advances in tar sands mining technologies, particularly the ever increasing scale of the trucks used to move the oil sands from mine to processing center, have reduced costs, leading to rapid growth in this segment; advances in steam-assisted gravity drainage (SAGD) are allowing further growth in the development of deeper oil sand formations (table 6.2).

Most important, as explained in chapter 1, the energy industry is subject to long cycles. When prices rise in an up cycle, a new set of opportunities become economic, and investment growth accelerates. When they fall in the down cycle, companies are obliged to rein in their investments and reduce costs for survival. In the view of the authors, recent high prices are signaling the need for changes that will be more difficult to achieve than those of the 1970s and 1980s. Further, the rise of China and India to materiality as global economies and in associated energy consumption has radically changed the demand outlook. If this is true, this up cycle may last longer, and its down cycle may be deferred beyond the 13 years that characterized the previous cycle, between the 1973 start and its 1986 collapse.

Strategies that would not have been natural for companies with merely a favorable competitive position in an aging segment can be revisited if the segment is now in a growth stage. This calls for more imagination and leveraging of core competencies, rather than "hunkering down" as in the past. For example, at $2.50/mcf prices, few companies could build an economically viable business in LNG or tight gas. At $5/mcf and above, new growth opportunities abound. Similarly in the refining segment, slow growth in demand, stable feedstock quality, and spare global capacity suggested the only refining strategy available was operational efficiency (and bottom fishing acquisitions). Today the situation is radically different: no spare capacity, rapid demand growth in Asia, changing feedstocks, and newly economic technologies such as gasification. Refining now looks like a growth segment. What is important is to understand the cycle dynamics. The refining cycle may be shorter in duration than the resource cycle, and strategies need to be carefully paced.

Table 6.2: Maturity of Energy Segments

Indicator	Embryonic	Growth	Mature	Aging
Growth Rate	Increasing	>GNP	<GNP	<0
Product Lines	Basic	Proliferating	Renewal	Reducing
Role of Technology	Concept Development	Opportunity extension	Process improvement	Cost reduction
Competitors	Increasing	Shakeout	Stable	Declining
Barriers to Organic Entry	Low	Increasing	High	High
Supplier/ Customer Loyalty	None	Increasing	High	High
Cost Importance	Low	Increasing	High	High
Energy Segment Positioning	• Renewables • East Africa Exploration • India, China Commercial Energy	• LNG • Oil Sands • Unconventional N. Am. Gas • Heavy Oil Processing	• OECD Refining • OECD Power? • C-Stores	• N. Sea E&P • GoM Shelf E&P • Nuclear?

While growth rate is a primary indicator of a segment's maturity, it is useful to validate its positioning by reference to the other considerations listed on table 6.2.

Note, for example, that we have some doubts about the status of OECD power and nuclear energy. Up to the mid-1990s, OECD power was clearly a mature segment, showing many aging characteristics due to its low overall growth rate and regulated regional monopolies. However, deregulation in Europe and in parts of North America has changed the maturity of the segment by allowing new competitors, lowering barriers to entry, and eliminating enforced customer loyalty. Deregulation opened the segment to strategies that normally would be considered in the growth stage of a life cycle, and a shakeout in the number of competitors is occurring. This will be amplified in the U.S. by the repeal of the 1935 Public Utility Holding Company Act (PUHCA), which will allow consolidation among regional power companies.

For decades, nuclear energy has been targeted by environmentalists, effectively preventing growth or a normal maturity for several decades. Moreover, several European countries had planned to retire their nuclear plants. These events gave the industry a clearly aging appearance in developed economies. More recently, however, recognition that nuclear energy produces power without emitting greenhouse gases has led to a reconsideration of government policies. Should nuclear power revive, it would indicate that modified regulations, technology innovation, and change in public sentiment can rejuvenate industry segments and open up new strategic options.

The deepwater Gulf of Mexico is a segment that has been opened up by advances in technology. If we suspect that the deepwater Gulf is currently a growth segment, we can validate its position by reference to table 6.2 and note that several different geological plays are emerging (the upstream version of product lines). Secondly, technology is focused on expanding the number of drillable opportunities. Thirdly, a shake out in the number of competitors has begun, with EnCana and Spinnaker exiting by sale to Statoil and Norsk Hydro respectively. And finally, companies are seeking to develop long-term relationships with suppliers such as drilling companies and platform fabricators.

More ambiguous is the entry barrier indicator. By virtue of the lease provisions set by the federal government, incumbents are obliged to relinquish leases that have not yet been explored. Many of these leases will be subject to new bid rounds in 2007 and 2008, thereby reducing the barriers to entry by allowing new participants to bid on the relinquished blocks. In this way, the government increases the competitive intensity (which we will discuss below), extends the growth stage of the maturity cycle, and increases its receipts from bonus bids. Notwithstanding the availability of leases, there are still significant barriers to entry for newcomers in the form of understanding the petroleum system and proven ability to see below the salt layers that hide oil-bearing formations from conventional geophysical processing. Nevertheless, deepwater Gulf of Mexico qualifies as a growth segment on most criteria.

As we pointed out in chapters 3 and 5, we believe that the industry in aggregate is entering a new secular growth phase, with the opportunities to be found in developing non-conventional resources to meet societal needs for mobility and power and in new value propositions to resource-rich countries. Strategies need to recognize that, while some segments may be moving steadily from growth to maturity and aging, other segments may be rejuvenated by a change in commodity prices or by new technologies; the strategies that have been adopted as suitable for an aging industry may be wholly inappropriate for the segments entering a cycle of rapid growth. Different strategies will be required from those that worked well for segments at the end of the last cycle. However, organizations tend to be skeptical or even defensive if faced with the possibility that prior business models may be becoming obsolete. Scenarios can be a very effective way of conditioning an organization for change.

Using Scenarios to Broaden Strategic Thinking

As noted in chapter 4, scenarios are useful for characterizing different possible geopolitical and societal futures. They are also valuable for opening the minds of leaders to new strategic directions. Scenario development was pioneered by Herman Kahn of the Hudson Institute in the early 1970s, and applied most famously by Pierre Wack of Shell.[9] Through this exercise, Shell came to understand the importance of rising nationalism, and was better prepared than other major oil companies for the 1973 oil crisis. Shell recognized that the new scenario played well with its strong competencies in becoming a local company wherever it operated, and it strengthened still further the role of country chairman. Other majors' business models were locked into an integrated supply chain from Middle East producers to refineries in consuming markets, and they had considerable difficulty in disaggregating the supply chain and building profitable local businesses.[10]

The first step in developing scenarios is to identify the important drivers of future supply, demand, prices and industry structure and assess which are most uncertain and which are most important. Of the various drivers, there will be some that are quite certain. For example, demographics are quite predictable. Demographics are also important, but since they are predictable, they should be incorporated as forecasts into all scenarios. Conversely, drivers that are uncertain but not important do not merit consideration. Scenarios should focus on those drivers that are uncertain and important.

Future scenarios should definitely include consideration of technological advance, regulatory frameworks, globalization, resource access, and industry structure. A scenario is a "story" describing a possible picture of the future business environment. Scenarios are not "wild guesses" or minor variations on a "best guess" single-point forecast. Instead, each scenario describes a plausible, but qualitatively different future. Scenarios should be relevant, internally consistent and unique—specific enough to test alternative future courses of action. Scenarios describe a possible world in the future, and how we might have gotten there. It is important to set the timeframe and the drivers to be relevant to the challenges facing the company.

It is also important to ground the scenarios in reality through rigorous analysis. For example, it may be tempting to develop a futuristic scenario in which technology will halve our consumption of transportation fuels. But however desirable such an outcome may be, such a scenario would require draconian government intervention in markets to force consumer choice in the direction of rapid adoption of the new vehicles. It may still be an important scenario to consider, but it must be thought through in detail so that all the connecting parts hang together.

The authors believe that all energy companies should give serious consideration to a scenario including a secular shift in energy demand growth as China and India continue an expansion similar to that delivered by Europe in the 1950s and 1960s, overlapping with Japan's growth of the 1960s and 1970s. The big differences are that these two markets represent the aspirations of 2.5 billion people, and it is difficult to see today the supply analog of the expansion of Middle East and Soviet Union oil production that fueled post-World War II OECD growth. Scenarios are important when the future is unclear.

Being Objective on Competitive Position

Whatever the future, strategic options are always shaped by a company's *competitive position*. This can be assessed for an industry segment by determining the key success factors and ranking the various competitors on their strengths and weaknesses relative to them. Success factors in the upstream normally include such characteristics as: understanding the petroleum system, scale, financial strength, operational efficiency, access to new opportunities, relationships with suppliers, relationships with partners, relationships with permitting agencies, expertise in relevant technologies, and access to infrastructure. In the downstream the same factors are still important, but a number of customer and product factors are also important, including brand value, product distinctiveness, customer intimacy, site location, and customer service level. Embedded in the success factors are the differentiating competencies of companies and their competitors.

To the extent possible, it is important to quantify performance of all the competitors on all of these factors through benchmarking, surveys, or analysis of public data. It is then convenient to rank the competitors and describe their competitive positions:

- *Clear Leader:* Would generally need to have a market share twice that of its nearest competitor, and can sustain its position regardless of their actions
- *Strong:* Has considerable freedom to take a wide range of actions, but must consider competitive implications before moving
- *Favorable:* Can sustain its current position provided it makes few mistakes
- *Defendable:* Can continue in business, provided stronger competitors do not target its space
- *Weak:* Is unlikely to be able to achieve acceptable financial performance in its current state

Steve Lowden of SER believes that the IOCs are losing competitive position because of dis-economies of scale and because of unintended consequences from Sarbanes Oxley.

> And I think the other trend, which is more of a Western influenced trend, is that things like Sarbanes-Oxley and consolidation in the West have moved the key decision-makers and leadership away from the host government customer interface. And therefore their relationships don't count very much anymore. And yet their key clients, stakeholders, which are governments, national oil companies, local investors, are used to dealing top to top. They want to deal president to president, CEO to CEO, and director to director. They don't want to work four levels down inside the organization. So that's why we've suddenly seen this influx of new companies into this business, in this space where they care about relationships. And their senior guys are directly connected to their counterparts in the resource-rich countries.

Meanwhile, local investors and national oil companies in producing countries are gaining competitive strength, as are the INOCs (see chapter 2). If this continues, IOCs will need to be more selective about their strategies, find energy industry segments where they can be strong competitors, and choose innovative strategies to offset their relative weakness in other areas.

Competitive position analysis also needs to look at very specific business positions. For example, let's consider how the competitive dynamics of the deepwater

Gulf of Mexico have evolved, and how that has affected the competitive positions of the competitors. Shell Oil, the U.S. subsidiary of Royal Dutch/Shell, was the first company to establish a position in this play. In the 1970s and early 1980s, Shell Oil still had a portion of its shares owned by the public, with Royal Dutch Shell owning approximately 70% of the company, and had considerable strategic independence. It had found itself stymied for upstream growth opportunities within its designated geography and prohibited by its parent from seeking international opportunities. Necessity being the mother of invention, Shell Oil redoubled its efforts to find new growth opportunities in the U.S.

The company invested in understanding the Gulf of Mexico petroleum system and determined that ancient migration paths containing oil precursors extended out from the Shelf to the deepwater. They took a gamble and built a commanding lease position at a time when few other companies recognized the opportunity, and even fewer believed that discoveries could be economic. Shell then got to work with its suppliers to design deep-water drill ships and to extend technologies already being used in the North Sea to design deep-water production platforms. By the mid-1990s, Shell had developed the technologies that proved the concept could be economic (in parallel, Petrobras was making similar advances in deep water offshore Rio de Janeiro).

In 1994 and 1995, Shell successfully completed the Auger and Mars developments and was happily surprised by the productivity of the reservoirs, which resulted in a faster ramp-up of production and, therefore, a higher than planned DCF return on investment. Shell was at that point the clear leader in deepwater Gulf of Mexico exploration and development. However, Shell had decided at the corporate level that the risks involved were such that it should bring in partners. BP gladly accepted Shell's invitation and has by many measures now overtaken Shell as the market leader in this segment. BP in turn brought in BHP as a partner, and then BHP brought in Petrobras, so the number of competitors expanded rapidly as would be expected as the segment moved into its growth phase.

Currently, Shell, BP, BHP, Petrobras, Chevron, and ExxonMobil can all be considered strong competitors in the deep water Gulf of Mexico. Companies such as Devon, Anadarko, Newfield, Murphy, Hess, Dominion, and ConocoPhillips can probably be considered as having a favorable position with a sufficient basis of knowledge and experience to sustain their positions. Several other companies have defendable or weak positions and would be well advised to focus their efforts tightly by adopting selective investment to build expertise in a niche within the segment. EnCana had a favorable position, but moved early to monetize its investment by selling to Statoil. As the segment matures, more companies will find they need to adopt a selective investment strategy, and some will fail to build a defendable position and will be advised to exit.

As with prior elements in strategy formulation, the important thing is to have a productive and honest conversation about the relative positions of the different competitors in each segment. Through that conversation comes a solid grounding in the current competitive reality. The conversation also produces a shared understanding of the relative importance of the key success factors and where a company has relative strengths or core competencies that can be leveraged, and also where the weaknesses are that need to be addressed.

Reflecting on Competitive Intensity

The next aspect of a sound strategic position analysis is an assessment of the relative competitive intensity of the business segments in which the company competes. For this, we use the classic Porter's Five Forces approach (fig. 6.5).

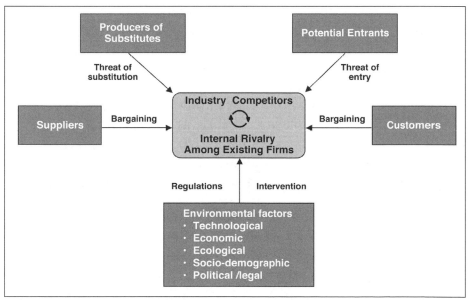

Figure 6.5: Competitive intensity is a key element in predicting industry profitability.
Source: CRA International, Inc., based on Michael Porter Five Forces model

Each segment must be reviewed for the relative power of suppliers and customers, for the potential for new entrants to cause havoc, for substitute products to emerge, and for the level of internal rivalry endemic in the segment. These five forces are conditioned by regulations, the business cycle, and environmental factors.

From a broad market perspective, we observed in chapter 1 that market shares and market cohesiveness are key factors in understanding the level of internal rivalry in a market segment. Additionally, when markets are growing and expansive, there is generally a lower competitive intensity than when markets are in a mature stage and companies are fighting for the same customers in a low growth environment. In aging industry segments, competitors rationalize and intensity tends to subside

The threat of substitutes is quite high for energy markets other than transportation, where fuels derived from crude oil are dominant. However, technologies exist today to increase significantly interfuel competition for transportation, using fuels derived from natural gas, coal, and biomass or ultimately, through advanced hybrids, from electricity. The other dominant factor impacting industry rivalry is the forward-integrating resource holder (supplier) entering and intensifying competition in the downstream and petrochemicals segments.

Looking at niche markets, the refining segment in California demonstrates lower competitive intensity because permitting for capacity expansion is very difficult. Hence, there is little possibility of new entrants. Moreover, substitutes for California's boutique gasoline are not yet on the horizon, although there are grand plans to promote substitutes. Customers have little power as individual buyers at the pump. And the feedstock comes from commodity market where no individual oil suppliers carry significant bargaining power.

By contrast, as we have seen, the Libya exploration segment has been intensely competitive, with many new entrants, a supplier with enormous bargaining power (the government), and few other international exploratory opportunities available as a substitute. In considering strategy, it is natural to favor the less competitively intense segments, though at this stage of the energy business cycle that is difficult to do.

Assessing Overall Strategic Position

We described earlier that there are generic strategies that seem appropriate for industry segments at different levels of maturity. Assessments of industry segment maturity and company competitive position can be combined in a matrix that provides some guidance on the degrees of freedom a company may have in its strategic choices (fig. 6.6). In this figure, we have illustrated increasing competitive strength of NOCs in a maturing conventional oil industry. By inference, IOCs may need to consider avoiding head-to-head competition in favor of more selective investments in conventional resources in traditional places, while focusing more intensely on the growth segments of unconventional resources and non-traditional places.

The third dimension of competitive intensity can be added by shading the segments. Companies operating in mature or aging segments need to have a strong competitive position in order to succeed, particularly in the most competitively intense segments. This is what drives consolidation in mature segments.

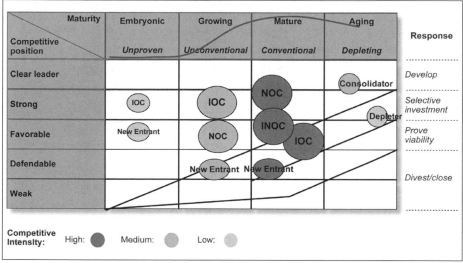

Figure 6.6: Strategic position analysis helps identify natural strategic responses.
Source: CRA International, Inc.

The overall matrix is useful in viewing different business segments objectively and leads to further productive conversations on where to focus the company's scarce human and financial resources. Gwyn Morgan's comments in chapter 5 illustrate how an honest internal appraisal of competitive position, segment maturity, and competitive intensity led EnCana to divest its Gulf of Mexico deep water, North Sea and Ecuadorian assets while further emphasizing its North American resource plays. Putting his segments into this framework reveals that EnCana had defendable positions in a growth segment in the Gulf of Mexico deepwater and in mature segments in the North Sea and Ecuador, but had a strong position in a growth segment in resource plays. By divesting his weaker positions in very favorable market positions, he was able to "double up" in the segment where he believed EnCana had truly distinctive competencies.

Developing Corporate Strategy

After determining the strategic positions of business segments, an organization can begin to develop strategic options for the corporation as a whole and for the various business segments. Corporate strategy development offers two challenges: 1) to shape the portfolio of businesses so that it can deliver a persuasive investor value proposition; and 2) to clarify the parenting advantage that will make the businesses worth more under the company's ownership than they would be with another owner. In this section, we will describe an approach to corporate strategy development and outline the major strategic challenges facing the individual energy industry business segments.

Making portfolio choices

Typically, energy companies will develop some simple themes at the corporate level to clarify the role of the business segments in the corporate portfolio. For example, Royal Dutch/ Shell CEO Jeroen van der Veer explained to us. "We are sticking to a very clear strategy of 'more upstream and profitable downstream'—investing to grow our exploration, production and gas businesses while refocusing our downstream strengths in oil products and petrochemicals." This simple message makes it clear that upstream should be looking for new investments, while downstream needs to work on its returns. Business strategies are generally left to the business segments, since they understand the markets, competition, and regulatory frameworks better than headquarters units. However, the corporate planning unit and top management have the opportunity to question or reject the business segment strategies if they believe that they will not deliver the intended results, are inconsistent with the corporate parenting advantage or are extravagant in their demands for resources.

Each business segment must identify the various strategic options and choose the most appropriate one. From a financial perspective, that would be the option that had the highest risk-adjusted net present value. However, it is often difficult to quantify risk and, by definition, impossible to quantify uncertainty around price and cost forecasts. Thus, leaders often use a set of filters based on attractiveness, "do-ability," and "fit" to narrow down the options. This reduces reliance on complex financial analyses and stimulates the right conversations on the merits of the different options.

"Attractiveness" refers to the fundamental economics of the business strategy for a generic competitor: is there growth in the sector? Will new options open up if we pursue this strategy? Is the competitive intensity moderate? Are margins reasonable? "Do-ability" focuses in on our own capabilities: can we get access to the opportunities that we have identified? Do we have the necessary relationships with partners and suppliers? Do we have the necessary information? Do we have the required core competencies? How will competitors respond? Can we finance the investments? "Fit" addresses whether the strategy is aligned with our overall investor value proposition: does it match our positioning on agility and focus, cost leadership and differentiation; is it consistent with our values? Is it complementary to the strategies of other businesses in the company? Even if we think we can get it done, can we sustain the strategy over the long term?

By using qualitative filters, the full array of options can be reduced to a manageable set that can be quantified in financial models and analyzed in depth for risks and rewards.

Assessing the corporate portfolio

Energy companies have a wide range of choices for creating value for their stakeholders. The primary choices are in the following areas:

- Which commodities (crude oil, natural gas, power, refined products, chemicals)?
- Which regions or global coverage?
- Where in the supply chain (discovery, transformation, monetize, customer services)?
- What business model will work best (integrated or not, insourcing, outsourcing, partnering, technology)?

It is informative for large energy companies to unbundle the corporate portfolio along these dimensions and reflect on whether the mix makes sense. If the investor value proposition is centered on returns and sustainability rather than growth, then it will be most important to see how the different portfolio elements are performing on those dimensions (fig. 6.7). This is a variation of the classic risk-return analysis performed by financial investors on stock portfolios. Algorithms are available to pick the portfolio elements required to produce a desired risk-return result. Risk can be expressed as the predictability of results, which historically can be seen in the variation of earnings from year to year. However, it is also useful to reflect on risk looking forward by making a judgment on future technical, political and commercial risk in each portfolio segment.

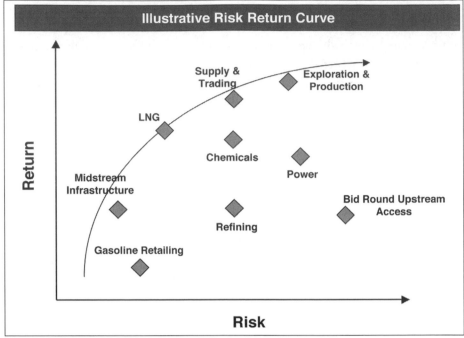

Figure 6.7: Investment shifts to difficult resources and difficult places.
Source: CRA International, Inc.

The choices are interdependent with the company's value proposition to investors. We have seen that the value proposition is crafted from specific positions on growth, returns and sustainability. Companies stressing their returns have to be alert to the commodity cycles in their segments, which will influence future returns and risk. Companies wishing to highlight the stability and sustainability of their results will most likely need exposure to several different business segments, since each of the commodity sectors experiences cycles in prices and margins that may be coincident or opposed. By contrast, companies with a value proposition centered on growth have generally chosen a relatively narrow focus of commodities and regions. These companies have to be particularly sure that they are downwind of the important trends that will propel their chosen segments forward.

Above and apart from the selection of a good portfolio, corporations must also address and define their "parenting advantage" (see chapter 5). They must be able to explain to their investors and to their employees why their assets are worth more in their portfolio than in someone else's. According to Michael Goold, in order to create more value than rival parents, a company needs:

1. A corporate strategy targeted on parenting opportunities that have not been seen by other rivals, or which are more substantial than those of the rivals, or in which the parent's depth of understanding of its role distinguishes it from rivals (value creation insights).
2. Parenting characteristics that are more suitable for realizing the opportunities it is pursuing than the characteristics of rivals that are targeting similar opportunities (distinctive parenting characteristics).
3. A clearer understanding than rival parents of the sorts of business in which its value-creation insights and distinctive parenting characteristics are most potent, and in which misfits with parenting characteristics that would lead to value destruction can be avoided (heartland businesses).

Shell's early move into deep water exploration, and BP's and ConocoPhillips' securing positions in Russia are examples of *value creation insights*. A parent can apply value creation insights through stand-alone, linkage, and functional/services influences. Stand-alone influences include picking the right people for senior positions, budgetary control, and strategy reviews. Linkage influences include lower transaction costs across internal divisions, benefits from coordinating strategies, pooling power with external groups (customers, suppliers, other stakeholders), improved utilization of physical assets, and better utilization of intangible assets. Functional/service influences can include the deployment of specific competencies in manufacturing, technology development, marketing, brand management, government relations, etc. Goold argues persuasively that a company should be able to articulate its parenting advantage along these dimensions, and this should be the foundation of its corporate strategy.

ExxonMobil asserts that its business model of technology deployment and capital discipline are *distinctive parenting characteristics* that will result in superior returns. Shell and BP have dramatically shrunk their exposure to petrochemicals to avoid potential value destruction, reflecting their perception that this business did not fit well with their parenting characteristics. EnCana divested its successful North Sea, Gulf of Mexico, and Ecuador businesses to focus on its North American resource plays, where it believes it has a true competitive advantage based on core competencies, and distinctive parenting characteristics.

The external environment of rapid demand growth in China and India coupled with peaking conventional oil and gas production opens up a plethora of growth opportunities in all energy segments, as outlined in table 6.3. Business segment strategies will be focused on managing the life cycle of existing assets for maximum value, through to eventual divestment and abandonment, and building robust processes for identifying, capturing and implementing the growth opportunities. We discuss these strategies in more detail below.

Table 6.3: Business Segment Growth Strategies

Exploration and Production	• Establish Arctic as a new exploration province • Continue growth in deep water • "Manufacturing excellence" in unconventional gas • Reduce costs and emissions of oil sands development • Further expand LNG supply chains
Refining	• Increase heavy oil capabilities (integrated to oil sands) • Participate in "refineries of the future" for Asian demand • Participate in GTL, CTL, and biofuels
Power	• Help design structures with role for private capital • Integrate power with refining, CTL, and GTL
Midstream and Trading	• Develop markets for new commodities • Arbitrage international imbalances • Provide customer choice with risk management products
Retailing	• Develop partnering strategies to access growth markets
Overall	• Improve partnering capabilities with resource owners and "gatekeepers" to emerging markets • Work on competencies to deliver large complex projects • Educate the general public and special interests on the role of energy in the global economy

Developing Exploration and Production Strategies

Strategy development for the upstream E&P segment of the oil industry focuses largely on three topics—securing access to areas known or believed to contain resources in commercial quantities, determining the types of assets to have in the portfolio, and assessing technology requirements and building supplier partnerships. Exploration for conventional oil in traditional places will suffer from extreme competitive intensity, so companies will need to be very selective and only compete where they have a clear advantage through existing infrastructure and strong relationships. The growth opportunities will be in developing unconventional resources such as oil sands, tight gas, and stranded gas, where the success factors will be the familiar oil industry competencies of resource capture, scale, technology application, project management, and integrated supply chains.

Resources

The dominant theme of exploration and production (upstream) strategy is the scale and quality of the portfolio of assets and of project opportunities, and the competencies that the company can draw on to exploit them. E&P strategy for the IOCs must increasingly take into account that most of the undiscovered conventional oil reserves are inaccessible to them.

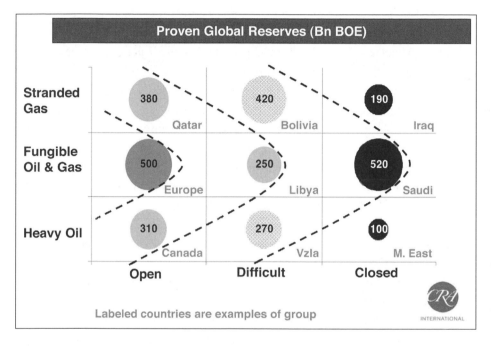

Figure 6.8: How can the individual value chain plays be creatively combined to optimize returns? *Source: CRA International, Inc.*

Global hydrocarbon resources can be pictured along two dimensions (Fig. 6.8). On the y-axis, we can arrange the resources in terms of their value: from low value oil sands and heavy oil, through high value conventional oil and fungible local gas, on to low value stranded gas. On the x-axis, we can consider the ease of access to the resources, from open access regions such as the Gulf Coast or North Sea, to more difficult opening regions such as Libya and India, and on to regions that are currently largely closed to international oil companies such as Saudi Arabia and Iran. While there are growth opportunities for fungible oil and gas development in open regions, these are intensely competitive. Consequently, companies are looking for growth opportunities by developing unconventional, lower value resources such as heavy oil and stranded gas in open regions. They are also looking for growth opportunities in regions that are more politically difficult to access. Even these are now proving to be highly competitive, as a recent bid round in Libya showed.

The IOCs also have available the final frontiers of E&P technology in ultra-deep water and in the Arctic. Don Paul, Chief Technology Officer at Chevron, gave an example: "We are at the edge of that with our joint venture with Exxon in the Orphan Basin, which is northeast of Newfoundland. We leased the entire basin, so if there is nothing, you get nothing. But if you do get something, then you are all set. The only problem is, you don't know how to produce it. You've got to have completely unpluggable production platforms of immense scale, or you've got to have really full sub-sea automation and long 1,000-mile tie-backs. But that's all right; ultimately if the resource is there, these challenges will be overcome."

Helge Lund has ambitions for Statoil in the Arctic also, and notes the technology challenges: "Our ambition is to take the role of an industrial architect in the Arctic. We have taken the first positions in Snohvit and Goliath in Norway, and we are working hard to qualify for Russian opportunities as well.

The Snohvit field is 140 kilometers offshore. With an onshore processing plant we are taking multiphase transport to a new level. We also remove the CO_2 from the gas and inject it back into the reservoirs. The challenge there is the well string, the distance, and to avoid hydrates. A field like Russia's Shtokman is 550 kilometers from shore. We need an advanced technology development to be able to undertake that, and also master the challenges posed by icebergs."

There will also be opportunities for inorganic growth by increasing share in mature areas, so that properties end up in the hands of the owners to whom they have most value. This should encourage regional consolidation, enabling companies to build economies of scale, increase their knowledge base and create efficient drilling supply chains. However, the strongest trade winds will be supporting the development of the more difficult resources such as unconventional gas and oil sands.

Upstream asset portfolio

Companies may choose to center their activities on pure exploration, on resource development, or on a mix of the two. Some small companies have been adept at securing access to acreage, devising and proving a concept, through some initial exploration, and then selling down their interest to larger players. An example was Triton, a small company that successfully found fields in Colombia and Equatorial Guinea before being purchased by Hess. Triton's history of volatile investor returns demonstrates that this is a difficult strategy to sustain. Steve Shapiro, CFO of Burlington Resources, characterized the challenge for us: "What we found is, exploration works but isn't scalable."

Most companies try to build a balanced portfolio of exploration and development opportunities (fig. 6.9[13]) in order to sustain and grow their business, and many companies renew their portfolios using inorganic (merger and acquisition) strategies as well as organic (exploration) strategies. The main focus for exploration-led growth is in deep water, and especially for resources beneath layers of salt, which makes imaging difficult. However, as noted above, companies are also progressively moving to lower value resources that require above ground transformation rather than below ground exploration skills. As such, the shape of the asset portfolio for most companies will be flattening in the future.

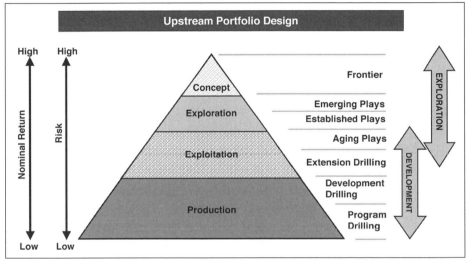

Figure 6.9: Upstream portfolio design. *Source: Dominion Exploration and Production, Inc.*

We find that it is often useful to imagine several E&P strategic themes consistent with the opportunities available and the companies core competencies, then build them out as hypotheses to be tested. Those that best fit with corporate objectives, that seem most likely to desirable results, and that are supported by the business unit's competencies can be selected for implementation. What does not work is to create a huge database of all available opportunities and rank them according to various political, technical, and economic criteria. This exercise almost invariably produces an inchoate set of unconnected opportunities with little chance of establishing competitive advantage.

Upstream technology

Technology in the upstream oil and gas business presents something of a paradox: no one doubts the value it creates, but it's notoriously hard to be confident that the money spent on it generates a good return. Technology has a central role to play in all parts of the business: improvements on existing techniques will drive economically marginal conventional opportunities, and new technologies are necessary to monetize unconventional resources. Additionally, access to business opportunities with non-traditional partners, such as OPEC and Russian companies, will be strongly dependent on technology excellence and reputation. As indicated in chapter 1, the authors expect to see a renewed interest in research & development strategies during this up-cycle.

Supplier relationships

One of the last important frontiers for exploration may be the deep water offshore the Gulf of Mexico, Brazil, West Africa, and Southeast Asia. For the past decade, exploration companies have been able to rely on there being surplus supplies of most of the equipment and services they needed and fierce competition among suppliers for their business. Since 2004, and exacerbated by damage caused by Gulf of Mexico hurricanes, this is no longer the case. There is currently a shortage of deepwater drillships and heavy-lift vessels, and constraints in the availability of fabrication capacity for production platforms. There is also a looming shortage of skilled personnel, not only in deepwater technologies, but throughout the technical underpinnings of the energy value chain. As a result, operating companies are making longer-term commitments to key suppliers, and deeper partnerships are emerging. Supply chain strategies have become critical not only in deep water, but also in "resource plays" that rely for their success on program drilling, and those that were previously based on reverse auctions and low cost bids are being revisited. Operators compete for the economic rent from exploration, but must cooperate to expand the number and value of opportunities. This demands a "co-opetition" approach of shared technology development and field experimentation.

Developing Natural Gas Strategies

The primary natural gas resource development strategies are centered on reducing the costs of extracting unconventional gas from tight sands and coal beds, and in building integrated supply chains for commercializing stranded gas—gas resources that are far from markets—through liquefied natural gas (LNG) and gas-to-liquids (GTL) technologies, as well as through long distance overland pipelines.

LNG

There are very strong trade winds supporting the monetization of stranded gas. Traditionally, LNG projects were put together piece by piece, in complex consortia of buyers, transporters and sellers, linked by long term, take-or-pay contracts, and backed by long term project financing (fig. 6.10).

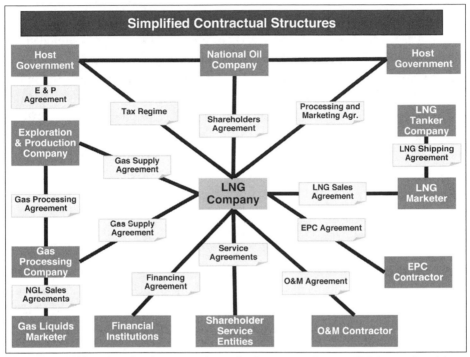

Figure 6.10: LNG projects involve a complex web of contracts. *Source: CRA International, Inc.*

The pioneering contracts were forged in Algeria and Indonesia. The Algerian contracts were between the national oil company, SONATRACH, and natural gas pipeline and distribution companies in Europe and in the U.S.[14] In Indonesia, the deals were between the oil companies with the natural gas discoveries and Japanese utilities and trading companies, with the national oil company, Pertamina, as a relatively passive partner. In these deals, Japanese companies received most of the engineering, procurement, and construction contracts as well as the orders for LNG tankers, so on a national economic basis they were quite favorable for Japan. Prices in both areas were linked to oil prices, though by different formulae. The deals were designed to be inflexible and bind buyer and seller together in a permanent symbiotic relationship. There were destination restrictions to force buyers to take the LNG to their home market, though there were so few choices at that time that these provisions were somewhat redundant.

Recent projects have been structured differently. The Atlantic LNG project in Trinidad and Tobago broke the mold. The project was designed to allow destination flexibility between the U.S. and Spain, and the investors (Cabot—now Tractebel, a unit of Groupe Suez, Amoco—now BP, British Gas, Repsol, and NGC) structured the project to be able to compete with the prevailing low gas prices in the U.S. in the late 1990s. They did away with the traditional construction gold plating and redundancies of the utility model, negotiated low costs for tankers, and achieved economies of scale by building a larger, single train. They involved the Trinidad government in the project as an equity participant through NGC and received strong support from the government. They priced the LNG on a netback basis, accepting the market price at the point of delivery and subtracting costs to arrive at the well-head value on which production taxes were paid. And then they got lucky, or their foresight was rewarded, because soon after commissioning the first train in March 1999, the U.S. gas bubble abruptly deflated in 2000 and prices rose.

LNG strategies still need to address the full value chain, since markets in tankers and regasification terminals are not as liquid or transparent as similar assets in the oil sector. Over time, LNG will come to resemble oil, but it would be a mistake to expose assets excessively to spot LNG cargo markets before these markets mature. Most recent projects in the Atlantic have been following the Trinidad model and are price "takers" from the prevailing market prices at the destination points. This assigns more price risk to the producer and liquefaction plant investor than the traditional model, and the allocation and management of risk has become an important managerial issue in LNG strategy development.

GTL

An important consideration for natural gas supplies to LNG plants is whether the natural gas would be better assigned to gas-to-liquid projects. We have seen earlier that LNG netbacks to the liquefaction plant entry could be in the range of $1.50/mcf (for the Middle East) to $2.50/mcf (for West Africa). GTL projects are being developed in both regions and, according to our calculations, will be economic if oil prices are above $35/B for the Middle East, and above $45/B for West Africa (fig. 6.11).

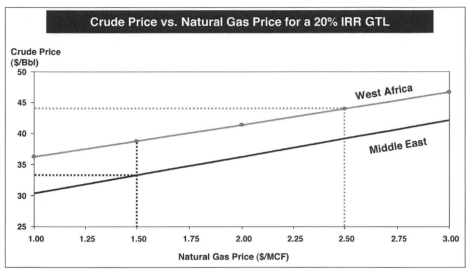

Figure 6.11: GTL could be economic if oil prices are in the $35–45/Bbl range. Note: 50 MBD plant with product netbacks from USGC. *Source. CRA Analysis*

Further, if we accept the premise that natural gas must in the long term compete with clean coal, then oil prices above $50/B will make GTL a more attractive use of stranded gas than LNG. Thus, the lock that conventional oil has on the transportation market would likely be broken if prices remain high through inter-fuel competition provided by advancing GTL technologies.

Non-conventional gas

The United States has an alternative or complement to LNG—to expand natural gas supply from North American resources that are more difficult to access, such as the Arctic, and from non-conventional gas resources. Figure 6.12 shows where the National Petroleum Council believes the resources to be (note that figure 6.12 shows expected resources for gas, while figure 6.7 showed proven reserves of oil).

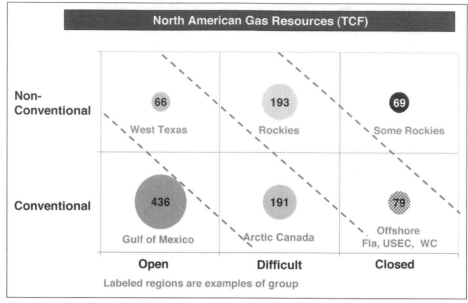

Figure 6.12: Investment shifts to dffcult resources and diffcult places. *Source: CRA International, Inc., based on NPC data*

Non-conventional gas include tight gas, where the gas is held in sands with low permeability that need to be fractured to release the gas from the formation into production, and coal bed methane, where the gas is found in association with coal. In the U.S., coal bed methane has benefited for many years from a government subsidy. These strategies are well described in comments by Gwyn Morgan, former CEO of EnCana, in this and the following chapters. Due to rising oilfield service costs and strong competition for acreage, the costs of tight gas and coal bed methane development have been rising. However, many deposits are economic at gas prices of $5 to 6/mcf, and companies are working hard to reduce production costs by strategies of customized drilling rigs and "Kangan" supply chains.

Developing Refining Strategies

We have described the strong drive toward increasing demand for transportation fuels worldwide. We have also described the opportunity for upstream companies to expand production of heavy oils. Finally, we note the rise of gasification technology that, if oil prices remain high, will provide a powerful incentive to upgrade all refinery streams, including petroleum coke, into electricity or, with Fischer Tropsch technologies, on to transportation fuels. These trade winds combine to create a favorable investment climate for refiners.

Heavy oil upgrading

On both sides of the globe, the imperative to upgrade heavy oil is again evident as it was after the oil shocks of the 1970s. In the East, the drive will be to upgrade heavy fractions currently produced from refining conventional crude oils as market outlets for residual fuel oil are eroded by competition from cheaper fuels.

In the West, the upgrading imperative will stem not so much from loss of markets for residual fuel, but from increasing supplies of bitumens from oil sands and heavy crude oils. Two-thirds of the world's deposits of heavy crude oil reside in the Americas (fig. 6.13). Venezuela holds approximately 270 billion barrels while Canada's oil sands reserves are estimated at as much as 310 billion barrels, compared with about 1,150 billion (1.15 trillion) barrels recorded for conventional oil reserves worldwide.

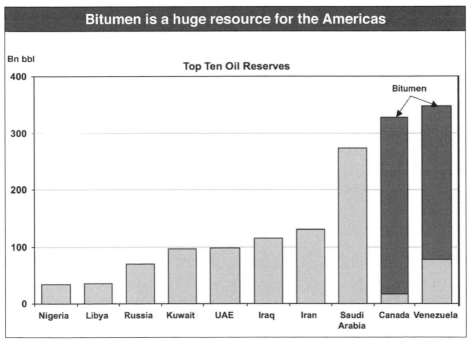

Figure 6.13: Heavy crude oil supply growth requires upgraded refining capacity.
Sources: BP Statistical Review, Oil & Gas Journal, PDVSA, EIA

In both regions, prices are signaling the need to upgrade residual streams: the price spread between residual fuel oil and sweet, light crude oil inflated to around $30/barrel on the U.S. Gulf Coast well before the damage inflicted by hurricanes Katrina and Rita. The implications for refining investment are substantial with investment requirements in the tens of billions of dollars. This is a huge growth opportunity, with economics underpinned by demand growth and by changes in the quality of feedstocks.

Consolidation

For the past 15 years, refining strategies have been dominated by the consolidation typical of a mature industry segment. Companies such as Valero and Tesoro in the U.S. and Petroplus in Europe were able to grow by acquisition, while the majors in aggregate divested capacity either voluntarily to rebalance their systems, or to respond to FTC directives following mergers. The industry's concentration in the U.S. has increased, but there is probably still room for some potential combinations that could add value. However, as the industry is rejuvenated by higher demand and the need to process new feedstocks, this strategy is less natural than it has been in the past.

Reliability

Commodity oil markets are volatile. Spot prices move up and down sharply in response to geopolitical events, unexpected demand changes, and supply curtailments caused by scheduled or unscheduled refinery shutdowns. Refineries have high fixed costs and low non-feedstock variable costs. They make money when capacity is fully utilized at times of high margins. In the final analysis, the most important refinery strategy is to assure reliable operations, to capture these ephemeral moments of profitability, and of course, to avoid the human and environmental distress caused by accidents. This is why almost every downstream company talks about a strategy of operational excellence.

There has been much discussion of a shortage of refining capacity as one of the causes of high prices for crude oil and refined products. After nearly 20 years of depressed margins, it is not surprising that companies have been cautious in their investment plans. As the realities described above become clearer, investment will increase and the spreads between light and heavy crude oil prices will decline. However, with rising demand for transportation fuels in emerging markets coinciding with the need to convert heavy streams, the prognosis is strongly favorable for refining for the first time in many years.

Broadening scope

In light of the likely pressure on feedstock supplies, the growing importance of non-conventional feedstocks, and the pressure to maximize yield and energy efficiency, we think it is time to develop comprehensive strategies to address the need for flexibility in feedstocks and to add further value by rethinking the refinery of the future. Much of the world's refining infrastructure was laid down in the 1960s and 1970s, when economists were advocating import substitution as the means to economic development. Experience from successful emerging countries has changed the economic recipe from small, sub-scale units to world-scale plants, whether the industry is textiles, electronics or agriculture. This raises the question of what a new-era world-scale refinery might look like (fig. 6.14).

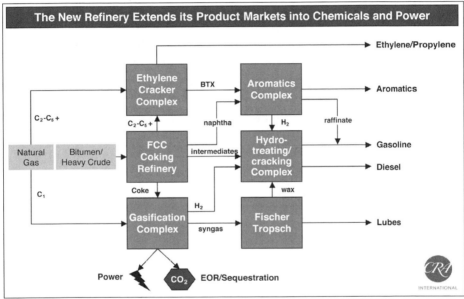

Figure 6.14: The refinery of the future will be more flexible. *Source: CRA International, Inc.*

In our view, such a refinery should expand its definition of feedstocks and of salable products. Feedstocks will include bitumens and natural gas. Salable products will include transportation fuels, petrochemicals and electricity. Emissions to air and water systems will be minimal. SO_x will be chemically reduced to elemental sulfur and NO_x formation will be minimized. CO_2 will be sequestered in geological formations, in some cases creating value through enhanced oil recovery in nearby oil fields. Biofuels and liquids from coal and natural gas will be blended in at refineries or terminals, and form a natural extension of the refining segment.

Such a refinery may be situated either at the resource location or the market location. The determinants will be the cost of construction, the availability of feedstocks, and the overall supply chain cost. The historical norm has been to construct refineries close to markets because it costs less to transport crude oil than products. However, the benefits of very large, complex refineries close to crude oil sources coupled to very large products tankers may in the future outweigh the logistics cost advantage of being close to market.

Examining Power as Part of the Portfolio

We have noted earlier the enormous importance of providing access to electricity for the billions of people who are currently condemned to use traditional fuels. This adds to the inevitable growth in power demand as economies develop and electrify. As we saw in chapter 3, electricity has the highest growth rate in energy consumption.

As often is the case in the power sector, the nature of the opportunities will depend on regulations. It is a sector that will show solid growth in the aggregate and higher growth in those regions with highest economic growth. It is a sector that will require massive capital investment. In the past, power plant investments were often financed by loans to national power utilities backed by government guarantees with the World Bank acting as final guarantor. There is more and more acceptance of the "moral hazard" of loans to governments and their nationalized industries, with the World Bank underwriting the loans. Recent problems in Argentina provide examples as described in chapter 8 by Ian Howatt of Total. Commercial banks have been forced to write off loans provided to Argentine institutions, and private investors have suffered losses due to government-imposed controls on energy tariffs.

Private sector financing

This and similar situations elsewhere have raised a question of how the investment required to supply growth in power demand will be financed if loans to government institutions are not available and private capital is not allowed an economic return. We tend to agree with former British Prime Minister Margaret Thatcher's TINA (There Is No Alternative) view that private sector investment will be required, and that it will only be forthcoming if a decent return is forthcoming. Most likely, the outcome will be most stable if foreign investors (e.g., IOCs) partner with local investors to form joint ventures, so that there is an influential local constituency with "skin in the game."

The World Bank has given much thought to the issue of making electricity affordable while allowing incentives to investors. "Macroeconomic policies should avoid discriminating against or favoring particular energy technologies. Price-distorting subsidies and taxes should be eliminated—though a need remains for well-thought-out, intelligently implemented subsidies that genuinely benefit the poor and yet avoid creating disincentives for energy companies." Investors may be well advised to engage international institutions in working with governments to design an equitable regulatory context for their investments. We return to this theme in the next chapter.

In a rational world, investing in international power projects will offer profitable opportunities. European utilities have made strong moves in this direction. U.S. utilities started to grow internationally, but in many cases they have abandoned this strategy as they were forced to address problems with their trading operations, with overbuilding gas-fired, combined-cycle, generating plants, and with the inevitable proportion of their international investments that hit bumps in the road due to tariffs and other disputes.

Deregulation opportunities

Deregulation of the U.S. power industry started in 1978. That is when the "Public Utility Regulatory Policies Act of 1978 (PURPA) stipulated that electric utilities had to interconnect with and buy, at the utilities' avoided cost, capacity and energy offered by any non-utility facility meeting certain criteria established by FERC." This encouraged large industrial users to build "cogeneration" plants where the steam and some of the power were used by their own facilities, and the surplus sold to local utilities at "avoided cost." At the same time, new combined-cycle technology radically improved the energy efficiency of natural gas-fired power generation.

The Energy Policy Act of 1992 allowed certain non-utility companies to access transmission networks. Then in 1996, FERC issued Orders 888 and 889, formalizing the procedures by which non-utility companies could gain access to transmission and requiring utilities to establish electronic systems to share information on available transmission capacity. This established the principles for wholesale competition in electric power, but the details were developed by state regulatory bodies. Enron and other energy merchants lobbied strongly for the changes and jumped on the opportunities as they emerged. However, a combination of poor market design, ethical problems, and miscalculations in the companies' risk management programs brought the whole edifice crashing down. The final beneficiaries in natural gas markets have been mainly the large oil companies, who had the competencies to step in and fill the gap left by the merchants.

It is an open question whether oil companies will make a major thrust into power generation. John Wilder of TXU believes the oil companies have the need and the competencies: "If I was running an oil company—I'm thinking I'm going to just get table scraps in Russia, West Africa, and Brazil, because they are going to disintermediate me. So where can I transfer my operational execution, discipline skill, and some excess cash flows? Look around and there is a commodity-looking business over here called power. I'd play around with it." Deregulation introduces conditions in which there will be a supply curve (as opposed to regulated cost-averaging), allowing the most efficient operators to earn above the cost of capital returns. This removes one of the fundamental arguments used by oil companies to reject power plant strategies.

Developing Midstream and Trading Strategies

Midstream and trading is the segment of the energy value chain that transports crude oil to refineries and petroleum products, natural gas, and power to market hubs. The segment used to be a function of the refining and marketing or power generation segment's business. The primary role was correcting supply imbalances through the distribution system. Natural gas and power markets were regulated, and this function was known as scheduling or dispatch. This has changed since deregulation, and the commercial function has proven its ability to create enormous value by adding risk management to the supply balancing functions.

The first key issue in midstream and trading strategy is determining the set of assets that supports the design and implementation of good trading strategies. Next comes the information platform that allows development of quantitative risk management products. This leads to segmenting customers into groups with similar needs to whom the products can be offered. And finally, the business model must balance the need for controls with the desire to foster an entrepreneurial culture.

The assets can be owned or secured by contract, though the "asset-light" approach of control by contract has come under question since the collapse of the energy merchants. There are many PhDs in applied mathematics at work on these topics, and the details are beyond the scope of this book. It is clear that trading is back from its collapse when Enron and other merchant players faltered. Renewed interest in trading is promoting sophisticated energy risk management educational programs, such as the one at the University of Houston's Bauer School of Business.

Providing customer choice

Some of the potential benefits of the risk management expertise that is being developed have not yet been fully realized because of ill-designed regulations that mask the true marginal costs of power and gas supplies from consumers. Without information on the marginal costs of supply, customers have no incentive to modify their consumption patterns. Advanced metering and control technologies are available that can harness demand solutions to resolve temporary supply problems. They would also encourage suppliers to offer innovative pricing products that provide customers with various options to select their preferred mix of cost and comfort. These products would improve the efficiency of transmission and distribution systems.

The potential to provide customers with choices that reflect the marginal costs of their purchasing decisions is enormous in terms of the savings that can be achieved by suppliers and passed on to their customers. These schemes are difficult to set up and require considerable customer education, but there is great potential for demand solutions to what may appear to be supply problems, if companies can find ways to share their cost-saving benefits with their customers.

Integration and network strategies

Liquid, transparent spot markets exist that allow conventional crude oils, natural gas, and power to be traded at major demand centers. Since the 1980s, exchange-based futures markets allow hedging and speculating on commodity prices and the spreads between different energy forms (e.g., the "spark spread" between natural gas and power and the "crack spread" between crude oil and transportation fuels). Therefore, there has been little value in forward integration strategies, since each commodity can be sold freely into these liquid markets, eliminating volume risk, while price and margin risk can be managed using futures markets. The same is not true, however, for heavy crude oils and low quality syncrudes, for LNG (yet), and for isolated regions where there are no liquid markets for certain products. In these situations, integration strategies are required until liquid markets develop. Creating these markets and finding arbitrage opportunities across them and across geographies provides an important opportunity, especially for the global super-majors.

Developing Retail Marketing Strategies

Retail marketing of transportation fuels has been revolutionized over the past 30 years by the rise of the "big box" retailers, who have over 50% of the market in some European markets and are growing their market share in the U.S. As one Costco store manager remarked, "We love gasoline. We can sell it at the same low mark-ups we apply to produce sales, and it's not perishable!" All that is needed is to carve out a piece of space from the parking lot, install a couple of tanks and some pumps, and they have added a new product and service offering. Their business model is highly focused on high-volume, low-margin sales and in cross-selling by offering "specials" on certain products (including gasoline) to attract customers. These companies are extending their reach globally (e.g., Carrefour recently announced the opening of its

69th hypermarket in China), and can be expected to include gasoline and diesel in their product mix wherever logistics and government regulations permit.

The stronger C-Store companies have been able to coexist with the "big box" entrants by providing a well-designed offering of convenience products and co-branding with fast food purveyors. Their business model is to use low-margin gasoline to attract customers to buy the higher margin convenience items and food products.

To a very large degree, refiner-marketer companies have encouraged their dealers to replicate the C-Store approach, but also to continue to sell the gasoline and diesel at premium prices on the basis of quality assurance and some differentiation in their additive packages that they claim improve engine performance. These companies also try to secure longer-term customers with loyalty cards. These business models are now pretty much set. There have been thoughts that internet-based services might emerge that would guide motorists to the stations with the lowest prices, and the large trucking companies do provide that information to their drivers; but it is uncertain whether these will emerge as strong influence in the consumer sector. To the extent that IOCs are allowed access, growth in this sector will be found in emerging economies.

Conclusions

The key lesson from this chapter is the need for honest and forthright dialogue on the various scenarios that may characterize the future, how a company is positioned relative to its competitors, and the various options open to it. Leaders must assess objectively for each business segment the stage in the business maturity cycle, the relative competitive positions, and the intrinsic competitive intensity. The energy industry is characterized by cycles of varying lengths, as shown in chapter 1. There are medium-term business cycles as supply and demand responds to investment patterns; there are longer-term secular cycles, as an inexpensive resource is used up, or a new major market needs to be satisfied; and there are very long cycles that presage transitions from one strategic energy form to another. The early part of the 21st century is showing signs of being at the start of a new, very long-term cycle, with huge new markets in India and China to be satisfied and a potential plateau of accessible conventional oil and gas production.

A change is needed in the process of providing price signals to consumers. Energy prices are formed by the shape of economic supply and demand curves. High prices signal that demand is greater than easily provided supply. High prices unfortunately do not tell us whether they are caused by a short-term business cycle that will be reversed when some new production is brought on stream, or long-term secular changes. Business leadership is all about making those calls. The authors believe that high prices this time presage a new long cycle that will rejuvenate the industry from a mature stage of development that characterized the 1980s and 1990s to a growth stage. As we explain in this chapter, growth industries require different strategies from mature industries.

Leaders must also reconfirm their vision in the light of the changed business environment. Vision, mission, and goals should be part of a persuasive shareholder value proposition and must ask: "Is the story that supported our value proposition in the past still credible for the future?" If it is not, then it may be time to start educating investors on a new value proposition, or on a new story, including a revised vision, mission, and goals that will be credible. Vision for a corporation is a long-term statement of intent, and should only be changed rarely. We agree with Kotter: "Vision creation is almost always a messy, difficult, and sometimes emotionally charged exercise," so it is not to be undertaken lightly.

However, when a new topography emerges, the corporation's vision may require adjustment. Even if the vision is unchanged, the mission may prove to be too restrictive if new business areas are going to be considered. For example a refining company may need to reframe its mission from "refining" to "transformation" if it believes that it may extend its business into power. A change in vision or mission will inevitably require a change in goals.

We have argued that the 1990s encompassed the mature and aging phases of the previous cycle, the growth portion of which started in the early 1970s. If we are really at the beginning of a new cycle, then the strategies that served the industry well in the 1990s are no longer appropriate. Further, the early stages of a growth phase are conducive to ambitious strategic choices such as:

- New value propositions for E&P access
- New technology strategies to develop difficult resources
- Extension of the inputs and outputs of refining
- New midstream business models
- Product line extensions into GTL, CTL, and biofuels
- Integration strategies for new resources
- Customer solutions allowing lower costs to be passed through as lower prices

New competitors, especially INOCs, are aggressively embracing different strategies, and they may have competitive advantages in certain areas. Traditional IOCs need to decide whether to try to "beat them or join them," and need also to enter into a dialogue with their home governments on how to combat mercantilism. Most important, they need to be ruthlessly honest with themselves about the key success factors in the new business environment and the competencies required, and adjust their strategies to take account of their new competitive position as well as the different maturities of the industry segments in which they compete.

We are convinced that there will be further changes in environmental and market rules, but would counsel companies to be wary of the Enron scheme of basing strategies on taking advantage of poorly-formulated regulations. Ultimately, thinking through the strategic choices carefully can make the difference in surviving and thriving or being lost in the annals of history books. Who would have thought that General Motors and Ford, the largest corporations in the world in their heydays, would be close to bankruptcy today? It happens.

Sometimes changing regulations or advances in technologies can result in the sudden emergence of a new "sweet spot," which allows companies to embark on new strategies in support of a new vision. Such changes can oblige companies to totally rethink the segmentation of their industry as new segments emerge first in embryonic form, then as growth possibilities, and the competitive position of previously strong players is eroded. It is often difficult for incumbents to accept that the world is changing, and this allows new entrants to capture substantial positions in the new segments before the incumbents have marshaled their forces to respond. Examples abound in telecommunications, where the status quo was overthrown by deregulation and cell phones, and in pharmaceuticals, where new entrants leveraged skills in genetic engineering to define new products faster and less expensively than the major drug companies. In the past decade, there have also been some abrupt changes in the energy sector that have presented opportunities for new business models. Sudden changes opened up opportunities for rapid success, but also for fraud and abuse in pharmaceuticals (Imclone), telecommunications (Worldcom and Adelphia), as well as in energy (Enron). This is why values are so important in underpinning all new strategies.

There will undoubtedly be further abrupt changes that will affect the energy industry, through regulatory changes, geopolitical events, and technology advances. These will create new opportunities for energy companies to build new businesses, some of them of enormous scale, in selected segments.

This chapter has suggested the steps that should be taken to prepare for developing corporate and business unit strategies and has outlined some important strategic challenges and opportunities in different energy segments. In the next three chapters we move on to review examples of strategies that energy companies have actually deployed in recent years.

1 Michael Treacy and Fred Wiersma, 1995. *The Discipline of Market Leaders*. New York: Perseus Books.

2 Jim Collins, 2001. *Good to Great*. New York: HarperCollins Publishers, Inc.

3 Philip Evans and Thomas S. Wurster. 2000, *Blown to Bits: How the New Economics of Information Transforms Strategy*. Boston: Boston Consulting Group, Inc.

4 Alfred Marcus, 2006. *Big Winners and Big Losers*. Upper Saddle River, NJ: Pearson Education, Inc. publishing as Wharton School Publishing, p. 307.

5 Shona L. Brown and Kathleen M. Eisenhardt, 1993. *Competing on the Edge*. Boston, MA: Harvard Business School Press.

6 Peter Drucker, 1973. *Management: Risks, Responsibilities, Practices*. New York: HarperCollins Publishers.

7 Peter M. Senge, Art Kleiner, Charlotte Roberts, Richard B. Ross, and Bryan J. Smith. 1994. *The Fifth Discipline Fieldbook*. New York: Doubleday, p. 302.

8 James C. Collins and Jerry I. Porras, 1994. *Built to Last*. HarperCollins Publishers.

9 Pierre Wack, 1985. "Scenarios: Uncharted Waters Ahead by Pierre," Harvard Business Review, September-October 1985.

10 Wack's successor, Peter Schwartz took his experiences from Shell, founded a scenario development consultancy, Global Business Network and published "The Art of the Long View" describing his learnings on scenario. development. Peter Schwartz, 1991. *The Art of the Long View*. New York: Doubleday.

11 Michael Goold, Andrew Campbell and Marcus Alexander, 1994. *Corporate Level Strategy*. New York: John Wiley & Sons, Inc.

12 Chevron's Web site notes its 2003 acquisition of a 50% working interest in 5 million acres in the unexplored Orphan Basin, which is located about 250 miles east-northeast of St. John's, Newfoundland, and 155 miles north of the Hibernia oil field in water depths from about 6,500 to 8,200 feet.

13 Reproduced with permission from Dominion Exploration and Production, Inc.

14 The U.S. deals collapsed with deregulation of the gas business by the Natural Gas Policy Act of 1978, which ended the pipeline practice of "bundling" high cost gas with low cost gas to provide a weighted average cost to customers.

15 World Bank. Energy Working Note No 4. Jamal Saghir. 2005. *Energy and Poverty: Myths, Links and Policy Issue.s*

16 The Restructuring of the Electric Power Industry, 2000, EIA.

17 John P. Kotter, 1996. *Leading Change*. Boston, MA: Harvard Business School Press, p. 79.

7 Setting the Direction

How do energy leaders discharge their huge responsibilities in leading their corporations? We interviewed more than 20 CEOs and senior executives to find out how they had addressed the challenges of the recent past in order to uncover lessons that can be used for new leaders as they address the different challenges of the future.

The Architecture of Energy Leadership

Our overall architecture of energy leadership is diagrammed in figure 7.1. In chapter 1 we provided a brief history of the commercial energy industry and concluded that the world economy appears to be on the brink of a "phase change" after which oil will no longer be seen as the strategic global energy source. In chapters 2 through 4 we completed a strategic assessment that highlighted the rise of national oil companies and associated loss of competitive position by international oil companies, and emphasized the continuing importance of geopolitics and societal expectations for the industry. We concluded in chapter 5 that prior shareholder value propositions based on high returns with limited growth are now obsolete, and suggested in chapter 6 that new strategies will be required.

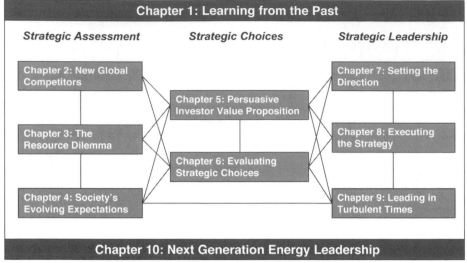

Figure 7.1: The architecture of energy leadership

We now begin three chapters that summarize the substance of our interviews with energy company leaders. These leaders have started the transitions that will be required by the phase change in energy markets, and provide important examples of forms of strategic leadership that have worked. In chapter 10, we will synthesize the lessons from prior chapters and draw up a charter for the next generation of energy leaders.

The topic of this chapter, Setting the Direction, links closely with the topics of chapter 8, Executing the Strategy, and chapter 9, Leading in Turbulent Times. These three chapters should be seen as iterative and interdependent rather than sequential. It is necessary to consider how strategies will be executed before setting a new direction in concrete. The primary leadership tasks are to set direction and assure that strategies are executed. The type of leadership required will depend on the direction that is set and the demands of implementation.

The examples we provide to illustrate how leaders set direction are obviously influenced by the circumstances facing these leaders over the past several years. The authors believe that, although the circumstances will be different over the next decade, their comments capture well the types of intervention that are available to energy leaders to reshape their firms. They will need to intervene in similar ways to meet the new challenges of the next decade. In chapter 10, we review how leadership models have changed over time. We then suggest how 21st century energy leaders can learn from these changes and from the examples provided in our interviews to respond to the new challenges of the transition to a new global energy portfolio.

From our interviews, we determined that leaders set direction for their firm through by taking on five important tasks:

1. Positioning downwind of major trends
2. Defining the corporate boundaries
3. Selecting areas of focus
4. Refining the business model
5. Building out the strategy (fig. 7.2).

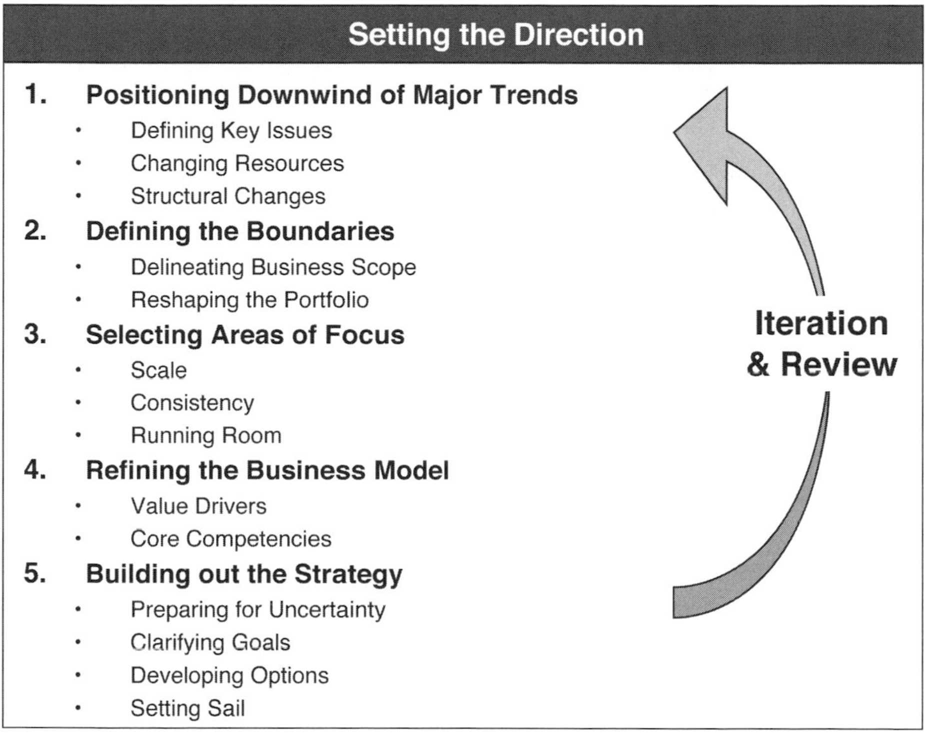

Setting the Direction

1. **Positioning Downwind of Major Trends**
   - Defining Key Issues
   - Changing Resources
   - Structural Changes
2. **Defining the Boundaries**
   - Delineating Business Scope
   - Reshaping the Portfolio
3. **Selecting Areas of Focus**
   - Scale
   - Consistency
   - Running Room
4. **Refining the Business Model**
   - Value Drivers
   - Core Competencies
5. **Building out the Strategy**
   - Preparing for Uncertainty
   - Clarifying Goals
   - Developing Options
   - Setting Sail

Iteration & Review

Figure 7.2: Setting the direction involves five tasks

In positioning themselves downwind of major trends, companies did some hard thinking to assure themselves that the trends were real and durable; but once they had confirmed what was working and what was not working, they were aggressive in pursuing their new course full-speed ahead. Some companies had learned well from adversity to avoid the frustrations of finding themselves upwind of trends.

Defining the boundaries is vital to ensure that companies are exposed to a full set of opportunities, yet do not get distracted pursuing backwaters. Leaders deliberately define the scope of the business, what is within the scope and what is outside; and continuously assess their portfolio, divesting pieces that are worth more to others and doubling down on those that represent a competitive advantage.

There is an inevitable tension between building a diversified portfolio of businesses and assets and focusing tightly on a single area. Diversification is expected to add to earnings stability, but the resulting portfolio may lack a clear parenting advantage and be more difficult to grow. Some of the most successful growth companies are clearly focused either regionally or on a particular type of resource. We describe how they came to choose their area of focus and the key factors that have contributed to their success.

Successful leaders refine their business model to compete effectively. Especially now, it is vital to prepare for uncertainty. What worked yesterday may not work in changed conditions tomorrow, and past business models must be scrutinized critically. Because energy is a commodity business, the over-arching consideration is achieving a competitive cost position, starting with a deep understanding of the supply options for each target market. However, it is important not to be frozen by nostalgia for an earlier, lower-cost environment. Costs only need to be lower than those of competitors supplying incremental demand; they may be higher than historical costs. Companies can then create information advantage and leverage infrastructure to sustain low costs and keep the advantage going by building on success. Our interviewees strongly believe that distinctive core competencies underpin successful business models and have built on them to create new business opportunities.

With a business model in mind, successful leaders build out their strategies using inorganic and organic means. Energy companies have created substantial value in recent years through inorganic growth, but asset values have escalated faster than corporate values in many cases, as we pointed out in chapters 5 and 6. The authors believe that we are at the beginning of a new growth cycle, and this stage of the cycle favors organic growth. By contrast, the late 1990s were the aging end of the previous cycle and favored industry consolidation. Organic growth is difficult, but successful companies are focusing on more difficult resources to find growth opportunities. Mergers and acquisitions still have a role to play, but they need to have a clear strategic or portfolio rationale. Several companies had found it necessary to redirect their companies in bold and interesting ways; they were adaptable, and were able to trim their sails if the winds veered unexpectedly. Some of the most successful companies demonstrated the openness and honesty to review their direction periodically in light of changing trends and competitive positions, and radically redirect their companies to better leverage their core competencies.

Thus, the system we describe is iterative in the best companies. We believe this system is probably applicable to companies outside as well as inside the energy industry. What is important, however, is that this is the system that best synthesizes the experiences of more than 20 energy company leaders. We therefore believe that it is uniquely tuned to the needs of the energy industry.

Positioning Downwind of Major Trends

We have identified three major trends: 1) the entry of new competitors into the oil E&P segment, 2) a focus on non-conventional resources and new value propositions for resource rich countries, and 3) changing geo-political orientations and societal expectations. These trends have transformed large segments of the industry into growth businesses, and together with altered competitive positions demand new strategies to address more difficult resources. Further, changes in regulations and the mega-mergers of the late 1990s restructured the industry, creating substantial new opportunities. Finally, increasing societal expectations invite new strategic responses.

Nick Butler, group vice president of strategy for BP, points out that the first imperative is to identify the challenges thrown up by the changing business environment and by BP's changing size and scope of activities. His summary is consistent with the trends we have identified:

1. "The first challenge is responding to a global market place that is changing rapidly, so that the areas of growth are not the areas we are familiar with. That's both upstream and downstream. The largest sources of new production for us are going to be Russia and West Africa. The largest new markets are going to be China and, if we find the right way in, India. None of those places are places we are familiar with either in our recent past or in our history, and they're not easy places to work in.

2. "The second challenge is we're going to have to get used to working in an environment where things aren't owned on a private company basis. The Anglo-American model of a private company owning equity barrels and standing separate, working on its own, is just not going to work in many of the countries I mentioned. We're going to have to work with state entities or quasi-state entities. They will be companies, some of them very good, some less good, whose welfare function in economic terms is different from ours. They exist for a different purpose. We will have to work together and find a way of balancing that. We can't expect things to be privatized. The energy industry is not going to be privatized in places like China.

3. "The third challenge, which is very different from anything we have faced before, is managing scale. We're now the sixth largest company in the world, which is quite different from when I joined in the 1970s. We work in 110 countries. We have 100,000 Russian staff, most of whom don't speak English. The UK is a great base, but accounts for only 12% of our business and is not growing as a percentage for obvious reasons. The growth is elsewhere. We have to be multi-lingual. The web site now runs in four languages. We have to understand you can't apply a command and control model to a 24-hour business, where many of the pieces have to respond to immediate circumstances in their own location and in the context of the local culture.

4. "The fourth challenge is slightly different in scale and is the environmental issue, where we are seen, probably with some justification, as one of the industries that causes some of the problems. We have to be in a position where we provide solutions, not just be a problem, and we have to do that in a context where there is no political agreement on what those solutions are. Climate change is just the most obvious example."

Challenges are the flip side of opportunities, and BP is developing strategies to invest in different places, in partnership with companies of different types, with a greater scale than ever before, while attending to society's demands at the same time.

More difficult resources

With higher oil and natural gas prices, resources that were uneconomic at low prices are now potentially commercial. EnCana saw the potential of resource plays as conventional resources become increasingly scarce and more expensive. Several of the leaders we interviewed had been early in identifying the growing importance of natural gas. Chesapeake learned from an earlier failure some key lessons on the changing U.S natural gas situation, and BG found success based on a holistic view on international gas, while Suntera hopes to position itself downwind of forthcoming changes in Russian gas. These are early examples of addressing more difficult resources in advance of the phase change that we anticipate in the global energy complex. Future leaders must internalize and further extend these initiatives.

Understanding the changing U.S. natural gas situation. Chesapeake was early in understanding the new North American natural gas trends. Faced with a collapsing stock price as the Austin Chalk play came to an end (see chapter 5), Aubrey McClendon and the Chesapeake core team learned from adversity and regrouped around two truths: their belief that natural gas supplies would be hard pressed to keep up with demand, so the market was bound to strengthen, and a belief that they had developed some strong competencies that could be brought to bear on the problem.

They made their call on the natural gas market just after the National Petroleum Council had issued its report concluding that North American supplies were endlessly elastic at low prices, so it flew in the face of the prevailing conventional wisdom. Also, their confidence in their capabilities was not at that time shared by the stock market: "Failure is probably a harsh word to describe us in '99. But you know when the stock price has gone from $35 to 63 cents it's certainly not a success! But it forced us to confront some basic truths that were emerging about our business and our industry, which were (as seen in 1999): gas is increasingly hard to find, depletion rates are accelerating because of technology and because of smaller targets, and consumers have latched on and have been sold this idea of a 30 TCF market by 2010 with gas prices not above $3.50/mcf and have invested tens of billions in equipment that might burn 30 TCF of gas by 2010. But we're are not going to be able to produce it!" So Chesapeake built its strategy on aggressive acquisition and development of U.S. natural gas and hasn't looked back since.

Seeing the potential of resource plays. Gwyn Morgan of EnCana made a radical shift in his portfolio, though he called it just "returning to our roots." After completing the integration of AEC and PanCanadian into the new EnCana, he and his management team took stock of what they had. They found that they had assembled "The strongest resource position for natural gas in North America, with the largest amount of undeveloped, unbooked resources. And a tremendous core competency in the best resource plays in North America, Canada, and the U.S." They had also been successful in building strong positions in the North Sea, Ecuador and the Gulf of Mexico.

But the trade winds were shifting. Gwyn took his successor Randy Eresman to a private gathering with some of the top oil industry CEOs and found that "around the table it was very clear that the number one issue was access to resources, and more and more of the world's oil and gas companies, both the independents and the majors, were getting rather desperate for access to those resources. So the value of our international assets was going up beyond our dreams." When they put the international assets in the context of the market conditions, they recognized that they would be worth more to a strategic buyer than remaining with EnCana, because EnCana had a massive inventory of opportunities in North America. "And when you do that analysis, you conclude that you are better off not to put more money into it; you are better off to sell it. Because your incremental rate of return there is dramatically lower than your incremental rate of return of your North America asset

Profitability

base. So it's a question of allocating capital to your best returns." They successfully sold out of Ecuador, the North Sea and the Gulf of Mexico and "can now focus on being the best that there is at two things only, North American gas, especially North American unconventional gas, and North American in-situ oil sands."

Clarence Cazalot, CEO of Marathon sees considerable organic growth potential in stranded gas, though he believes the stock market is slow in recognizing its value:

> A lot of our technology efforts are around gas utilization; we're looking at GTL as an alternative to LNG. There's a lot of discovered undeveloped gas around the world and that is one of our top priorities. If you look at what we're doing in Equatorial Guinea, the first LNG train is about 60% complete and we'll deliver the first cargoes in the third quarter of 2007. So that project is going very nicely, but it's only utilizing gas reserves from the Alba field. But around us is a lot of other gas, much of it currently being flared and some of it undeveloped. So we believe there is a great opportunity for a second train expansion that makes tremendous sense. So much so that the Nigerians have looked at their Eastern Delta gas, and said that from a government to government point of view, we want our Eastern Delta gas to go to Bioko Island, Equatorial Guinea, for their second train. So there's a double benefit of eliminating flaring and capitalizing on an infrastructure that we've already built with the potential of becoming a significant LNG hub in West Africa.
>
> So I think a lot of opportunities will be gas related on the upstream side. I think the integrated nature of those projects, where you can develop the gas, liquefy it, ship it, bring it to market, be in the full value chain, is something we're going to be engaged in as we advance our integrated gas strategy. Equatorial Guinea will be the exception because we've already entered into a long term LNG offtake agreement with BG. But going forward, we're in a better position to go after these opportunities by virtue of our downstream capabilities. The person who led our work on the condensate and LPG phase was Ken Woodward and he came out of our downstream operation. We've got a number of our downstream people there now, and their skills are important. This is a clear example of how our integrated structure can provide a competitive advantage.

Matching resources to markets. Frank Chapman, CEO of BG Group, believes that his company has been able to spot major trends early because they are focused on markets as much as on resources and see the opportunities from the point of view of the full natural gas value chain:

I think we coined the phrase first—"gas value chain player." And we started to put this notion together of working value along the chain. We start by saying, "We want to supply customers in gas markets." Because it stems from our heritage. That's a good thing to want to build on.

And it's distinctive because we are a large integrated player, notwithstanding the fact that, because of the way these chains work, we end up with two-thirds of our business upstream. We think about the business in terms of markets. Now we're not talking, when we say "customers," only about retail customers. Because in many cases we are talking about large industrial customers, power generators, aggregators; so in many places, like in the UK, we play the role of wholesaler in some form or another. But this notion of supplying markets and wanting to add value, from that flows everything that we are today.

Because we then say, "Well, what is it about the market that attracts us?" It's usually one of two kinds of market. It can be a market that is growing very rapidly. Brazil and India are, for us, the best examples of this kind of market. They are characterized by very rapid underlying growth in the economy itself, and a large population growth in some cases, as well. And there is a correlation between that and energy demand, clearly. They have to be accessible markets, where you can own infrastructure, and you can remit dividends. You can earn a return, and there are the flexibilities to repatriate those earnings.

I know that in Brazil, owning the concession, I can develop all the infrastructure in the way I want to. I can anchor it with large industrial customers and power customers. I can then start building out for my light industrials and commercials and residentials. And I have a program of 15 or 20 years of developing the network, aggregating demand and supply in that market. That's a tremendous grow-grow-grow opportunity to contrast with an E&P business, which from the first day it is started up gets smaller.

The other markets we are interested in are not necessarily growing, but are undergoing some fundamental change. Now for us this is really Italy, which is small for us at the moment. And the

UK, very big—a hundred billion cubic meters per annum of gas needs to be imported by 2015, which is almost the whole size of the market today. So that's the sort of change that is very interesting for BG, because that's a place where we can play a role in changing the supply dynamic.

Then we saw the U.S. as the world's largest and most liquid gas market, a perfect opportunity for BG, being champions of liberalization and really able more than a lot of other players to function in a fully liberalized market, reveling in it, actually, having been the architect of it over here. What was quite interesting about the U.S. market, of course, was that, in fact, they had so much gas at one stage that they were actually exporters of LNG. In fact, BG exported the first experimental cargo of LNG from Lake Charles to Canvey Island in 1959.

So we started to think about LNG and about the U.S. market in particular. One of the first things we did was unprecedented. We went to Trinidad and said, "We've got no gas reserves,"—although we had acquired prospective acreage as part of the Tenneco acquisition years ago—"but we want to participate in the LNG plant." So we participated in the discussions with the government and negotiated for ourselves a position in that plant, and actually went on to lead the change in the design of that plant from an air products process to an optimized cascade process. And we built that plant at that time at a completely unbelievably low cost.

Now the interesting thing about that, of course, was that this plant bought gas and sold LNG. It was a midstream profit center and immediately put us on the LNG technology curve once again, and made us an LNG manufacturer for the first time. And this started a whole train of things in terms of prospecting with the sole purpose of finding gas that could subsequently be transatlantic LNG, and that story has since grown.

Taking a position in Russian gas. Moving further into the international arena, Steve Lowden of SER has picked a specific trend:

Another trend I see right now is Russia has gas prices between 40 and 80 cents a million BTU. And it's got 65% of the world's gas resources. And it is going to export a huge amount of it if the global energy equation is even remotely right.

Russian gas is going to go overseas, so it's a good business to be in. So we'll invest in natural gas. And we'd like to invest alongside Gazprom for sure. And I think the Russian vehicle we'll invest in will be a good partner alongside Gazprom. But, you know, when you work alongside Gazprom, best to let Gazprom be in charge!

There are many things we can bring to Gazprom. I think Gazprom will be receptive to Russian partners. And the vehicle we will have will be a Russian partner with international connections. Gazprom is not only the second or third largest energy company in the world by lots of metrics, but it's also very bureaucratic, and execution is a huge problem. And so execution is what we can help them with.

Capitalizing on structural changes

Structural changes can often create opportunities for huge value creation and destruction. Deregulation of energy markets in North America and Europe resulted in a huge shift of competitiveness. In Europe, there is a gradual move towards a few continental gas and power players, in a highly political environment in which countries support their national champions. In the U.S., a new competitor group of "energy merchants" emerged, grew rapidly and then failed spectacularly, leaving the gas and power marketing field open to large oil companies. Other important structural shifts have been seen in the consolidation of the refining sector, and further shifts will be seen if greenhouse gas controls become constricting.

Coming change in the upstream industry structure. Steve Shapiro of BR sees that difficulty in finding economic organic growth projects will result in structural change for the E&P segment, and reinforces Clarence Cazalot's assessment that integrated companies will have an advantage over "pure" exploration and production companies:

Today what I think the majors are doing, risk-wise, is substituting geopolitical risk for geologic risk. And then what you are seeing is large E&Ps—and we were a big part of that trend— actually reducing risk because what we found is, exploration works but isn't scalable. And then you have small guys, where exploration is scalable because you start so small that you continue to have innovation at the micro- or mini-cap kind of level. And to me what that drives you to is an industry structure that has two or three very large upstream E&Ps and a bunch of small guys, with nobody in between, and an integrated model with the different levers to pull with the international geopolitical skill set to work those projects. But even that is becoming more difficult because competition from host countries is increasing. Returns are coming down, so they can't replace their assets with the same returns they've had, which drives them really back to OECD. ConocoPhillips hasn't said this, but I have to believe that is part of the equation—getting back into Libya doesn't do enough, return-wise.

Sensing opportunity in the U.S. refining industry. CEO Bill Greehey positioned Valero to take advantage of a shifting trade wind in refining. As we have explained in chapter 5, large companies became disenchanted with refining in the 1990s, so were prepared to sell off refining assets at a fraction of replacement cost. This trend was amplified by the mega-mergers, after which they were obliged to divest certain refineries for anti-trust reasons. Greehey talked about the strategy originating with his top refinery employee at the time, the late John Hohnholt, and himself: "If two people were really involved in the strategy, it was John and I in selling the gas pipeline, and then really concentrating on refining. And thinking that we would be successful in buying assets at lower than replacement cost and then envisioning the environmental movement [would lead to a premium] in processing heavier sour (crude oils)." We will describe in chapter 8 Valero's "secret sauce" in executing this strategy. The result has been that the structure of the refining industry in North America has been radically changed. Whereas the integrated oil companies used to be the largest refiners, with a group of small, undercapitalized independent refiners trying to survive in niches, Valero is now the largest U.S. refiner, overshadowing its independent refiner rivals Sun and Tesoro, and followed by the super-majors. Most of the previous integrated oil companies have been absorbed (Gulf, Texaco, Getty, Arco, Amoco, Unocal, Kerr McGee), dismantled (Tenneco, Sun, Tesoro, Diamond Shamrock), or merged (ConocoPhillips, ExxonMobil).

Preparing for new environmental standards. It is also important to think ahead on important changes in societal expectations, such as the potential for controls on greenhouse gases. Rick George of Suncor is very conscious of this issue:

We're in a warming trend now. What's causing it, I really don't know. But I really don't like engaging in that too much. Here's what our obligation is. (This goes back, again, to how you view the company.) We have an obligation to search out even new technologies to reduce our environmental impacts on air, water, and land at every step we can. In fact, we don't invest in any technologies or businesses that don't do that. That's our obligation.

And so it's why we invest in ethanol [Suncor has completed the largest ethanol plant in Ontario]. Why we invest in wind power. And, by the way, I've explained that to shareholders and they buy it. It's the right thing for that hundred-year model. You get to the moral obligation. Listen. If we do the right thing consistently, it'll pay off.

George continued:

We're working hard here in Alberta on a CO_2 sequestration and re-injection pipeline. We'll need some support from the government, but both levels of government have been very supportive. I actually think that this will happen. Now what I can't tell you exactly is the timing. All I can tell you is that Suncor is taking a leading position trying to push this through. And in my mind, it is not necessarily connected to Kyoto.

Opportunities in deregulated power. There is no doubt in our minds that deregulation will bring greater customer choice and lower costs in the long term, while regulated systems stifle innovation and encourage higher costs. Similarly, environmental objectives are best achieved through "cap and trade" systems. Progress along those lines has been interrupted by trading abuse, rising commodity prices and resistance from incumbents. However, progress will resume and successful leaders will plan accordingly.

John Wilder, CEO of TXU is acting decisively to lower power generation costs so his company can be a winner in a deregulated industry:

When do you desperately need a market solution to allocate resources? It's when resources are scarce. That's when we absolutely need to make sure we are putting in the best generation technology whether nuclear, clean coal, gasification, or ultra super critical boilers. But the regulated companies are saying, "Don't worry about any of that. We're going to give you relatively inexpensive electricity from here until kingdom come. And we'll build some expensive new stuff, and we'll just roll it in and you'll still have relatively cheap electricity."

But regulation brings a tremendous amount of soft cost because of the governmental processes put in place to justify an allocation of capital, are buried into the tremendous amount of non-customer value added. So, take a long time, lots of hearings, and lots of debates. No one wants to take any risks, so checkers, on top of checkers, on top of checkers, on top of checkers. And you get through one of those processes, and you analyze how much productive capital went into the asset that is going to generate a product for the customer? Our Comanche Peak nuclear plant would be one of the more dramatic examples. It cost $16 billion, but has only about $200, $250 million dollars worth of physical cost. What's the difference between $16 billion and $250 million? Lawyers, engineering study on top of engineering study, NRC, triple, quadruple inspections.

We announced the largest coal power plant construction program in the United States. 10 billion dollars, in fact. It rivals the China coal build out program. We've had Bechtel engineers, Fluor engineers, TXU engineers, TXU traders, a big integrated team working this for about nine months. We've broken down the construction process, and we've taken the regulated coal plant model, which started at $1,500 KW cost, and right now we are at $1,000 per KW. And we are going lower.

Customers and TXU shareholders will both be the beneficiaries of Wilder's aggressive moves to provide lower cost power.

In each of these cases, thoughtful analysis of changing structural conditions led to a change in direction for the company that was still consistent with the original value proposition to investors and, with good execution, has created enormous value.

Defining the Boundaries

Good strategy often lies in deciding not to do something in order to concentrate resources on the most prospective business areas where a company can build distinctive core competencies and competitive advantage. The core competencies can then form the basis for growth through incremental extensions of the business into related industry segments. Although companies of all types and sizes need to define boundaries, the task is more complex in larger corporations. Smaller companies often focus on a congruent set of opportunities. Still, they have decisions to make about the regional span and types of their activities.

The largest companies have broader degrees of freedom. In the 1980s, oil majors plunged into unrelated diversification. Several companies acquired coal and metals mining operations, Exxon bought Reliant Electric and tried to enter word processing, Gulf bought the retailer Montgomery Ward, BP entered animal foods through Ralston Purina, and so on. In retrospect, these unrelated moves destroyed shareholder value, weakened focus on the core business, and had to be reversed by new management in the low-price 1990s. The super-majors have no intention of making the same mistakes again.

Business scope

In a period of important transition in the energy industry, we should expect to see shifting boundaries between different resources and products and different business models. There have already been significant changes, such as the emergence of regional European power and gas giants, and the shift of North American natural gas supply and trading to oil companies from the pipelines. However, we found that the oil industry is quite cautious about expanding petrochemical operations or moving into power generation. This should open opportunities for others to reshape these parts of the industry. But we did find growing oil company interest in biofuels and greenhouse gas sequestration. It is our belief that the business environment is changing faster than many companies have recognized, and that the future will be won by leaders more open to business scope extensions than the current generation has been.

Caution on power. Nick Butler, group vice president of strategy for BP, is cautious on extending the scope of BP's business mix:

> For the next years we don't see a shift in our product scope. However, we are announcing this week that we are going to establish a green energy business, which will combine all our work on renewables and put a lot more money into it. We see that as a set of fuels that will grow over the next decade but won't make a material difference for 10–15 years. That's coming. But for the next decade, it's oil and gas: finding it (we mostly have it found), developing it in some of these complex areas, and managing the environmental issues around development and consumption.
>
> Power will be part of the next strand. It will grow in the next 10 years, we'll invest in it in the next 10 years, but you won't really see the impact on the global energy market until later. The IEA predicts that all of renewables will only be 3% of global energy demand by 2015. It then grows after that, and we want to position ourselves for that growth. But it's still quite minor. The challenge is sequencing.

CEO Dave O'Reilly is also cautious on extending Chevron's business into power:

> We're a long way along in building our global gas business as an addition to our oil business. Power is a bit more difficult, because I think it primarily depends on the opportunity and the attitude of the government. Because of course, power is very strategic. It is important to know what the philosophy of the government is: is it deregulated? In most places, power is a highly regulated business, and as long as it is highly regulated we don't belong in it. There are specific opportunities where we can make an investment in power that makes good sense, and then we will do it. But it's not a priority for our company until we see more of an opportunity to invest independently. We are in some power businesses, and our experience has been mixed.

Don Paul elaborates on Chevron's position: "I think one of the key challenges to the whole power generation world is that the merchant power market is a cost-of capital-business. It doesn't have much variance, which is the good thing; but still, you are not going to generate any serious returns in that business." Though this has historically been true, innovations in pricing and metering could begin a new trend that will allow the low-cost producer to earn higher than cost-of-capital returns.

Retreat from chemicals. BP, Shell, Chevron, and Phillips have all been conservative in their chemicals businesses. In 2004 BP announced:

> Later in 2005, BP will begin to exit the Innovene (BP's former olefins chemicals) business through an initial public offering (IPO), subject to market conditions and approvals. In order to realize its full potential as an industry leader, BP believes Innovene needs the freedom to operate differently from the rest of BP Group. As a stand alone company, BP believes Innovene can respond better to the increasing competitive demands of different markets.

In fact, BP sold Innovene to Ineos in December 2005 and redefined its business scope to exclude olefins.

Preparation for sequestration. Helge Lund does not see power as a part of Statoil's core business but is interested in the business potential of sequestration:

> My philosophy related to this, is that Statoil needs to concentrate and develop along or within areas where we can be among the best. Moving beyond the core business and competence may weaken a company. Over time the definition of "core" may change, but it has

to be closely related to the fundamental capabilities of the company. The things you do you have to do good, otherwise I don't see the point. Doing something halfway with your left hand will increase risk and erode value.

But of course we have experience in CO_2 sequestration, storage. There might eventually be a business there. There might be some learnings, some competence that we can bring, for instance, into the coal part of it, which is facing many of the same issues.

BP is also interested in sequestration and has a power project in Peterhead in the UK that will inject CO_2 into the declining Miller North Sea oil field; and, as we heard earlier, Suncor is also preparing to participate in sequestration.

Experimentation in biofuels. Shell has recently reaffirmed its commitment to biofuels, as Jeroen van der Veer told us:

Shell has an established position as the world's largest marketer of biofuels, as well as a leading developer of advanced biofuels technologies. Biofuels are fuels derived from biomass such as plant crops like oil seed, or plant wastes like straw. They can be used either pure or blended with standard automotive fuels dispensed at today's filling stations with the potential for much lower CO_2 emissions.

In partnership with Iogen of Canada, cellulose ethanol biofuels are being successfully produced from plant waste. By producing biofuels from plant waste instead of food crops, the potential stress on the food chain is alleviated. The Iogen process produces a fuel which can be used in today's cars, cutting CO_2 lifecycle emissions by 90% compared with conventional fuels. Shell recently announced a Memorandum of Understanding with Volkswagen and Iogen to explore the economic feasibility of producing cellulose ethanol in Germany. Shell Canada has been working with Iogen to develop a viable commercial framework for a facility in Canada. These projects complement Shell's existing partnership with CHOREN Industries of Germany. CHOREN have a patented biomass-gasification process that converts biomass—such as woodchips—into ultra-clean synthetic gas that can then be converted for use in diesel through Shell's Gas- to-Liquids technology. CHOREN is preparing construction for the world's first commercial biomass-to-liquids facility in Freiberg, Germany.

Both Shell and BP have extended the scope of their business to cover renewable fuels.

We noted earlier that Rick George of Suncor has made initial steps into ethanol manufacture. Chevron has also invested in biofuels, so there is interest and some movement by energy companies into biofuels development.

Rebalancing the portfolio

Once the scope of the business has been defined, there remains the question of what portfolio of businesses within that scope will best deliver the shareholder value proposition in terms of growth, returns, and sustainability. When Clarence Cazalot was appointed CEO of Marathon Oil Company he faced an immediate portfolio challenge:

> The key to an upstream business is growing your reserve base. As I explained to our organization then, it doesn't start with reserves, it starts with resource. And there are only two ways to add new resources to your company: successful exploration or acquisition of resources. This could include buying companies, it may be entering new projects, or it may be securing stranded gas, but you have to access that resource first, then commercialize it. At Marathon, we were milking the set of legacy assets we already had—largely U.S. and UK—but the great hope was Sakhalin. Sakhalin was where we were putting a lot of effort, technical resource, a lot of money was going into it and at that time it was an oil producing operation only about six months out of the year.
>
> Marathon was justifiably proud that we were the operator of that operation, we'd put special structures out there to withstand the ice during the winter freeze-up, and we were producing at that point and starting to move ahead with an LNG project, which was going to be the big piece of the project. At that time it was estimated it would be an $8–9 billion project, of which we had a 37.5% interest—it was a big expenditure for a company our size with a lot of single project risk, geopolitical risk, and compared to the rest of the company, we were betting the farm on that one project. Sakhalin was viewed as the major project for the company and was on the cover of every company publication, and people were very proud, justifiably so, of what they had done. But it was just as clear as the nose on my face that this project was just too big and too risky for us to stay in.

I paid a visit to Shell and met with Phil Watts in the spring of 2000 and by the end of the year we had the deal done. What we were determined to do was not simply sell our interest. Our need was for new production, new resources and so the trade we did was for all of their interest in the Foinhaven and East Foinhaven assets, in the Atlantic Margin, and an over-riding royalty interest in the Ursa field in the Gulf of Mexico. Suffice it to say that we go back every year with our board and we do post-expenditure reviews on all our major investments or divestments in this case. This one looks like a gem, because of both the increased costs of the Sakhalin project and the excellent performance of the assets we acquired.

What we worried about was: we were operating the physical project; Shell was responsible for securing the markets for the LNG. At the time, this was our one big LNG project. It was number one on our list, but we knew that Shell had all these other projects, and we didn't know where it stood in their queue. But Sakhalin was our one big LNG project: what we thought about was if we got hung out there, with our share, say $3–3.5 billion invested, and we were to incur a one or two year delay it would be disastrous for us financially. For Shell, they would take a little bit of a hit but nothing significant. So for me it was an easy decision. We just couldn't afford the project.

The portfolio lessons: overweighting on one project is too risky and results in loss of flexibility; divergence with a partner can be a problem, unless you're operator for the entirety of the project including construction and marketing. If you're driving the project, then you can assure yourself of getting it done, but in this case the responsibilities were split and that was a clear problem. Then there are the temporal aspects. You need to cover short, medium, and long term needs for reserves and production.

Clarence also favors a mix of assets:

The point I try to make about our portfolio is when you look at a pure E&P company, they are all on a treadmill. And the treadmill never stops; in fact it runs faster now for all of us because we've become so good at developing assets and producing them, with new completion technologies, horizontal wells that expose us to more of the reservoir and give us higher production rates. All we're doing is exploiting that reservoir and depleting it faster. From an economic point of view, that's what you want. But again, if you're not replacing your resource and your reserves every year, you're liquidating the company. That's becoming an increasingly difficult challenge. But if I do a refinery expansion and build a coker, I'll have the higher capacity five, 10, and 15 years from now; I'll have to put maintenance capital into it but I'm not having to rebuild it every year. And it's the same for our LNG project in Equatorial Guinea. LNG projects stay flat for 30 to 40 years and have a steady stable cash flow. So from the standpoint of our CFO, there's a pretty predictable cash flow stream that she can plan on every year. The E&P business sits on top of that so I think it's important to bring those components together.

But some investors in the market today take the view: I don't need to invest in an integrated company like Exxon, ConocoPhillips, or Marathon. I can go buy Valero and buy Burlington or XTO and I'll build my own. In terms of putting the two pieces together, that's a reasonable strategy, but my contention would be that in the world that we're in today and that we'll continue to be in, access to resource will get tougher and tougher. The world is rapidly becoming bifurcated between the resource holders and the market holders, and if you've got market, the resource holders need you. So I think there are integrated deals to be done between the resource holders and the market holders so that we can capture resource opportunities by using our markets to do so. These opportunities are not accessible to the pure E&P companies.

We're about 50% downstream. We're the most heavily downstream leveraged of the integrateds with ConocoPhillips just behind. One of the worst things I ever did was about three years ago; I said to the market that our intent was to be 60% upstream, 30% downstream, and 10% integrated gas. That was really based on a view that our downstream business had limited growth opportunities, because of our geographic focus in the Midwest. But I think that the quality of downstream opportunities today is as good as or better than the upstream opportunities, with far greater control and far less risk. Take the proposed expansion project we've announced at our Garyville, Louisiana, refinery for $2.2 billion. It's a full upgrading expansion to include a coker so we'll have the ability to handle any number of crudes from medium sour to heavy sour. We'll make this investment decision based on mid-cycle margins, not current high margins and it's a good return that we can control. We've got the land; we can build it at a cost we can predict; and it'll be a great performing asset for a long time with no decline curve.

But on the upstream side it's really getting tough. Exploration is getting more difficult; the prices people are prepared to pay for assets today are through the roof for things that just don't make sense. The competition for assets is just so strong today, coupled with the rising costs we're seeing throughout the business, and the issue is not growth but profitable growth in the upstream and this is a big, big challenge. The market has not recognized that it's profitable growth that matters.

We'll all see in March/April when the companies' 10Ks come out and the analysts will all run their reserve replacement rates and finding and development costs. But even for a company that's replacing 130% of its production, if the margins on the barrels that you're bringing in are half the margins of the barrels that you've just produced, the value of that 130% may not replace the value of the barrels that you just produced on a similar price basis. So it's really about value, creating the value streams in the future. I find today that investors want to judge integrated companies just on the basis of their upstream. Even Exxon takes a stock price hit if they miss a production target.

Somehow we're going to have to get the investment community to understand that it isn't about whether an income stream comes from an upstream producing asset, or from a refinery, or a methanol plant or an LNG plant, it's really about the quality, repeatability and sustainability of the income stream. And yet, the market will value a dollar coming out of gas assets in Oklahoma more than they'll value a dollar coming out of our refinery in Garyville, Louisiana, and it doesn't make any sense because there's a high decline rate on the gas production. So there's some fundamental changes going on in the business that will cause people to look at which businesses make the most sense, and how they allocate their capital.

Creating a robust international growth portfolio. Helge Lund took over as CEO of Statoil at a time when the company had suffered acute embarrassment from a alleged bribery scandal involving Iran. He quickly established a new set of directions to rebuild momentum:

> I came in after the Horton affair. The previous chairman and CEO had left their offices and in a way Statoil was an organization in disbelief.
>
> Before I took office, I wanted some months as an outsider and therefore only started in August. I spent the period between March and August researching the company, talking with people inside and outside the company, in order to be able to move quickly once in office. Then we changed the corporate management team after two weeks, and the next layer of management teams after six weeks. The new team then designed a new strategy that we presented to the capital market in December. The purpose of this approach was really to have people forget about this one incident, and focus on moving forward. In this respect speed was essential.
>
> There were a few specific areas that we needed to attack more forcefully: firstly, people had apparently talked themselves into a position that the Norwegian Continental Shelf (NCS) was short of opportunities. The dominating perception was one of decline. This affected the organization and therefore our priority one was to rejuvenate the assets and the opportunities in Norway. We now believe we can produce the same level from our Norwegian assets in 2010 as we did in 2004. People in the organization as well as in the general public in Norway now have a different mindset. A part of this rejuvenation was also a strong offensive strategy in terms of opening up new acreage in the northern part of Norway and identifying partnering options in Russia.
>
> And secondly, there had been a discussion internally, and partly externally, around Statoil's international involvement. In some camps there was deep skepticism toward Statoil pursuing and developing business opportunities outside Norway. I firmly believe we have an obligation, not only to our shareholders but also to our employees, to capitalize on the vast experience and technology development that we have developed over the last 35 years to put these to use internationally as well. This is no longer an issue: Among Statoil employees, confidence in our international strategy has never been stronger.

Leadership + Organization [handwritten margin note]

Thirdly, I believe a distinctive profile of Statoil is our gas position. We have launched an ambition of doubling our total gas production from the current level of 25 BCM per year up to 50 BCM in 2015. This is globally. This will entail not only a much more aggressive approach on the NCS but also take Statoil in a more forceful way into global gas plays and LNG.

And finally, we need to really capitalize more on the opportunities stemming from Statoil being an integrated oil and gas company: we have broad mid- and downstream competence and a very strong position in oil trading as the third biggest net crude oil seller in the world. I sincerely believe that the competence we have as an integrated oil and gas company is part of our competitive advantage as we go forward.

These were the four business-related themes. In the last two years we have also made an enormous effort to upgrade the work on our corporate values, corporate culture, leadership development, succession planning, and general competence development.

Portfolio risk management. Duane Radtke of DEPI recalls the evolution of their portfolio within the context of a gas and power owner's corporate portfolio:

Dominion started with the idea, "We are going to have a gas portfolio." When Dominion started in the early 90s, it was tight gas in the Appalachia, coal-bed methane in the Black Warrior Basin, driven a lot financially by the tax credits that came with that. But if you think about those types of assets, and then moving into Michigan, there was almost no risk. The issues are how you manage, how you squeeze cost. But the way we might look at E&P assets, it certainly was almost zero risk.

When Dominion bought CNG, it did not do it because of the oil and gas assets. The oil and gas assets just happened to come with it. At that time Dominion already had a very significant position of low-risk long-life natural gas. But when they acquired CNG you had opposite ends of the risk portfolio—CNG was almost a pure exploration and development company—real exploration in deepwater and the Gulf of Mexico shelf. And so now you had both ends of the spectrum on the assets.

Dominion acquired CNG in February of 2000. I joined them in April of '01. Tom Capps [Dominion's recently retired CEO] is extraordinarily intuitive and smart. I showed him two graphs at the time. I said, "Here's the five year plan that we have. And here's the identified production. Here's the identified exploration that we have. And here is a model of what we expect to be able to do." And 70 or 80% of the future production was unidentified. I said, "No one can take that kind of a risk, let alone Dominion."

I said, "We have two choices. We can sell all of the Gulf of Mexico and just become an onshore player," because we did have some other onshore assets. "But we will be very fortunate without acquisitions to just maintain production. It's the nature of the onshore. But if you look at the historic returns, as to what the Gulf has done, the other option is making onshore bigger, so we can afford the risk of the offshore. And let us change the profile of what's identified and unidentified."

So when we bought Louis Dreyfus, that rebalanced the portfolio (fig. 7.3). Now we could afford the risk. We had a much larger onshore asset base. It doubled our position in South Texas; obviously Sonora was the key; and we paid $2.3 billion dollars for all of Louis Dreyfus. It is safe to say that probably Sonora right now is worth two to three times more than we paid for all of Louis Dreyfus. Prices had a lot to do with that, but we have also almost doubled production.

We had the capital—because of the parent—to do a lot more with the assets than Louis Dreyfus had. We always felt there was a lot more opportunity there. But what it really did for E&P within Dominion was give us enough size onshore to be able to rationalize the risk that went with the offshore. Now we had the opportunity for real growth through the Gulf of Mexico.

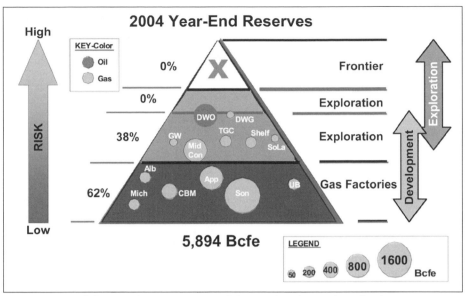

Figure 7.3: Portfolio overview. Source: *Dominion Exploration and Production, Inc.*

Taking larger risks. Duane Radtke always said that DEPI did not do frontier exploration; but with a stronger, larger portfolio, he has extended a little in that direction:

> We actually have a small slice of the ultra-deep test on the shelf, Treasure Island. We have 7% of it. One of the super-majors is the operator. And people have asked why.
>
> And our guys wanted to take it. I said, "Well, okay, but you need to explain the logic to me. What are we doing? Where is the program?" Well, these are huge structures, and there could be huge reserves. It starts at 28,000 on to 32,000 ft. There is a whole range of targets. It is a massive structure. You can see in the seismic. The question obviously is does it have anything in there? Is the reservoir preserved? Do you have any porosity. And, can you even get the well down? Because mechanically it is very challenging: there are all kinds of risks with temperature and pressure.
>
> So I said, "Okay. I understand what we are doing. But what are the options? There is another feature across a saddle—who has it? Well, the same super-major group." So we wound up with an option on the second structure. But at 10% interest. still promoted. But think of what you've done from the risk side. It is an option, not an obligation, to drill at a higher interest.

Then we had two big features locked up. But the other thing it did, it allowed us with our data and with our acreage, at least one year and maybe two years lead time on what we would consider our competitors in the area. I think we would be the preferred partner on some of the other things with the super-majors in there. They are not going to want to continue to share technical data with a wider and wider field. I think it puts us on a preferred list. We are big enough; we can afford to do that for an option on something that could create significant value.

This idea of fitting the risk exposure to the overall portfolio shape and scale is echoed in the comments of Helge Lund of Statoil: "I think size matters in several ways. For a company of our size, there are limited big bets that you can take in order to have an acceptable risk-reward balance. You must make sure not to overstretch when taking on several major development projects in different regions at the same time. That could put quality and precision in execution at risk. Striking the right balance between risk, reward, and capacity is a key leadership responsibility. In terms of capacity to take on multiple complex projects at the same time, there are arguments for being bigger than Statoil is today."

Selecting Areas of Focus

There is an inevitable tension between diversifying to mitigate risk and specializing to achieve clearer focus. There is an undoubted strength of shared purpose, like a rowing crew in perfect harmony, in a company with a single business model pursuing a set of congruent targets. Specialization fosters the accumulation of information, knowledge, insight, infrastructure and relationships that improve effectiveness and efficiency and open up new opportunities for business extensions. However, there is also the risk that the chosen area of specialization will attract excessive competition that will drive down margins, that the available targets will run out and make the company's competencies functionally obsolete, or that the market will change in a way that challenges the validity of the business model.

On the other hand, it is difficult for diversified companies to reach the same level of expertise as the specialists. The leaders of diversified companies are often challenged to compare investment proposals for a "gas factory" operation drilling up to 1,000 wells per year with a single deepwater frontier well off Africa or a refinery expansion project. The technologies are quite different, the competencies needed for success are different, the performance metrics are different, and the risk is different.

For this reason, diversified companies present value propositions to their investors that are usually centered on sustainability, while specialized companies generally emphasize growth.

Jack Welch changed GE from a conglomerate with many different businesses to a powerhouse by demanding that all his businesses be leaders in their industries. The energy industry is, contrary to popular belief, quite fragmented, and there is still tremendous opportunity for industry consolidation, particularly in the upstream. The challenge for the largest companies is to select the right portfolio of businesses and assets and to recognize that different asset types require different business models. The challenge for specialized companies is to make sure that there is sufficient "running room" so that they do not suddenly become cramped for lack of opportunities.

Resource plays such as oil sands and tight gas seem to provide sufficient potential for long-term growth, and new resources such as coal and biofuels have even longer term potential. In selecting areas of focus, companies should ensure that the segment allows for the following: 1) development of scale economies so that low costs can be achieved; 2) consistency over time so that learnings and competencies retain their value over time; 3) a focus that can help build employee motivation; and 4) sufficient running room for a consistent strategy to be durable.

Scale economies

Duane Radtke of DEPI is contemptuous of unfocused efforts in the upstream: "I call it 'putz' drilling. You can't do 500 wells in 500 places. And I don't care how good the prospects look technically. The numbers just will not work eventually. That's a loser's game. If you drill a hundred wells in five places, you are a whole lot better off. And obviously at the end of the deal, you can't drill all of your wells in one place. You are not going to control the world in that one basin. But I mean, that's the scope of things if you look at it. It's finding the right mix in between." Companies that achieve "scale in basin" can achieve lower costs and greater effectiveness than unfocused players, but they need new basins in which to grow.

Consistency over time

Rick George "tested the waters" carefully before deciding on oil sands as his area of focus for Suncor:

> I'd say that only after we saw our return on capital numbers come up, by 1994–95 did we make the oil sands the center of our forward thinking. That's when we really developed what is the strategy of today. If you look back, though, one thing that's always helped us was the ability to have a strategy that is basically contained

on one sheet of paper. And in an oil company sense, it's a pretty simple thing. We have a bitumen production business. Today we're now both in bitumen mining and SAGD; I like both technologies, we're putting large R&D money in both of those. We're moving to be North America's largest upgrading complex, and then moving that to the downstream markets. And then we have a natural gas business. Each one of these businesses is an energy consumer of gas. So you've got a natural kind of hedge on the gas side. And so you're able to paint that picture.

And you can look back on one of our annual reports at least 10 years. And, okay, we've changed. We added assets, and the thing keeps getting bigger, but the strategy has not changed dramatically from that moment. You know, some of the best-run companies, in my opinion, are ones that have a strategy and pursue it dogmatically. When we started this thing, nobody really knew much about oil sands. There were only Syncrude and us, and they were owned by 10 oil companies, and weren't pursuing an integrated model. It is kind of a form of flattery, but now everybody's dying to have our kind of a model.

It's the dogmatic pursuit. But that's often what happens in the oil patch, in my opinion. If you look at Bill Greehey at Valero, you know. He started buying refineries when they were 10 cents on the dollar. People said Bill Greehey was crazy. Now his logic is clearly solid. So sometimes the trouble with this industry is it runs as a herd.

People are a huge part of this, getting the people's confidence in the strategy. I think one of the big strengths at Suncor has been the ability of the leadership team in general and the employees to understand the strategies. We find we spend less time with politics. We spend less time with "where is the company going?" Because with a clear vision and strategy about where you're going, I think it's easier to get people to line up.

Rich Marcogliese elaborates on the benefits of consistency for Valero, compared to a diversified company such as ExxonMobil, where he spent 26 years:

What has been great at Valero, in contrast, is we have a different view of the downstream business and the refining business. You know, our refining business isn't competing with other worldwide business segments. Our mission has been to acquire depreciated assets, upgrade them, and grow the business; and make our system more competitive. We don't have a lot of internal competition from power plants in China, or drilling for oil off West Africa.

Running room

EnCana's Gwyn Morgan is also a believer in focus:

Randy Eresman has been with the company 25 years. I've been here 30. All of those things we did were based upon our philosophy of long life reserves, applying technology, and taking time to continuously improve recovery, our whole resource play and unconventional focus was embedded in this company. We are better at it than anybody else of our size.

And we have the best resource position. We examined our risk-adjusted incremental return expectations from developing our North American asset base, which we knew had a very long life. Like, one of our oldest fields in Southeastern Alberta which is hitting production records 30 years later. The rest of them are all only 10 years or younger, some of them only two or three years. So they are on this early part of the curve. And they are huge. And we know we are going to be able to do what we have done before: keep driving down costs, improving recoveries, and increasing production.

What happened was we built these assets. We owned the land. We have the resources. We paid the original upfront price building our position at a time when gas prices were still low. So now the incremental returns on what we've built in North America are so high, and the life looking forward is so long, that we said, "We can grow this company organically faster than virtually any of the larger companies in North America or internationally, with a country risk and a rate of return that is better than anything else."

Employee motivation

One of the benefits of focus can be improved employee motivation. Valero's Bill Greehey emphasizes the advantage of focus to the employee base: "And if you're a chemical engineer, and you look at the industry, where would you want to work? From one refinery we have grown to 17 refineries in seven or eight years. The promotional opportunities here have been just outstanding."

Gwyn Morgan of EnCana echoes the same sentiments in his focus on tight gas: "If you're a geophysicist, geologist, or whatever you do, you are able to apply some of the most incredibly technology-intensive approaches—much more technology-intensive than conventional stuff. Geophysicists working these micro-fractures in this rock to figure out which way we should drill these wells, and then drilling them and finding out what it is like, they're in heaven. They are applying the most complex, detailed technology. It's getting people inspired about what it is that they are doing." Having made the course correction and unloaded parts of the cargo that were worth more to others than to EnCana, Gwyn is generating enthusiasm around his company's core competencies to build momentum and move farther, faster in the chosen direction.

Refining the Business Model

A successful business model combines assets, knowledge, and competencies to create competitive advantage. The assets and knowledge must align with the value drivers of the industry segment. The competencies must enable the company to sustain growth and a competitive cost position.

Understanding value drivers

Energy is a commodity industry. Little differentiation is possible in the product: an electron, methane molecule, or even gasoline is pretty much the same, whoever provides it. There are opportunities to differentiate in customer relations and on pricing and service models, but not so much that a company can ever forget the importance of a competitive cost structure. Indeed, the best differentiation enables low costs for access and supports sustainable low capital and operating costs. Most other parts of a successful energy business model are subordinate to developing a competitive cost position.

Low costs are, of course, relative. They must be measured against the appropriate competitive supply options for a particular market. If supply options are insufficient, prices will rise to levels required to reduce demand, which raises the allowable cost for all suppliers. Nevertheless it is dangerous to adopt a strategy that locks in costs that are higher than those of competitors. Competitive costs are achieved by putting together assets, competencies, and insights into a business model that can sustainably provide incremental supplies to an individual market at low costs.

Starting with the market

Low costs have to be defined by reference to other ways of supplying a market. In the energy industry, companies need to focus on the relative cost of incremental supplies required to meet growth. Frank Chapman explained that BG pays careful attention to the industry cost structure in specific markets. He elaborated on how BG recognized the opportunity in changing UK natural gas supply costs:

> Now that opportunity, I believe, doesn't readily appear on the map for resource led players in the same way. In BG's world, our response to the recognition that there was going to be this big change in the way the market was supplied was to say, "Where can I get gas from economically? Let me understand what the cost profile of supplying this market is from Norway, North Africa, Russia. Let me understand the way these cost patterns are going to change in time. And let me understand, therefore, what the long run marginal cost of supply is for this market. And having established that, let me understand to what extent I can access supply that can be competitive against that framework.

> So BG's response has been more North Sea drilling to capitalize on the infrastructure that we already have installed: making deep, high-pressure, high-temperature gas condensate plays economic through existing infrastructure. Something we call play extensions. We're only just starting to talk about it in the world. After working on it for several years, we have accumulated several hundred new prospects. Quite small in size but they can be tied back to our infrastructure. It's going very well. We've had a couple of very nice discoveries doing that: four, in fact, last year.

Chapman continued with another example of market-led thinking:

> We did a bit of pro bono work putting together a gas marketing plan for Egypt. We're good at that. Some players wouldn't know where to start. It's part of our whole fabric. So in Egypt, when we made our first discovery in Rosetta, we went straight to the government and said, "We know that we are at the bottom of the list of half a dozen [development opportunities]. But if you develop our field first, you'll save X amount of money on pipeline investment for the infrastructure, because over this side of the Delta, you are not using your infrastructure. And over there, it is all full. So why don't you make us first instead of last?" They did. So immediately after

that discovery, we got sanctioned on the project. And we could have been waiting there for years.

Now we also knew that if we were successful in the Delta, we'd have too much gas in the country. So before we drilled the first exploration well, we understood the value chain to Europe. We understood it would be LNG. So from the date of spudding the first well, to the date of export of the first LNG, was just six years. And in the process we wrote the law for exporting LNG. Twenty-three volumes, I think, which were passed through the People's Assembly into law, covering all of the fiscal and legal arrangements required to export LNG from Egypt. Our lawyers and commercial people wrote the lot. And, that's what we are good at, dealing with huge commercial, technical, but also legal complexity, which characterizes these chains.

Aiming for low costs

As we have described on previous occasions, energy is a commodity business, subject to price cycles of uncertain timing. It is vital for long-term success to secure a competitive cost position. Frank Chapman, CEO of BG Group reinforced the need to think ahead as he expanded on his Atlantic LNG business in Trinidad:

You assume that in the limit, ships and terminal capacity and LNG liquefaction will become commoditized, and the rent will shift out to cost-effective supply and customers. What customer propositions have you got? How do you get it to your customers? So when people say to me, "What happens if they build 20 new terminals?" I say I don't worry, because I've got lowest cost. We set about building the lowest cost structure in the industry: lowest cost with LNG supplied by the lowest liquefaction operating cost; some of the most competitively priced supply contracts; way the lowest costs of regasification in the U.S.; and every time we build another tranche, we are driving the cost structure down. This has made our supply portfolio robust against a U.S. market price of $3/MMBtu.

Leveraging infrastructure and knowledge

We presented earlier Aubrey McClendon's view of the advantages Chesapeake gains in focusing disproportionate resources on a play relative to the competition in Oklahoma. He continues on the critical, differentiating result of this focus:

The final and most important one was information. If you ask any business man around the world today what separates success from failure in your business, I'm hoping they will say that it's information gathering and decision making, decision implementation, and then execution of the operating strategy; but the very core of it is information. So what we found out is all of the sudden we were collecting 20–30% of all the drilling information being generated in Oklahoma today, 60% in a few months, and this was the second largest gas producing state in the country, so we were seeing everything. It was like this veil or film was lifted from our eyes, and we had all these guys trying to solve all these geological issues, we had all this information coming in over the transom, and for the first time in the history of the business, one company was gathering 30, 40, 50, 60% of all the industry information; it was a huge advantage.

Frank Chapman of BG believes there are still opportunities for companies to differentiate themselves through superior knowledge and insight. However, the differentiation he refers to is mobilized in support of developing low cost access to resources:

Norway and the UK provide an area where you can compete on technical capability rather than the amount of money you're prepared to put in play, because you have competencies you can leverage. And you become, therefore, an attractive player to a host government. In so many areas around the world today, you license on terms of financial commitments, and that is a game which we don't want to play because it is a winner's curse. The fellow that wins is the fellow that has paid the most, and therefore is least likely to make an economic return.

So this year, for example, as a result of that type of strategy, we were able to add 40% to our exploration acreage portfolio for the cost of 130 million pounds. Some players have paid several times that for a single block. This is absolutely based on competencies. It has something to do in certain locations with access to existing infrastructure. The type of play one is pursuing will sometimes be a play that is accessible to you because of the existing cost structure you have. We are able to look at things other people are maybe not looking at. But also we are able to convince the government on the basis of a combination of market and upstream skills.

Why wouldn't a government always go to auction? Let me give you a practical example. You've got a piece of acreage, and you've got one anticline in that piece of acreage, and let's just say it's not a huge one. If you drill the anticline with a single well, it is either going to be a success or it is going to be a failure. If you bid 10 wells on that one anticline, the government knows after the first well if it's a failure. If it is, what are you going to do with the other nine? You're going to ask the government to release you from it. So it's meaningless. It's no good the government forcing you to drill the other nine into the same structure. It's throwing money away and providing tax offsets to somebody in the future. So the government knows this is nonsense, and it's not a game that they encourage. So what they are really interested in is, "What can you see in this play in terms of concept that other people can't see?" You establish a play that, if you are successful, will encourage others to want to bid in later licensing rounds. Of course, in the process it also creates wealth from that particular venture.

Then it takes us to Norway. What is over the median line from Norway that can be tied back to our infrastructure? Given the fact that, if you do a map showing the stratigraphics of all wells drilled, particularly into Lower Jurassic and Triassic plays, UK versus Norway, you'll find a dense block of dots on the UK side. Blank on the other side. Now the Good Lord didn't organize it like that. So there must be Jurassic and Triassic prospectivity on the other side. Under the Norwegian fiscal regime, that is not likely to be attractive and, therefore, players haven't gone there. The new infrastructure that would be required. . . . How does this compete with a 40–90 TCF Troll, a 50 TCF Ormen Lange? How does it stack up against that? Not very well.

But when you realize that it is within spitting distance from your own infrastructure on the UK side of the median line, and there is the Frigg Treaty, and then a new UK-Norway gas treaty is signed. Well, then things start to look a little different. So that's a second strategy, to go and explore in Norway. We have acquired fifteen licenses in two years, with eight of them operating, and some of them 100%, which is a remarkable achievement in Norway, which is quite a conservative country in terms of the way they license new acreage. They normally tell new entrants that they should be investing partners in other consortia operated by Statoil or other established players.

Building on success

Chapman finished by explaining how BG had combined its market insight and belief in the importance of infrastructure to extend its UK business model into LNG:

The third thing we are doing in the UK is to bring LNG in. So the response is the Milford Haven "Dragon" plant. You know, most people come at this and say, as we did ourselves in '96, "The North Sea is declining. Where else are we going to go to?" We say, 'Well, hang on a minute. The North Sea will decline, but the market still has real growth potential." It was one of those moments in time—it wasn't quite a "Eureka!" moment, but it was that sort of moment where we started to say, 'Well, who are we? Where did we come from? Doesn't this market still have growth potential? If it has growth potential and we are experts in gas, what does that mean for us?' And what it starts to ask you is, "How do you supply this market? Where do you bring it from economically?" And buried in those questions are opportunities.

These examples develop a business model which is market-led, anchored in low costs, leverages infrastructure and knowledge and continuously builds on successes. There are lessons to be learned from this model for different regions and for different commodities.

Building Out the Strategy

Once positioned downwind of the major trends, with a clear sense of the firm's boundaries, areas of focus and business model, leaders must decide how far they want to take their companies and how broad they want the advance to be. The first lesson is that of humility: companies must be prepared for uncertainty. Our interviews suggest that clarifying goals is the critical next step after setting a new direction. Leaders next have to develop options for future growth. Many of these will be found in new frontiers. BG was able to assemble a successful business from a series of discrete moves, each of which enabled new moves that would not otherwise have been available. The key is to capitalize on core competencies to build competitive advantage.

It is clear that inorganic moves and major restructuring can still create value. We found powerful examples of new directions in transforming non-OECD scope and scale (Total), building scope and scale in North America (ConocoPhillips), and engineering a turnaround (TXU). Leadership is about change, and the examples below are transformative in nature. Future energy leaders will find themselves taking a hard look at their firms and making similar decisions about whether the current scope and scale of their organization is right or wrong for the emerging business environment.

Adapting to changing circumstances

Capitalism was famously described by Joseph Schumpeter as "creative destruction." In this respect it can be seen as a natural system, subject to evolutionary pressures. Companies, like species, evolve in order to prosper in changing environments. Those that fail to adapt to changing environments die out. But in adapting, species and companies build on what they have rather than totally reinventing themselves. Those that can take existing strengths to new habitats or add new strengths to compete better in existing habitats will survive and prosper.

The three ingredients to successful adaptability seem to be recognizing uncertainty, learning from adversity, and experimentation. Smart leaders recognize that the future is unknowable, and that plans that depend on the actions of others may have to change. The key is to move promptly with a new plan if the prior initiative is frustrated. Especially when technology is involved, there is a need for experimentation to test market receptivity and to establish a successful business model. Also, failures can often yield insights that can form the basis for a new direction. The adaptive behaviors described here will be relevant to companies wishing to evolve successfully in tomorrow's changing business environment. The specific moves can't be copied, but the approaches and attitudes can.

Recognizing uncertainty. Nick Butler of BP reminds us:

> I think you always work in conditions of uncertainty. We're very conscious that we're 0.01% of the world economy. We're quite big, but the world moves without paying much attention to us. You have to be very flexible. I think if you have your broad objectives, then you have to apply your best efforts, but without any sense that you can plan a future for yourself. It's like the old line: "What makes God laugh? People who say they have plans."

I think, let's go back to the strategy we did 10 years ago, when we decided we didn't want to be a niche player but become one of these global players. The first step we took was to try to take over Mobil. We had intensive discussions over a long period of time, and it would have been a great deal. But in any event they decided they wanted to retain their independence. There was no ill will—they recognized the potential benefits, but they didn't want to be absorbed. So we had to go for a second option (Amoco and then Arco), which was good but perhaps not as good.

Learning from adversity. Things do not always go according to plan, and a company must be able to adjust. Few companies have done it with more flair than Chesapeake. Aubrey McClendon talks about Chesapeake's first growth cycle in the Austin Chalk in the 1990s, which was very successful for a while. It established two themes that are still key components of Chesapeake's value proposition: geographical focus and leveraging technology. But when this play reached maturity, McClendon and his team started thinking that their experience might be typical of trends throughout the Lower 48. These were that U.S. natural gas discovery and development were becoming more difficult, while demand was growing strongly. Putting the two together, they realized that prices were going to have to rise, and that they could take their competencies in geographical focus and drilling to other plays with greater resource potential than the Austin Chalk. They redirected the company to go after Mid-Continent natural gas resources and have shown extraordinarily profitable growth since that moment.

Experimentation. John Browne of BP summarizes his view: "So you can't do strategy on the basis that all will fall in line with what you want. I think you need the right direction and some idea about the means, in this case the use of mergers and takeovers to get us into different areas of the world and different businesses. You then need a degree of experimentation, and to be prepared to fail, which is how we did the mergers."

Clarifying Goals. John Browne emphasized the importance of clear goals as the vital complement to setting direction: "The most important thing for a leader is to set a direction, because nobody else can do that. You define the goals and the means of getting to those goals, and then establish the resources that aspiration will need and make sure they're in place. Perhaps most important of all is to set a standard by what you talk about and, therefore, what you care about, what you spend your time doing, how you behave when others see you because you're a very visible leader, how you behave with people outside the company—the manner and tone of all this.

Through these behaviors you set the boundaries of what is acceptable and what is not acceptable." In this brief statement, he covers vision, purpose and goals (setting a direction), and values (the standard). He then makes sure that resources (human, financial, technological) are available to develop and implement the strategies that his businesses formulate to meet the goals he has set.

Nick Butler elaborated about the goals for BP and the thinking that went behind their "extreme makeover" of the 1990s:

> It's not just being Number One in terms of size. That's not a great driver. We would want to think we have a good strong growth position in some key areas:
>
> - Access to available resources of oil and gas.
> - The ability to get those resources to market in terms of being positioned in the infrastructure.
> - Positioned in the downstream growth markets around the world.
>
> We wouldn't want to be third or fifth in any of those areas. We don't absolutely have to be first in any single one, but the aggregate of where we would want to be does mean we would think we should be first or second overall. Otherwise, this is where we were 10 years ago, when we were fourth or fifth in the industry and we could see Shell and Exxon, in particular, moving ahead. We felt we would either become a niche player with a problem of declining assets in the North Sea and Alaska, or we had to make some big steps to be working on a global level, which is where the growth was and where the new resources were. That was a conscious strategic change we made 10 years ago.

Gwyn Morgan of EnCana calls it strategic leadership:

> The next part of the foundation, in my mind, is what I call strategic leadership. Making sure that you tie together what you want to accomplish, be sure you're heading in the right direction. Inspire people about that. You know, when we first started, I remember, people used to say, "Well, these resource plays, they're not as exciting as going out and drilling big boomer wells." But then our leaders would talk to the staff and say, "Look, we're not going to have lots of discussions on what is more exciting: drilling a big boomer well, and it's all over and it's all downhill from there, or working over time to realize the full potential of a complex structure."

Developing organic options

Adaptability is about preparing for uncertainty, experimentation, learning from failures, and building on successes. The purpose is to grow the business organically, by uncovering and investing in new business opportunities, or inorganically, through mergers and acquisitions. In both cases, the critical first step is to create options by originating and capturing opportunities.

John Lowe of ConocoPhillips understands:

> I think what our shareholders want now is primarily the ability to generate organic growth. I think that most people would recognize we do have a competency at making acquisitions and making them work, but I think the driver and where they want the emphasis to be is on the organic growth, and I think there are only a handful of companies that can compete for the big projects around the world.

But organic growth will not be easy. John Lowe recognizes the difficulty:

> The competition for assets these days has gotten completely out of hand. We talked a little bit earlier about the heavy oil sands, people are bidding $1.00 a barrel for undeveloped resources that they don't know will ever be developed, and if they are then its going to cost billions of dollars to develop them. It's almost as if they use that as the rationale: once we spend $10–15 billion developing these, then in won't matter whether we pay $1 billion for this billion barrels of resource, or $500 million or a billion and a half, and I think they just talk themselves into bidding these numbers. We don't have any engineers running the company, but we do understand that if you expect to give your shareholders a 15% return, and you spend a billion dollars this year and you don't have anything developed for 10 years, then that just cost you a billion and a half dollars of value over that time period. We still do IRR's and NPV calculations, and it doesn't do too well when you spend billions of dollars and don't get anything for 10 years.

In the 1970s, oil companies found growth in frontier exploration in areas requiring new skills, such as Alaska and the North Sea. We have heard earlier from Clarence Cazalot and others that there are good opportunities in stranded and unconventional natural gas. In addition, there may still be new frontier exploration opportunities. The other source of organic growth is more systemic, building on existing assets and competencies and capitalizing on successes. It is this systematic process for developing growth opportunities that we would like to highlight in this section.

New frontiers. Don Paul of Chevron thinks there are still large-scale exploration targets to pursue:

> I don't know if I agree with you that exploration won't play a role. I think the Arctic is still clearly there. Most of the Arctic has not been explored. All of Greenland is basically open. I mean, anything that is more than 500 miles north of Newfoundland is open. I don't think there is any question about it that there will be hydrocarbons there; the geology is definitely prospective. And in fact some recent academic resources indicates it is every bit as favorable as Alaska. So I think there are more Prudhoe Bays.

But Paul recognizes that the technology does not yet exist to recover the hydrocarbons, so this is a long-term bet with great risk attached. Ian Howat of Total agrees that this is an important role for super-majors: "That's something that we pay a lot of attention to, trying to get acreage where the technology is not yet working but we think it will at some time."

Don Paul continues:

> So if you step back you see that in the end, the real issue is that this business has always been about the large resource bases that are yet to be developed for some reason. They are not politically accessible, or they are not technologically accessible. So where are those resource bases? I think that most people would say there still is a substantial amount of opportunities in deeper water. There is also in the Gulf of Mexico, potentially a substantial amount of really deep gas at 35–40,000 feet. Then you go to the really big resource bases. You've got the bitumen basins: Athabasca; Orinoco; then coal in the U.S. and China, and India to a lesser extent.

And ultimately, along with unconventional gas, these resources are where the organic growth will come from, and we would expect more and more companies to develop organic growth strategies based on them. The challenge will be for companies to develop the technologies that allow these resources to be developed at acceptable returns in a cyclical industry. But this is no different from the 1970s, when the companies did not know at that time how to develop North Sea and Alaskan resources economically at the then prevailing $12/Bbl price. By the mid 1980s, the view was that the North Sea needed a $30/Bbl price to be economic, but by the early

1990s, companies had totally revamped their business models to reduce costs such that new fields could be economic at prices below $20/Bbl. So the issue is to view these new resources not as single projects but as programs with the learning-curve opportunity to lower costs over time.

"Asset accretion." Frank Chapman, CEO of BG Group, summarized the overall approach he calls "asset accretion" to create new organic options from each strategic move. He described how BG had capitalized on its success in Trinidad with the Atlantic LNG project (ALNG) to build out a complete Atlantic Basin LNG business:

> While we were building ALNG1, and we moved to ALNG2 and 3, we were thinking about what was happening in the Atlantic Basin. And we had formed the view that the Atlantic Basin LNG trade was going to grow six times faster than the Pacific trade. Because there were markets that were liberalized and/or intended to liberalize in this arena, it would be less closed than traditional point-to-point trades with Japan and Korea.
>
> And so in 1998, I think it was, Martin Houston came to the board with the proposal to take four LNG ships. I was on the board and I gave him a lot of support for this, and we got it through the board. And we got these ships at $32,500 dollars a day. The shipping market was really long on ships then, but within a year or so the shipping market went really short all of a sudden. That's when the LNG plants began to come on stream. And we started to charter these ships at $160,000 dollars a day. Four ships earning $100,000 dollars a day is not small beer, but it's also not a company-making process.
>
> The real reason we did it was to see how we could leverage the ships into achieving supply and developing market positions. Now, again, this gives you a little insight into the way our minds are working. It is not being driven by resource. This is being driven by, "What is it in the gas chain which is currently the bottleneck?" And by owning or accessing those short commodities you open up opportunities either side.
>
> Now this has absolutely played out and has caused BG to become the largest LNG player in terms of trading cargos in the Atlantic Basin today, and also the largest supplier of LNG to the U.S. market. We supplied 37.5% of the market last year in the U.S. And that came about as a result of having the ships.

It was then seen that the multiple facilities built to bring in LNG to the U.S. would have to start to operate again because the U.S. indigenous supply would start to decline. We went there and we got all of the Lake Charles capacity. Now that Lake Charles capacity was very much sought after. But the reason that we got it was, we said, "We'll have it all. The whole lot. Hundred percent for 22 years. We made a $1.2 billion commitment to capacity without having a single cargo of LNG.

That 4.5-million-ton Atlantic Basin idea has since grown from 2001 to today to a 40-million-ton idea, half of which looks to me to be really quite secure. Now that's a big business. We turned over one-and-a-half billion dollars last year in the U.S. LNG profits last year totaled something like 175 million pounds, and we grew 18%; and that will continue to grow very powerfully. We are going to grow volumes something like 20 to 25% per annum between now and 2012.

This business is going to grow—and it stems from a sequence of things. Having ships when nobody wanted them. Using the ships to get terminal capacity. Using the terminal capacity and having the market access to be the only person in the market that could buy LNG unconditionally. That allowed us to do EG and Nigeria; enabled us to trigger ELNG Train 2 in Egypt; enabled us to get into Brass Island LNG recently. We are still the only guys that can bid unconditionally because new regasification plants have yet to get built. We could probably take about five million tons a year without batting an eyelid. And that's a deliberate strategy to be market long. So having the supply, we then can build a customer base because we've got volume certainty. So we now have 100 to 150 major customers, and we are spreading those away from the hinterland in the U.S. So ships become terminal capacity, become cost-effective supply, become customers.

This is a concept that we call "asset accretion." How can we use what we've got now, and find a way of either gaining an entry into something else, or leveraging those assets to create value in a new venture? So our experience in Egypt and in Trinidad, and the fact that we were open for business in the U.S., made us an interesting counterparty for NLNG in Nigeria. So we bought cargos from NLNG on a long-term contract to supply the U.S. That meant that for the Nigerian National Petroleum Company, BG was now a customer. So when we walked through the door, we weren't another guy banging on the door to request a license for some prospective acreage. We were actually a customer.

So we went to Nigeria, and the idea of OK LNG was born. And BG, having no position in Nigeria, was actually welcomed by NNPC and Chevron as a catalyst to get this thing done with a new type of approach. We are now acquiring upstream assets which can supply that facility. So it's another example of having market, going back into the midstream, and then using that to leverage entry into the upstream. And the farm-in is because the acreage is 50 kilometers from OK-LNG. So you are leveraging backwards.

Leveraging core competencies

Competencies execute strategies, and they also open up opportunities for new strategies. Respect for competencies attracts new talent in a virtuous cycle. Responses we received emphasized the importance of competencies in the commercial, technical and project management domains. EnCana's Gwyn Morgan is of the opinion that a set of differentiating competencies is, with superior assets, the primary route to competitive advantage. Of all the lessons in this chapter, the importance of core competencies is perhaps the most universal.

Commercial competencies. Core competencies can be focused on a variety of areas. John Lowe believes that ConocoPhillips has particular strengths in collaborative deal-making: "So as a core competency there, I think it is a willingness to be open to new ideas and new ways and not be completely set upon 'one formula works for everything,' and recognizing that everyone needs to be a winner on this thing long-term for you to succeed. It just doesn't work for one person to be a winner and one person to be a loser."

Steve Lowden is building a company on a related core competence—the ability to forge productive relationships:

> This new venture is a relationship business. At the center it is differentiation. Any way to bring energy to the market and develop a resource in the energy space, frankly, is in our focus. Our focus is not geographical or geological or a basin. It is a business model focus. And it's about a relationship with privileged access.
>
> And so we focus on connections and relationships. And we'll focus in countries where we know a lot of people, where we are very comfortable, where they know us, where they feel

confidence. We are going to be recruiting a bunch of people who have good relationships. We are already about 30 strong, and we will be 100 strong by the end of the year. And half those people will have extensive relationship skills and track records from past relationships; things you carry with you.

Many companies think they have a relationship, but they never do. Relationships are with the person, not the company. So people will come onboard with existing long-term relationships. I'm sure the super-majors could argue that mostly they don't need relationships. They are big enough, and the brand is so strong, they can go anywhere. I'm sure that the big brands can carry them a long way. But I'm also sure that is no substitute for relationships. I am convinced that the relationship between John Browne and Vekselberg and Fridman is what keeps TNK moving. Not BP's brand. So I think for all company sizes and shapes, relationships have become more and more important.

So the brand, basically, helps get the door open. How do you get the door open without a brand? Where we have the relationships, they open the door easily as fast as a brand does. But also, you remember that the people that are coming into our company right now all have execution and track record with brands. So they have the past brand that they can carry with them. They carry the brand, and they carry the relationships. The company can't keep the relationships without the people.

So Steve is building a business model on a singular core competence in relationship building in the energy space.

Technical competencies. As Chesapeake extends its geographic reach, it is focusing on specific competencies. CEO Aubrey McClendon asserts:

We want to be known as the guys who have the greatest knowledge about black shales. We're in the Barnett; we're in the Fayetteville; we're in the Caney in Oklahoma; we're in the Devonian in Appalachia. Nobody else is in more than two of those plays. We should be able to learn more about shales and to transfer technology across divisions. Some companies, you actually sit with your asset team, and I do see value in that engineer, geologist and landsmen sitting side by side. We do it a little differently: we put all our geologists together, all of our engineers together, all our landmen together; so you work on an interdisciplinary team, virtually through e-mail, and then physically together in meeting rooms. But then when you go back, I want you to be with your professional colleagues in your own discipline because I'm hoping that the challenges that you're facing in this problem over here have already been resolved on that problem over there.

Gwyn Morgan is a strong believer in the importance of core competencies and continuous improvement to hone those competencies:

The classical resource play, every year you are going to drill more wells. And you are going to drill them for decades and decades and decades. And every year you can find out more. You can adapt your equipment. Today, after 30 years in our oldest project, the nominal dollar cost to drill a completed well is roughly the same. And you know what has happened with the Canadian currency and service sector costs. We are drilling faster and using completely different equipment, that has been tailored and designed to optimize results. We use endless tubing; drilling rigs and frac right through the same equipment. Then we get it all done and move on to the next.

The merger of Pan-Canadian and AEC reinforced these competencies:

Pan-Canadian's major traditional base, and its largest asset, was adjacent to our big gas field. Almost all of their Canadian position was tight gas. And these two companies grew up focusing on that kind of stuff, mostly with their North American base. We also had our in-situ oil sands. We had the two best positions in that. But, again, they were companies that basically built their whole culture and their technology around recovering gas and oil from difficult reservoirs—unconventional reservoirs, in-situ oil sands, tight gas and CBM.

Project management competencies. Frank Chapman highlighted another area that he believes is distinctive in BG:

There is another very important difference. BG is investing three or four times its depreciation rate in capital investment. Now if you do that, you have to do it well. If you don't do your projects well, you will destroy your company if you are investing three or four times your depreciation rate. So we have to be very good at execution.

If we're not good at execution, we die. And that's much less of a problem for a company that is only investing 1.3 times its depreciation rate. And that's caused us to value engineering execution skills in a way, I think, that some of our competitors haven't. So we've become a very good executer of projects, because it's a core skill for us, just the same as exploration or geophysics. And does this affect our ability to attract good people? I think it does, because it is a position that is esteemed in the company.

Attracting personnel. Helge Lund of Statoil believes intensely in the importance of building competencies:

I can only see one path that I am 100% certain of, and that is competence will be more important over the next 10 years. Therefore, I have moved HR up into the corporate management team, and we are more or less reshaping all our programs in this area. This will be completed by 2006: leadership training, leadership development, leadership succession planning, project academies, knowledge academies, business schools in business units. All over the organization, we have initiatives to address this area. Statoil has been the most attractive employer in Norway among graduate engineers for nine years and we have had the same position among MBAs for five years.

We really have to be successful in bringing that to bear internationally—to have an equally strong recruitment position in those markets that we would like to grow. I was extremely pleased with this year's record application to our corporate trainee program. And I think 50% of the applicants were outside Norway. We had several thousand applicants of which a substantial part was non-Scandinavians. That makes me optimistic that our plans are working.

Valuing deep expertise. Reflecting a similar respect for competencies, John Wilder of TXU was dismayed when he came to TXU by the need to reaffirm respect for specific expertise:

What I'm about to announce is what we are calling a TXU Fellows Program. It's an orchestra example, right? The great musicians are deeply skilled in their own unique area. And with just some coordination through the conductor, music can be fantastic. But the world is shifting. Much like a university setting, much like a hospital setting, corporations are shifting to a deep set of skill specialists. And our culture had, and still to a certain extent has, a zero respect for the specialist.

And so we started this Fellows Program to make the first-of-its-kind corporate recognition of talent at the specialist level. And so we are in a senior team meeting. What does it take to be a Fellow: to have testified on a technical topic before Congress; to get frequent invitations to speak to audiences of 300 plus people at technical conferences; to have multiple patents? No one could think of anyone.

And I said, "We've got a guy that is actually very talented at transmission. He knows the whole Texas transmission system literally in his head. What about that guy buried in our trading group that knows every generating facility in Texas. He knows when it gets dispatched, what kind of fuel it uses. Have you guys noticed who I call up, who I talk to, who I e-mail, who I ask to come see me and talk about things in the company? It's these kind of guys." And so we are going to announce some TXU Fellows, and we are going to go off for a weekend, like to Barton Creek, invite their spouses. Tell them how important their unique knowledge is. We are starting with a half dozen, we just don't have many. We're going to name them a Fellow, and we're going to give them some stock.

Building competitive advantage. Gwyn Morgan of EnCana continued:

If there is one other thing that I would say that I've been fixated on, because of my way of thinking about this business ever since we began, it's core competence. Competitive advantage is created through core competency and superior assets. So when somebody would ask me about mining projects, I'd say, "Okay, now, you tell me one thing an upstream oil and gas guy, who knows how to drill wells, and put stuff onstream, and sell it at the plant gate knows about mining. Where did we learn that? Second thing is: What do we know about refining? Where did we learn that? What's surface oil sands plant? It's mining and refining."

That's why we focus on underground reservoir engineering. Most of us are petroleum reservoir engineers. Randy Eresman is a petroleum engineer. I am. We've got, of course, geoscientists as well. But we look at it from the point of view that what we do; is to drill into reservoirs—mostly challenging reservoirs—and apply technology, time, and focus to maximize the returns. And so that's why, when we came to look at the oil sands years and years after our initial thrust in tight gas, we obviously gravitated towards drilling for it, not digging for it.

We noted earlier in our discussion of the benefits of focus on a segment that both Bill Greehey of Valero and Gwyn Morgan of EnCana highlighted the advantages to professional staff of working for a company that provides continuous opportunities to apply their skills. This further reinforces the strength of the competencies. Morgan can draw on the experience of a large number of petroleum engineers who have the skills to detect fine fractures in difficult rock, and others who can design drilling rigs that are perfectly adapted to a specific play. Greehey has attracted the foremost experts in every refining process technology, and they can continuously improve the performance of Valero's refineries.

Frank Chapman emphasized that BG has built a business model that would be difficult to emulate, with distinctive competencies in both commercial and execution arenas. And he also believes, with Gwyn Morgan, that competitive advantage comes from the combination of superior assets and distinctive competencies:

Now this growth idea and doing things, having a flux—is very motivating for the staff. They are highly energized when you are going into new places and doing new things. So that's part of the mix. People are always onto me about how others are going to challenge us, how others are trying to emulate our strategy and how they are trying to do what we are doing—and, of course, one has to be alert to this. But when people think about doing the same as us, the truth is that some of these things can be emulated, and some of them can't.

What people try to emulate when they look at BG is what we have. And they see that we have supply and markets on both side of the Atlantic. So they try to acquire supply and markets on both sides of the Atlantic. In itself, it's not an easy thing to do. But let's assume they are successful in doing that. Where are they then?

Then they haven't got BG's strategy. What they've got is a replica in some form of what BG's assets are.

What they don't have is our mindset. And our mindset is that we are, first and foremost, marketers. We are interested in supplying markets, and we have a skill set that can enable us to play anywhere along that value chain. We can be in power. We can be in distribution. And it might be because that's the part of the chain that needs completing. That's the part of the chain that has the most value in it. It might be all along the chain, because we are the architect.

Draw a chart of the industry. On the manufacturing side you'll find Exxon and Shell and Total, all these people—and BG. On the purchasing side, you'll find Osaka Gas, Tokyo Gas, Tokyo Electric, GDF—and BG. BG was until recently the only company on both sides. We are both a purchaser, a marketer, and also a very good manufacturer.

Positioning for inorganic options

Mergers and acquisitions have played a large role in the growth of energy companies in the past decade. Going forward, the authors believe that mergers and acquisitions will play a less prominent role, but those with a clear strategic purpose and with unrecognized value will continue to be important. The key success ingredients are the ability to understand optionality—the opportunities that will become available after an acquisition that would not have been available without it—the honesty to understand that your company may not be positioned to be a future winner, and the ability to structure win-win outcomes. Mathematically, win-win is possible because an asset can be worth more in one portfolio than it is in another portfolio. By transferring the asset, both portfolios can be enhanced. Acquisitions also allow a company to reach material scale quickly, without the costs, risks and time required to build a position organically.

Understanding optionality. We met John Lowe the week before the Burlington Resources transaction, and he explained the attraction of acquisitions:

When it comes to deal making, I would say there is certainly an element of risk taking in there, but it's very informed risk taking. Each of the transactions that we've done, if you have quality assets and a viable industry, it's highly unlikely that you're not going to get a return on those assets. Maybe the return is not going to be as great as you would like, but when you get high-quality refineries in the U.S. market, is that going to be a 15% return or an 8% return? Right now it's 35 or 40% return, which we wouldn't have expected. But is it reasonable to expect a 10–15 % return, and then if things don't go very well maybe you get an 8% return or 6% return? Those are bets that Jim Mulva is willing to take and a lot of other people aren't, because it's a lot easier to do nothing; then you're not second guessed and criticized and yelled at by your shareholders, and bad things are not written by analysts.

Then there is the optionality in acquisitions:

The Alaska deal for example, you run NPV13 numbers on Alaska, and in our calculations we put in zero for the gas. But just at Prudhoe alone, there's eight TCF of known gas: non-conventional gas is selling for $3.00/mcf, just crazy prices, so how much is that gas worth today: $20 billion? More than we paid for Arco Alaska, so we're definitely aware of the optionality, and the options and upside potential that are created by the transactions. Just like anything else, you try and get the asset without paying for it, but it's getting much more difficult in this environment.

Visualizing future industry winners. Steve Shapiro of BR believes that the large independent E&Ps have some important competitive disadvantages relative to the integrated companies in this next cycle that make consolidation likely:

They will essentially become trust-like vehicles that the market will accept as an $80–$120 billion E&P that has slower growth, better yield; so from a market point of view, it's a pure play on the commodity, which markets like, combined with higher yield. And so the value creation is capital allocation, not exploration, and exploitation skills, leverage on the cost side, etc.

On the other hand, Shapiro sees interesting industrial logic, but considerable risks, in consolidation through the merger of equals:

You look at Burlington and EnCana, a powerhouse in North American gas. Or us and Oxy, a free-cash-flow machine with a Middle East position. I mean, just huge free cash flow and a Middle East position. Or us and Devon, room for operating efficiencies. I mean, you can go around, and there's industrial logic in each of them.

But the execution risk was also very large in all of them. And your ability to deliver on the industrial promise gets in the way of doing the transactions. The execution risk is in joint management. You all agree on this global concept of what you are trying to achieve. But the way you operate your business day to day is very different. And because you can't pay a premium to get the deal to happen, you don't get it executed because the risk of executing after the deal appears to exceed the theoretical value that could be created. No one side has control. You have shared management. It's executing after the deal. It's delivering on the promise. You lose the people. You lose the momentum. You lose the skill sets that the companies need to develop. It's risky. And boards tend not to want to go there, and that's understandable. It's hard to get to one culture.

Nevertheless, large mergers continue to take place including the merger of almost equals Anadarko and Kerr McGee, confirming Shapiro's original prediction of a few, very large independent exploration and production companies.

Constructing win-win outcomes. We asked how it can be a good deal from point of view of both sides, and Shapiro drew the general lessons based on his experience with the acquisition of BR by ConocoPhillips:

There's two ways. One is, of course, different perceptions of future markets, which is a guessing game, and I don't think anybody would want to say that is the basis for a deal. The real reason I think it could be good for both sides is because of the different portfolios and what each side is looking to accomplish. For us, what we were faced with was how to migrate into an asset base of long-lived projects to take some of the pressure off the day-to-day stuff that we do. We found that very hard to do for two reasons. Number one, we didn't have any competitive advantages in terms of those kinds of projects. Nor necessarily the culture to take the big leaps, nor the portfolio size to justify taking the risk. For them it was almost the opposite. They had a very large slate of large international projects. Tremendous option value on those projects, but perhaps a riskier portfolio was evolving. This allowed them to look for a little more stability, a little more OECD.

You know if you look at ExxonMobil, if you look at BP, Amoco, and Arco, a lot of the value there was based on the cost savings in the R&M, in the retail business, in the overheads of the companies. In our business, so many of our costs are at or near the wellhead. And although there are efficiencies in the field—I don't want to deny that, and we will work hard to get at them—generally they don't drive the transactions. There's about $375 million of synergies. On a $36 billion deal, it's not what drives the deal. It's generally portfolio, I think, that drives the transactions.

From the point of view of BR:

This didn't come out of portfolio weakness on our part. I mean, I think the drivers for us were a couple of things. Number one, we didn't know where we were in the cycle. The fact is, it was hard to imagine margins growing at the pace they had been historically. So, margin growth at best was going to slow down, or at worst was going to be a flat line. Also, we never felt we were getting paid in the marketplace for the margins we were generating today. But with the premium that came with the deal, we traded at the upper end of the multiples of our peer group. So maybe the whole group comes up, but ConocoPhillips essentially gave us a multiple-year head start with the premium that was in the deal, rather than the uncertainty of the marketplace down the road. The second thing was, we were trading half the value of the company into ConocoPhillips equity that was valued at the low end of their peer group. So we gave our shareholders a chance of just getting up to the average there, which could be a 30 or 40% movement in their equity.

Gwyn Morgan echoed the importance of win-win value propositions as EnCana was divesting of selected international assets:

We knew when we were selling both the North Sea and the Gulf, that the buyer would most likely be a company that was looking for a strategic platform. In other words, it is very hard to go into the Gulf of Mexico with no land, and no reserves, and no staff, and start. You need to have a critical mass. And if you've decided you are going to be there, and that your ultimate investment is going to be much larger than the entry investment, then you are prepared to pay, quote, "too much," for the initial piece. It gets you in the game. And that is strategic.

So we knew that we were selling a strategic platform in the Gulf, and a tremendous position in the North Sea with a great discovery in Buzzard. And of course, as Nexen is telling me, everything we said is working out as we said. I said to Charlie Fischer, "If we make this deal, all I can tell you is that the way the project is going, our current estimate of cost, how we see the reservoir, I don't know of any problems. We're selling it because of a strategic decision." And as far as I know, they have been saying it is even better than they thought. So that's good. Then it's a win–win. But, for them, it transformed them. They were looking for some near-term growth and for another platform, and it gave them a place outside of places like Yemen and Colombia and so on, where the world is a little more difficult. Statoil now is going to be a player where they want to be in the Gulf. So, when you have those situations, there is a tremendous win–win strategy. But I've found that if you can sell something to a strategic buyer, you'll almost always do very well.

A jump to scale. One of the strategic buyers was Helge Lund of Statoil, and he believes that well-chosen acquisitions are an essential part of strategy:

I think maybe size is part of the apparent conflict between growth and capital discipline in the majors. Statoil is growing from a much smaller position, but maybe with the skills of a major in many areas: our experience as an integrated oil and gas company comes from almost four decades of work with large and complex developments in a harsh operating environment. Statoil's success with complex oil and gas developments, increased oil and gas recovery, and acquisitions with growth potential and good strategic and industrial fit have given investors confidence.

You mentioned the EnCana deal: I admit that when we announced the deal, I was uncertain how the shareholders would react. The market's general reaction was positive, which I believe was driven in part by the good oil and gas price environment. Since then, I think confidence in the deal has increased. And why did we do it? We did it because we believe that we can add value to the deepwater assets in the Gulf of Mexico based on the experience and technology we have. We see similarities between some of our biggest legacy assets on the NCS and fields in the Gulf of Mexico. The deal allows us to put our increased oil recovery skills and competence into play.

The EnCana deal is a good example of a decisive move based on technology, skills, and experience. Previously, in 2003, Statoil

Profitability

Portfolio

made the major acquisition in Algeria. This deal also had a strong basis in our skills set, but even more so in our gas strategy because it supplements our supplies from the NCS.

With our big position in Shah-Denis in Azerbaijan we already had the possibility to supply Europe from the East. By moving into North Africa, we strengthened and diversified our gas position with assets from the South. This move was primarily driven by our gas strategy for Pan-European supply. I believe our investors acknowledge the logic. They see that rather than going on a big spending spree, we are looking at acquisitions that really fit with our key competencies and core strategy.

Growth

My philosophy, in simple terms, is that long-term exploration is the key route for long term growth and value creation. Then we use our skills and NOC position to team up with likeminded partners in resource rich regions, as we are doing in Algeria, Brazil, and Russia to mention a few. From time to time you will see Statoil make acquisitions.

But you should not have to think long and hard about why we did these acquisitions. You immediately recognize the logic and strategic rationale. In the Gulf of Mexico, we farmed into the Tiger well with Chevron in 2003, and subsequently they made a discovery. Then we made the EnCana acquisition, and subsequently co-bid on a few licenses in this same area with Exxon. At present Statoil is among the 10 biggest deepwater license holders in the Gulf of Mexico and we are starting to build a material position. We have the potential to produce more than 100,000 barrels per day after 2012, which really makes a difference for a company our size.

In the context of the political and fiscal risks we face in other areas, I believe the transaction makes a lot of sense in terms of balancing the international position of Statoil. We paid a good price, we like what we see, and I think the organization is even more happy today than it was at the time of the transaction. And I think our colleagues in EnCana are also very happy.

Setting sail

With clear goals established, and an understanding of the options for organic and inorganic growth, companies can set sail on their new strategic course. In times of change, the new direction may involve substantial transformation. Three examples follow.

Transforming non-OECD scope and scale. Ian Howat of Total described the transformation of his company through acquisitions:

> The current Total Group is clearly the result of the two mergers we did, firstly with Petrofina and then six months later with Elf at the end of the 1990s. I think that in doing the mergers, we were driven by two things: in the downstream business you don't have to be everywhere, but you have to be strong where you are, and when we started, outside their home countries, Total, Fina, and Elf were all at 3–4% market share everywhere. And of course, the market was now mature, so it was very difficult to grow organically. So that was the first driver for the mergers, that it got us up to roughly 10% market share in all the European countries, and Total is now the number one player in Northern Europe; and now in Africa we've overtaken Shell since we took on some of Exxon's business. We don't have to go further in these areas, though we may want to expand in Asia, but we now have a very comfortable position of being the number one player in the markets we're in.
>
> The other driver was that we perceived that we needed to bulk up in a world where Exxon was merging with Mobil and BP was merging with Amoco. The logic was to really create out of these three groups [Total, Petrofina, and Elf] a super-major. The super-major model really does have a meaning. One of the principal meanings is that you have enough scale and enough financial muscle to be able to take very big shares in very big projects, so that hopefully you can become operator, in a lot of different, very risky countries; because of course, most of the growth will come from very risky countries, and you don't want to put all your eggs in one basket. The fact that you need a lot of geopolitical risk dispersion means that if you want to do a lot of big projects with big shares in a lot of difficult countries, then you have to be quite a big player. And we think that the company we have created is now big enough.
>
> In fact, if you leave aside the OECD countries, and we recognize that's a big thing to leave aside, where we don't have the production of some other super-majors, outside the OECD we're already in absolute terms a bigger company than Exxon or Shell or Chevron. We're not now quite as big as BP because of TNK-BP,

though before that we were also bigger than BP. And when you think of all the hot areas, we're in the Gulf of Guinea, we're in the Caspian, we're in heavy oil, we're in LNG, and we're a big player in all of those things. So in all the things that are going to be big in the oil and gas business over the next 20 years, we're there punching well above our weight in terms of an oil company with a market cap of $160–170 billion compared to Shell and BP of around $250 billion and Exxon of close to $400 billion now. When we present ourselves to host governments, we look a much bigger company than we are, with a very big presence in those areas.

So we think the mergers have worked quite well, and that the super-major model does work very well. And of course, in many of those countries we have a long history. We're used to working with a lot of these governments in difficult countries. We're an extremely patient company. We found that some of the American companies, some of whom have since disappeared, when they came out of the U.S. and tried to do business around the world, didn't seem very clued up on how to handle the geopolitical issues. We have just launched the Yemen LNG project, which, depending on how you define it, took somewhere between 12–15 years to get going from its inception. The Nigerian LNG deal took something like 25 years, so we're a patient company and we're very used to dealing with these host governments.

So Howat thinks that they have achieved the critical mass and competitive strength that Total needs, and he is optimistic that their core competencies in working with difficult countries will propel them to a bright future. And the authors concur as long as Total can continue leading its peers in production growth.

Building OECD scope and scale. John Lowe describes the powerful change of direction that Jim Mulva brought to Phillips Petroleum that has resulted in the creation of ConocoPhillips, which is now either the smallest of the super-majors or the largest of the integrated companies, depending on your perspective:

Jim Mulva took over as CEO that summer, in July 1999, and he gathered key members of his staff or organization together and asked, what are we going to do? The fear was of becoming irrelevant. We said we really have done okay at Phillips, but with all the mergers and big transactions, even back then, you could see the advantages of scale: ExxonMobil needed 100 people in the center to run their company which was 10 times our size; we needed the same 100 people in the center to run our company. So you could just see the efficiencies.

Also, there were a lot of questions around commodity prices, and where we had been competitive in businesses before, we felt that that may be slipping away. Even in the midstream business, where we'd invented the business and been the leader for decades, we found that Duke Energy Field Services had actually gotten larger than us. We looked at chemicals, where we invented polyethylene and polypropylene, and yet we saw people kind of going by us. So the objective was "how do make sure we stay competitive in everything we do."

So we said we need to do something in all four business lines. How do we do it? We can't just go borrow money; we don't have the story to issue a bunch of equity. We agreed that we wanted to stay integrated financially and physically, and we thought we could achieve that by doing some transactions. So the leadership team agreed: what's most important to us is to have strong exploration and production; we'd really like to have a stronger refining arm because we think that that integration is particularly important; and we don't want to lose the physical or financial benefits of midstream and chemicals. But we don't have the financial capacity to do everything.

So we put the midstream business in with Duke Energy Field Services, and we pulled out $1.2 billion of cash; we took a 30% interest, but we got some financial flexibility. $1.2 billion may not sound like that much now with the size of company we are, but it seemed a lot then. Right on the heels of that we announced our 50/50 chemicals deal with Chevron; we pulled $800 billion out of that, so that got us a couple billion dollars of financial flexibility to work with. In December 1999, we announced the midstream deal with Duke and closed that in April 2000; we announced and closed the chemicals deal the second quarter of 2000. At the same time we announced the Arco Alaska deal right around February 2000. So that put the $2 billion to work quickly, and we were working all three of those simultaneously as well as trying to figure out what to do with our refining and marketing!

I believe the reason Phillips is still around is because of the Arco Alaska transaction, it's a world class asset, which doubled our reserves and doubled our production; it was a watershed moment in the company that said we were going to be around for the long haul. We still had to fix refining and marketing, and that one was tougher because we were really small in that segment. When we talked to potential joint venture partners, we said we'd always have to have the ability to move our crude into these refineries, and that's difficult because they say, "Well, but the venture needs to be able to optimize the venture's income and not to optimize your position in upstream."

Ultimately, the Arco Alaska acquisition changed our mindset on what we could be as well. We thought maybe we could acquire refineries and get that into good shape, but we were highly leveraged. With the Arco Alaska deal we were up over 60% debt to capital. Then we looked at the Tosco assets and said, "Here's an opportunity to issue $7 billion of equity that's not dilutive;" basically it was neutral to earnings, so $7 billion equity issuance solves the balance sheet. And they've assembled real refining assets. Tosco wasn't Tosco assets, these were major company assets: Bayway refinery was an Exxon asset, and Trainer and Alliance were BP assets, the West Coast were primarily Unocal assets, Wood River was a Shell asset. Not all of them had been maintained throughout the years, but they were major company assets. These were refineries that were all going to be survivors in the long term.

Then all of a sudden, we're a pretty good sized company, and we really have positioned all four of those business lines to compete. So that was kind of the first phase of why get larger—just to be able to compete and stay relevant and be able to go in credibly around the world to people who own the large resources, and say we're going to be a long-term player. We'd said that before, but people didn't believe us, so that was important.

Then we saw the opportunity to fill in with the Conoco merger and essentially double in size and add excellent people; the people resources were a significant issue with oil companies, even back then, to get the right people to run these assets, to further get efficiencies. We drove out close to $2 billion in efficiencies in these two companies, which created significant value for everyone and created a lot more financial flexibility for the company to be able to do these large projects which we think are the most attractive way to get access to long term resources, and which ultimately drive the value of the company. Conoco had just acquired Gulf Canada the year before, so now we have sustainable assets that can be built upon in most areas around the globe.

Since then, ConocoPhillips has added further North American natural gas reserves and production through its acquisition of Burlington Resources to become a very strong North American competitor in oil and gas.

Engineering a turnaround. John Wilder of TXU took his company through a classic turnaround:

So this 90-day process develops. Probably about 60 days into it, we started crafting kind of a forward view strategy and vision of each of the businesses. And we then communicated that, both internally and externally. Here are the businesses we're going to be in. Here are the critical success factors of each of these businesses. Is this a business we want to grow, or is this a business we want to just maintain? Here are our core four businesses. Here's why we think we have a structural advantaged position. What was unique about that business that made it a business we believe we can compete in over the long term? What kind of people and capabilities are required for that business? What kind of technology drives that business?

So we painted a picture at about 60 days of what each of our businesses would look like. And we have refined it over the last couple of years, but we stayed fairly true to it. It was a pretty decent piece of early work. Like in our power business, we painted that as being a business in which we were going to drive ourselves to high performance levels around production and cost.

We had a three-phase turnaround. First phase was portfolio rationalization, which is, get the 'where and how to compete' right. What business do you want to stay in? And then we had a whole bunch of financial stuff going on. We had these weird hybrid equity instruments, and all kinds of waste built into the capital structure. So we sold $15 billion worth of businesses, and we bought and sold $15 billion worth of debt and equity. We did all that in 60 days. So we were a big hedge fund over here, and we were kind of an asset manager and risk operating company over there.

So we did all that and said, "Now here's the core in terms of the capital needed to run this capital intensive business here that we want to be in." And then we said, "Okay. Next phase is performance improvement phase. And that phase will be zeroed in on driving each of these businesses." We measured a little crudely, but a decent benchmark. The nice thing about regulated businesses is there is much public stuff you have to send to all these agencies. So with a little bit of work, you can get some pretty quick high-level benchmarks. And we are kind of median to bottom quartile performer in a lot of dimensions. And when each business gets to top quartile—our ultimate goal is top decile—performance, then that business has earned a right to grow.

So until then, no capital. Performance improvement, period. And our vision, for our power business was, if we could drive these performance levels in terms of production cost against all other coal-fired power production across the country, it's $50 billion of value.

And as performance improved, he was able to start the third growth phase of the turnaround, the capital program described earlier.

Each of these examples represents an "extreme make-over" that goes beyond the portfolio reshaping moves described later. These make-overs yield companies that look and feel quite different from their antecedents, and they have created tremendous shareholder value in the process.

Conclusions

Ultimately, leaders make decisions. Strong leaders recognize that each strategic decision serves to adjust the course of the company. Leaders think deeply about the major trends that are emerging in the energy industry and position their companies downwind of them. We described the successes of Valero, Chesapeake, Suncor, British Gas, and EnCana in setting sail on the right course, then adjusting as they captured more information. We also noted the significant redirection of companies such as ConocoPhillips and Total as they substantially increased their size and scope through acquisitions.

Leaders clearly set the boundaries of the business to avoid distractions, while encouraging innovation and expansive thinking within the boundaries, and emphasize the need for continuous portfolio review. Within the portfolio there will be areas of focus that have the potential to reach scale, that can benefit from a consistent approach, and that have enough running room to warrant a sustained effort. There is general agreement that scale matters, that markets matter and those markets potentially can open up access to the more difficult resources.

Leaders refine the business models that will bring competitive advantage. In the overall corporation, this is known as the parenting advantage; for each business it will, in the energy industry, demand clarity about how to build and sustain the ability to supply incremental energy at lower cost than competitors. In each case the business model must address the value drivers for each industry segment and build the competencies required for sustainable, profitable growth. Both EnCana and BG emphasized that it is the combination of distinctive assets and distinctive competencies that creates competitive advantage in this industry.

We discussed how to build out the strategies and outlined two different points of view on organic growth—that difficulty in access will constrain growth, and that there are very promising growth opportunities in unconventional resources. It is certainly more difficult to find growth in conventional resources in traditional places. However, this reminds us of an observation by Dwight Eisenhower: "If a problem seems insoluble, enlarge it." Companies need to look beyond their traditional areas to find growth, and when they do, the challenge will be transformed into a suite of opportunities presenting a variety of strategic choices.

Acquisitive companies have created huge value for their shareholders by "bottom fishing" and acquiring valuable assets at low transaction prices. This stage of the cycle is over, and assets are now fully valued. Mergers and acquisitions with a clear strategic purpose and the potential for tapping unrecognized value will happen but will be difficult to find. Mergers based on exchange of stock avoid the price risk in cash transactions, but still present considerable post-transaction execution risk.

Ian Howat of Total remains optimistic for the super-majors' ability to adapt to changing conditions:

> I think the big groups, all the super-majors, have been very good at reinventing themselves since 1986. The oil companies worked very hard for the 15 years after the price collapse to get costs down so they could produce oil profitably at $15 per barrel. When you look at these very asset-heavy companies, there is a tendency to see the assets as the essential element of the company. I don't

think that's the case at all. I think the essential element is the human resource, and the ability of the human resource to be imaginative and to keep rethinking the model to adapt to a changing world.

I don't know if there will still be five super-majors in 10 years, but I do think that if there are, the things they will be doing will surprise us. We're going to have to keep reinventing the model. I am still passionately attached to the super-major model, the scale, the enormous range of skills that are available to you so you do have the ability to move with the times. But we're making it very clear to our people now that we will have to reinvent ourselves, and the way we have been thinking of things is not necessarily going to work in the future. The world is changing in ways that we did not predict. We will have different relationships with host countries; there is great uncertainty on how fiscal conditions will evolve and what oil prices will do; and then there are new competitors coming in. It's going to be interesting—in the sense of the old Chinese curse!

8 Executing the Strategy

Many books explain how to develop strategy, but few offer an exemplary model of strategy implementation. Execution is a difficult topic because it is complex and the details are different in each case, so that it is hard to extract lessons of general applicability. There are always many moving parts: leadership, the organization, its people, tactical decision making, operational choices and performance, customers, partners, suppliers, shareholders, community, government, and, not least, competitors trying to preempt or frustrate execution of the strategy. For example, Arco knew for a decade that it needed new sources of growth to build on its successful integrated strategy to produce, transport, refine, and sell Alaskan crude oil through am/pm convenience stores. Arco also knew that the most attractive growth opportunities lay in international exploration and production. However, the company was unable to gain access to the right opportunities, deploy the right resources, and organize effectively to find or develop new oil and gas fields to execute an international growth strategy. Ultimately, their leadership failed to secure the commitment of the right people internally and was unsure of the commitment of its shareholders to the new strategy. Unable to present a persuasive growth story, Arco was acquired by BP in April 2000.

Execution is also challenging because of the competitive interplay. Other companies do not stand still and watch while strategy is executed: they are busy executing their own strategies and competing vigorously for access to resources, to customers, and to partners. The theoretical basis for thinking about complex options in the face of competition is found in game theory. "In many cases, game theory can suggest options that otherwise might never have been considered. This is a consequence of game theory's systematic approach. By presenting a more complete picture of each business situation, game theory makes it possible to see aspects of the situation that would otherwise have been ignored. In these neglected aspects, some of the greatest opportunities for business strategy are to be found."[1] Barry Nalebuff encourages role modeling to understand the interests of all the parties engaged in business situations and to discover how to collaborate to create value. The essence of his message is to first figure out how the parties can collaborate to increase the size of the pie, and only then worry about how to divide it later.

Executing strategy in the new energy business environment will require collaboration with others. Partnering with resource rich countries and their NOCs can help gain access to resources (see chapter 2). There are important choices to be made about partnering with suppliers and technology providers. And society is increasingly demanding tighter collaboration with communities and civil society (see chapter 4).

Five Tasks for Executing Strategy

Following our "rule of five" as in chapter 7, we found in our interview process five key tasks behind successful strategy execution (fig. 8.1). These are

1. Securing commitment
2. Creating opportunities
3. Allocating resources
4. Organizing effectively
5. Addressing stakeholder concerns.

We set stakeholder concerns last not because it is an afterthought, but because it is the natural place from which to start a new iteration. Companies that address outside stakeholder concerns well will be able to present a more persuasive shareholder value proposition, will be a more attractive place to work, and will gain access to more opportunities.

Executing the Strategy
1. Securing Commitment • Internal (leadership team and employees) • External (shareholders) **2. Gaining Access to Opportunities** • Access to International Opportunities • International Partnering • Access to North American Opportunities • Executing Mergers and Acquisitions **3. Deploying Resources** • Capital Allocation • Defining the Extended Enterprise • Leveraging Technology **4. Organizing Effectively** • Accountable Business Units • Functional Excellence • Accountability and Control • Energy Supply Chain Integration **5. Addressing Stakeholder Concerns** • Public Awareness • Community Relations • Incident Management

Iteration & Review

Figure 8.1: Executing the strategy

There is a saying around kitchen tables: "The chicken contributes to a bacon and egg breakfast, but the pig is truly committed." We need to differentiate commitment from contribution, cooperation, and compliance. If strategies are to be successful, then those charged with implementing them must be committed, and not just complying with instructions. Similarly, major investors must support the strategy. If they withdraw their support, the stock price will plummet and it will be difficult to finance the strategy.

Strategic progress in any direction is impossible without access to opportunities, and this is becoming increasingly challenging as traditional areas mature, resource nationalism restricts access in other areas, and INOCs compete with increasing intensity in the remaining international areas. Competitive intensity is also high as energy companies seek to deploy their investments into resource plays in North America. Mergers and acquisitions are more difficult to justify as market values approach or exceed mid-cycle asset values. In our view, a critical capability in this next cycle will be to enlarge the problem statement, identify all the stakeholders with an interest in its resolution, visualize the ideal outcome, and then accommodate the claims of the various stakeholders. It will be very difficult to implement strategies if companies adopt a win-lose posture and view every interaction as adversarial.

In order to capture opportunities and realize their value, companies must deploy resources effectively. Capital allocation is at the heart of the choices that energy industry leaders must make as they execute strategies, and may, in fact, be the single most important strategic business process in the energy industry. Strategizing, evaluating capital projects, and budgeting should be, but often are not, seamlessly integrated. In the current environment, leaders will be called upon to challenge their organizations to create value from an increasingly difficult set of opportunities— to "make lemonade when presented with lemons." They will need to move from screening projects out to screening projects in.

Companies also need to make important strategic choices about how to organize internally and where to set their priorities in developing and nurturing competencies and in managing such enterprise extensions as suppliers, business partners, technology providers, and customers. We heard from many of our interviews that availability of skilled people and specialized equipment were a significant constraint on their growth. The superordinate goal should be to expand the availability of these limiting factors worldwide, and this requires looking at the full supply chain that determines capacity for each scarce element. The internal issues include the roles of outsourcing and insourcing, determining which capabilities to retain or strengthen internally, and accessing and developing needed technologies. These choices must be shaped by a common philosophy and clear understanding of the value drivers for each business segment. The same ideas will also inform the organization required to execute the strategy: whether to be centralized or decentralized, integrated or functionalized, insourced or outsourced.

Finally, and this is a recurring theme of this book, leaders must interact positively with all stakeholders in the strategy, particularly the public and local communities. Leaders are especially called upon to take the helm at times of crisis. This industry operates facilities that transport and transform hazardous fluids that are often toxic, corrosive, highly pressured, hot, and explosive. Accidents can harm people, property, and the environment. Most companies have a goal of zero incidents, and many of our companies start each meeting with a "safety moment" that is almost akin to morning prayers. Yet the leaders recognize that despite their best intentions, accidents still happen. When they do, our leaders must personally take the helm and address the concerns of all stakeholders.

Securing Commitment

Strategy implementation requires the commitment of two key stakeholder groups, employees and investors. Leadership is about change, and change is uncomfortable for both groups. Change threatens employees. It may involve exiting businesses in which employees have skills, making them redundant; it may demand entry into new areas, requiring employees to learn new skills, or exposing them to competition from new hires. Change also threatens investors, who may have bought the corporation's stock to fill a specific need in their portfolio that it no longer meets; investors may have to search for a substitute with similar characteristics, or even rearrange the whole portfolio. This section addresses both these issues in turn, drawing on comments from our interviews.

Internal commitment

After deciding on a direction described in chapter 7, the conversation now turns to delivering results by executing the strategy. The key issues in this section are alignment of the strategy with employee personal beliefs and goals, applying the principles of change management, and overcoming the obstacles often presented by organizational inertia.

Enlisting people

Any conversation about change strikes deeply into the personal beliefs of the executives concerned, and conversations can (and really have to) become emotional. Enlisting people is about satisfying their needs in a way that drives their behavior to achieve the organization's vision. If the company sets a stretch vision, with a BHAG (Big Hairy Audacious Goal), how will that affect the balance of time between work and family? If the company decides to mold our business model into new shapes and forms, can the existing management team make the changes needed in their behaviors? If the company has been proud of its consideration for employees and their need for long-term job security, how will it deal with those who do not make the effort required to create the value posited by the new vision? If the company has been rewarding managers for controlling costs and aiming for high returns, how does it compensate someone for taking on a risky new-growth venture? And how does it respond if a manager runs up a lot of costs and fails? If the company believes that the most attractive growth area is in West Africa, but that it will require the best talent on site to be successful, is it prepared to divert this talent from existing projects, and will the talent be prepared to go?

We introduced the concept of "the left hand column" in chapter 4 as a useful tool in understanding conversations between companies, NGOs, and governments on the environment. It is also a vital tool in assuring productive conversations internally on these touchy issues, since the leadership must be aligned in order to achieve difficult goals. David Simon, John Browne's predecessor as CEO of BP, once wrote to the Harvard Business Review, "In my view, more wealth can be created by aligning purpose, process and people within an organization than most of us realize."[2] He goes on to say that "strategy is not about pie charts showing a pre-determined distribution of assets. Strategy is a continuous process of choices. . . . It is iterative and interactive, with management setting the challenges and employees on the front line responding with the ideas." To be effective, there needs to be agreement on vision, purpose and goals, and the conversations need to be open and honest between corporate management and the "employees on the front line."

Overcoming organizational inertia

The discontinuous change in the business environment that is currently in progress should be infusing energy companies with a sense of urgency for change. Organizations the size of the super-majors, however, must overcome the inertia of the present course and accumulated ballast before change can occur. In this section, we address the process and activities that need to take place to overcome organizational inertia and initiate productive change.

Commitment to change does not just happen; it requires attention and hard work in the field that is generally referred to as change management. John Kotter proposes an eight-stage process organizations can follow to get past the inertia and into needed changes:[3]

1. Establishing a sense of urgency
2. Creating the guiding coalition
3. Developing a vision and strategy
4. Communicating the change vision
5. Empowering employees for broad-based action
6. Generating short-term wins
7. Consolidating gains and producing more change
8. Anchoring new approaches in the culture

To execute the strategy, companies will have to pay full attention to Kotter's stages four through eight. People will be committed when the change vision is persuasive, when they are empowered to contribute to implementing the change and when the short term wins give confidence that the change is working.

In communication and empowerment, there is a wide array of tools that leaders can use that go well beyond written communications and town hall meetings, as set out by Joan Lancourt and her co-authors.[4] At the top of the hierarchy, leaders must demonstrate their personal commitment by role modeling the behaviors and actions required by the new direction and induce the desired change in others by projecting confidence. Clearly, change will always require persuasive communication using a variety of forums and media. The more this can be two-way, the more the messages will be internalized.

Gwyn Morgan described how EnCana's decision to exit from international businesses received the full commitment of his management team. "The management team was very much on board. In fact, interestingly enough, our leaders in the North Sea were saying to us, 'Look, we've sat here and we've looked at what you can do in North America. We can't match that. We can't ask you for more money for the next play because we can't deliver, can't compete with the other opportunities.' And these guys were well taken care of." The decision came out of an honest, productive conversation about the future, and was founded in trust that everyone would be fairly treated. Consequently, those responsible for selling the properties were fully committed to a successful outcome.

Companies may also restructure the organization to align better to the new strategies and modify the performance measurement and reward system to encourage the desired behaviors. Kotter's empowerment stage should include participation, deliberately involving a broad range of people in working through the details so that they feel ownership of the results. In addition, if new skills are required, there is an obligation to provide training so that existing employees have the opportunity to grow into their new responsibilities. Kotter rightly is critical of many efforts: "Training is provided, but it's not enough or not the right kind, or it's not done at the right time."[5]

In any organization undergoing change there will be four camps: those that are truly committed to the new direction and want to make it succeed, those that are generally supportive and are pleased to be engaged in the process, those that are accepting, and are waiting for more detailed instructions, and those that are truly opposed. Carrots generally work better than sticks for the first three groups; but there may be employees who simply cannot come to terms with the new direction, as John Wilder, CEO of TXU Corp., found.

Wilder lamented the difficulty of enrolling middle management in his transformation program at TXU:

> It has been a really hard process. We started cleaning house
> at the top. We designed the new organization, widened the span
> of control, empowered the people and so forth. And we ran a

systematic severance process, using the same kind of standards as earlier programs, but yet the culture still isn't driving performance.

For example, we know there are employees who are counterproductive to our organization. They aren't on board with where we are going. Their fellow workers know it. So let's identify those employees. We ran through that process and only identified one employee. I told my management team, "This is ridiculous. I know 5% of our employees fit that characterization. Are you telling me that out of 14,000 employees we could only find one?"

So the transformational change is going to take a long time. It's about changing the behaviors of thousands of managers. I know this gets studied a lot in academia, but that concept of the resistant middle is real. I get letters from employees all the time. "Thanks for changing the culture. Thanks for making the *suits*"—that's one term they use—"making the *suits* work."

Creating stretch plans with performance feedback

Finally, what drives performance is having well-defined plans that stretch the organization to achieve the vision. People like clear goals that they can relate to in their work activities and that correspond with the strategic direction they are hearing from their leadership. At the corporate level, this process culminates into a short-term operating plan and capital budget for the year containing key targets that the leaders commit to achieving with their board of directors. This is the well-known process of management by objectives, but creating goal alignment both vertically and horizontally within the organization is easier said than done. Larry Bossidy[6] of Honeywell insists on the importance of regular personal contact between leader and staff to discuss performance goals and how to achieve them. The reason the plans need to create some stretch and challenge is that performance tends to reflect the quality of the goals. Mediocre objectives achieve mediocre results at best. The only way to achieve high performance is with high performance goals, but these goals need to be meaningful. Goals that focus on key value drivers supporting the strategic business model lead to successful strategy execution.

Nick Butler talked about how to encourage performance from BP's far-flung business leaders: "You have to set boundaries, but within quite broad boundaries they have to have trust and clear responsibilities to be able to deal with events and circumstances. Other things that can help are clarity of expectations, clarity of their leadership. So you give people space, but also a line up if they do need help;

an understanding of where they fit in the company. So they understand the overall strategy and which pieces are needed now and which later. And then give them praise, which is also very important."

A strong performance assessment process causes goals to become real and secures commitment throughout the organization. People become much more committed when they recognize they will be held accountable for the results achieved. The performance assessment process should focus on learning and driving improved future performance rather than simply going through the motions of comparing plan to performance. Leadership becomes the crucial ingredient in developing a dialogue that brings out the "brutal facts" when necessary, but focuses on improvement rather than punishment.

External commitment

Any change in direction requires the support of the investors. It is risky to change direction because investors who bought into the previous direction may sell before the new direction has proved its value enough to attract new investors. The result will likely be a short-term stock price decline. Managing shareholder expectations and communicating likely future actions in advance without violating SEC rules or giving away competitive information requires a complex set of skills and competencies. It is noteworthy that the largest companies are quite frugal in the information they provide; but the information becomes more and more granular in the smaller companies, with new IPOs often providing detailed information on specific drillable prospects.

A number of CEOs seem quite contemptuous of the financial community and seek to provide as little information as possible. They believe (rightly) that it is dangerous to promise results that may not be delivered, and (in our view wrongly) that a full exposition of the persuasive shareholder value proposition is not necessary or would provide useful intelligence to competitors. Burlington Resources found a good intermediate position.

Burlington Resources (BR) did well by its shareholders in choosing a clear value proposition, setting a firm direction of moderate growth with sector leading returns, and finally selling out to ConocoPhillips in a deal that provided a cash return and continued exposure to the sector. They also benefited from a very thoughtful and well structured investor relations program (see box). Ellen DeSanctis was in charge of investor relations (IR) and strategy at Burlington Resources. She explained her approach to the job:

Objective

Simply put, the job of an IR practitioner is to lower the firm's cost of capital by making equity expensive. That's really the main role of IR. But you can do it a lot of different ways; and as I matured in the job, it became more and more obvious to me that IR sat in a unique position in our firm—part communicator, part chief of staff, part strategy advisor.

The basic premise of efficient capital market theory is that markets are rational and information is perfect. IR people likely believe this to be valid in the long run, but not necessarily the short run. However, a huge aspect of our IR strategy was to provide reliable information, set realistic expectations, and then deliver on both. The more IR can have a role in setting expectations, the better. IR can greatly accelerate value recognition and drive multiple expansion by managing expectations and effectively communicating the performance against expectations.

Values

You have to have a consistent story to all investors. At BR, IR was managed very much in accordance with our values about integrity, about personal accountability and responsibility, even about morality. To me, the essence of corporate morality is being consistent and fair in all situations. It would have never felt authentic, defensible, or genuine to tell one story to a growth investor, and another to a value investor, and another to a hedge fund.

Positioning

At BR, we did not try to be the perfect or the only energy investment. I only wanted to be the best wherever we competed. I recognized that there are many ways to own energy. I aimed to be the stock within the energy sector that people choose if they wanted low relative reinvestment risk; if they wanted modest growth but relatively high returns; if they wanted a more consistent, stable performance, and if they wanted a gas-prone story. We didn't offer people a lot of exploration growth. We didn't offer people a lot of international optionality. I wanted to be the "sleep-well-at-night" E&P stock. For a while, we dominated that space.

Signaling changes

If there are changes in strategic direction planned, you want IR in the loop on these discussions. Managements can benefit by getting input about how the analysts and investors will react to strategy shifts. Internally, organizations can benefit by having an early warning system that says, "What we've been doing may not be sustainable, or is at risk for being out-competed." With strategy shifts, you want time. Fast direction changes are the death of IR.

The real power of having time and having good relationships with the marketplace is that when you do a large transaction or implement a strategy decision, it's possible to have a whole bunch of people who actually take credit for thinking it was their idea. You've enlisted them early by asking their input, their feedback. You have people saying, "Well, I understood that BR was going to do this." People taking credit for thinking this was the smart thing for us to do. I wanted the market to trust our every move.

Fair disclosure

I think every IR person will tell you, the biggest single change in the IR field over the last decade is the emergence of regulations for disclosure. Reg. FD was implemented in very late 2000 to eliminate selective disclosure. Before FD, analysts called and you could say, "Hey, your model looks a little out of line. Why don't you go through and tweak it and let me know." You can't have any of those conversations now unless you have them publicly. You can't say anything material and non-public to anybody.

Internal alignment.

This might be the key to it all! There is no doubt in my mind that BR was a stand-out company because we got the message right for this company, we got the assets lined up with our capability, and we got the leadership style right. Our models, our accountability systems, the compensation systems all lined up with the assets, lined up with the capability of the people that run the assets, and lined up with our values. This alignment was absolutely critical.

Not making IR a central part of the alignment process is a big mistake, in my view. In fact, one of the best things IR practitioners can do is get employees on their side. In BR, we worked hard to make all 2,200-plus employees think and act like IR ambassadors.

Investor segmentation

Traditionally, there were two kinds of energy investors. "Indexers" (market weight) had to have energy whether they needed it or not—these folks generally wanted relatively low risk, safe, good returns. And so their investment decision was around, "What relative returns do I need? What yield do I need?"

Then there were about 50 people, we used to call them the "Band of 50," who were money managers with particular expertise in energy—these folks were more discriminating energy buyers, willing to be above or below market weight based on their view. They were cycle players, and they had very deep knowledge of the sector.

And then the last five years, we had a new class of investor come very heavily to the sector: hedge funds. These folks made decisions on many different kinds of technical metrics, a lot of them around classical arbitrage: "Where are the relative values? I like gas. I'm going to be long the company that trades at a discount to BR. I'm going to be short BR, who trades at a premium." I actually spent a lot of time with hedge funds over the last few years. I wanted to know how they made their investment decisions. Obviously, with a mind toward avoiding being the company that was the short side of their trade.

Process

I do think technology, conference calls, Web teleconferencing, video conferencing, may change some of how IR is practiced. But fundamentally, I ran IR like a very strong customer relations function. I had a whole process for it.

My formula for IR was based on the acronym CUE. My job was first to *clarify* who and what we were. Why did we exist? What was our business purpose? Was our business authentic, defensible? Then it was important to *unify*. As I said before, we had to get the message right, get the communications working, and get alignment internally and externally. And then we had to *energize*. We made the annual report bright and colorful and bold. We got the charitable programs going. We got the people out into the communities wearing shirts that said "Burlington." We held hundreds of meetings with investors.

So, *clarify, unify, energize*. That worked very well for BR. And then we made sure that our board was informed through management.

Creating Opportunities

Executing the strategy is about making things happen. With a committed organization behind the vision and strategy, the next step involves tactical maneuvers to open doors and to create business opportunities. In the energy business, many opportunity domains can be pursued, but one of the biggest arenas remains getting into energy rich resource countries.

Accessing energy rich countries

Steve Lowden, CEO of SER, emphasizes the growing competition for resources from inside the producing countries themselves and from the emerging giants of India and China:

> I guess the first and most obvious trend is the changing access to resources, and the trend that the owners of the remaining resource today want to play a bigger part in investing in those resources. Not just the owners as the government, but their people. Their private enterprises want to play in the business. And really what has been holding up Project Kuwait is just this. The Kuwaiti private investors are saying, "Why can't we invest in this project? Why do we need to have the oil companies? Or why can't we invest alongside them?"

> And of course the longer oil price and gas prices stay high, the more wealth those countries generate, and the more they want to do just that: the individuals in those countries and the governments. I see that trend accelerating. And the longer the access becomes difficult, the higher the oil prices are going to get. That means more money is going to end up in those places making them more motivated to invest in their own business. So this is a trend that is inexorably going in that direction.

> And being ahead of that is very much about Russia, India, and China. To play in those countries, you have to play with those governments, and with those private investors. You see it in Russia. A few years ago it wasn't clear. Now it's pretty clear. And you can see it in India. It's very, very clear. Some of the biggest companies

in the world are now Indian. The biggest refiner in the world is about to be Indian. The biggest steel company is Indian. And a very significant part of my capital is coming from India. So I know this firsthand, as it were!

The other interesting trend is that the available resource in the developed world is very clearly in steep terminal decline. But it is difficult to find growth from remaining new resources in the developing world. Those countries have become more confident. They want to do it themselves.

With the intense competition, energy companies need to tailor their approach to the needs of the stakeholders in any country where they seek to gain access to potential resources. One way to address the issue is to map separately the hierarchy of needs of the national oil company, the host government and the citizens. There is a case to be made that energy companies should be prepared to deepen their engagement in the economies of host countries in order to respond to their needs. Certainly, the competitive intensity for access to international resources is increasing, putting pressure on margins for conventional resources in traditional places.

Understanding NOC needs

In our experience, national oil companies evolve a hierarchy of needs as they become more capable (fig. 8.2). In the beginning, an NOC just wants to survive. Then it seeks to wrest some control from the ministry so that it can take shape as a quasi-autonomous entity and build an independent culture, and the employees can begin to feel secure in their jobs. When all that is achieved, the NOC will start focusing on its primary functional goals of expanding hydrocarbon production capacity and securing markets for the hydrocarbons produced. Then, with a bit more bounce in its step, it will likely look for diversification opportunities into refining and petrochemicals.

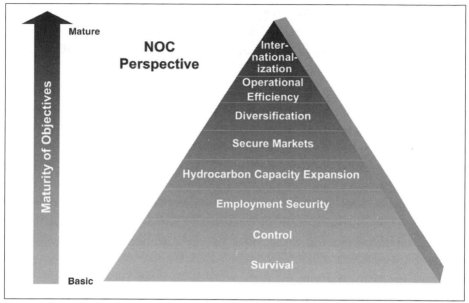

Figure 8.2: National oil company hierarchy of needs. *Source: CRA International, Inc.*

Generally, a little later, the company will look for benchmarks to test its operational capabilities and will start measuring itself against IOCs. As confidence in its abilities grows, it will tolerate the idea of competition and opening some of its acreage to foreign bids in exchange for more freedom of maneuver. Finally, particularly if the indigenous resource is limited or non-fungible, it will internationalize. As it progresses, it will behave increasingly as a commercial corporation and less as a government bureaucracy.

Each NOC is different, and it is important for energy companies to understand where the NOC stands on these dimensions to tailor a partnership value proposition. As outlined in chapter 2, INOCs must be considered as equal business partners (and competitors) by IOCs. National operating companies (NOCs) may need specific technologies, or access to international downstream markets. National supervisory companies (NSCs) may be most interested in developing their own competencies through open alliances and training. In all cases, personal relationships are of paramount importance.

Understanding government needs

The government has a different hierarchy of needs (fig. 8.3). In most countries, it is wise to network among energy, finance, and other ministries to assure a full understanding of the government's needs. The issues start with national security;

then the imperative of internal political stability is layered on. National prestige and image are early additions to the government's need hierarchy. Beyond these basics, the serious business of government must be addressed. First, there is assuring that jobs are created; and here the lure of populist, socialist dogmas can be a siren call. Next comes accessing stable sources of revenue; as noted previously, in an oil-rich country, the separation of revenue generation from the population is potentially a "moral hazard" for government. With stable revenues, government moves to address social welfare needs such as health and education. At a more sophisticated level, governments look systemically at economic development and create an environment in which the private sector and the public sector can work together to enhance economic welfare for the country. Finally, and regrettably it is often thought of as a luxury, environmental protection becomes a core value in the community. Energy companies need to be sure that they are aligned with the real priorities of the government they are dealing with.

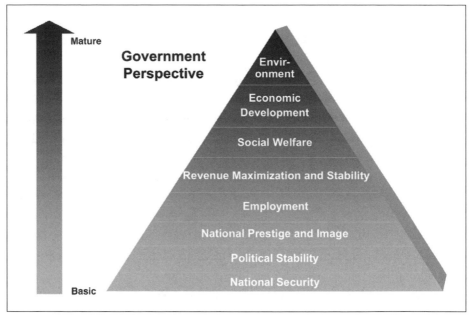

Figure 8.3: Host government hierarchy of needs. *Source: CRA International, Inc.*

Balancing all the needs

So energy companies wishing to do business in resource rich countries have to respond to the potentially different value hierarchies of the government and the NOC, and also be aware of the needs of the citizens and local communities. NOC management wants to create an effective institution; the government wants to use the NOC as a means for satisfying national policy needs; and the citizens follow the established Maslow hierarchy of needs described in chapter 1. Regardless, energy companies will be expected in many difficult countries to provide what governments have failed to deliver. The new INOCs appear to be further along in understanding this than many IOCs and other shareholder driven companies. Balancing the needs of these different stakeholders is always a challenge. It is difficult to match the INOCs in terms of infrastructure development, but IOCs can offer education and training with greater credibility than INOCs, can also help develop the capabilities of local private firms, where that is a government objective, and can extend their investments downstream into pipelines and power if such investments are economic.

Developing winning value propositions

Total accepts that they will need to change the value proposition offered to resource owners. Ian Howat said:

I think one of the things we're very conscious of right now is that host governments are becoming more demanding. Obviously they're going to become more fiscally demanding if the oil price remains high. But also more demanding in what they are going to want an oil company to do. They're not just going to want us to do E&P. Some governments are now saying it would be fine if we could do some refining, perhaps some petrochemicals or some power plants—we've done power plants, for example, in Abu Dhabi, which is one of our core countries. So our attitude is, why not? We are an integrated oil company, and we have all these strings to our bow. Obviously we'll look at the overall economics of the total package, but we're not dogmatic about going into this country or that country just to do E&P. What we're trying to do is to see what the host government is looking for, and to try to see what the fit with our aspirations might be, to find some kind of win-win situation. So I think that's going to be a big development over the next five to 10 years; we're going to see companies getting much more involved in the framework of the host country, rather than operating outside the local economy, apart from taxes paid on production.

Of course that's going to create a lot of risk that's going to have to be managed. Our investments in gas transport and power stations in Argentina, for example, cost us a lot of money because there were changes in government regulations there. It's a less comfortable existence. You're often dependent on the health of the local economy, and if the local economy comes a cropper, as it did, the government is likely to react. And you don't have the comfort of just producing oil offshore, putting it into a ship and going off with it, and not being subject to the risk of not being paid. So there are some quite different issues to be handled.

Ultimately, companies are also going to be associated with their home country, and local foreign policy will play a part. Chevron CEO Dave O'Reilly observed: "Strategically, this is a key point; we're still identified as an American-based company, and there are advantages and disadvantages. That parentage has value in a global business environment that we have to think carefully about. In the same way, BP is very definitely a British company. It is British Petroleum. Total is a French company. That doesn't mean that these companies don't have major operations outside their home countries. Of course they do; but anyone who thinks they are not primarily identified with their home country is being unrealistic."

Howat of Total saw both sides of this: "We all wave our home flag when it is convenient, and keep it furled when it is not."

And this brings up another important example of "the left hand column" outlined in chapter 4, as evidenced by remarks attributed to Vladislav Surkov, deputy head of the Kremlin administration: "Sometimes [Washington's] words and thoughts do not coincide. For example they talk to us about democracy while thinking about our hydrocarbons."[7]

O'Reilly draws some bright lines on the scope of corporate engagement:

> In thinking about company engagement in resource rich countries, you have to separate the economy from politics. You have to stay out of local politics. Let's be clear about that. I think the issue for us is to play a constructive role in the economy, but to be very mindful of what our capabilities are. What we do is invest; by investing we create jobs; creating not just direct but indirect employment by the fact that local people or companies can build businesses around supplying the key products and services that we need.
>
> I recognize that historically, much of the supplies and services came from offshore. But more and more this is coming onshore, because the capability is increasing. One of the roles we can play is try to help that capability increase in these countries, and not be blind to the fact that there are opportunities. For example, we just approved an appropriation request yesterday for a facility

offshore Angola, for which the jacket will be fabricated in Angola. And it makes good sense and good economics to create these kinds of opportunities. We have been giving more and more of the basic facility manufacturing work to Angolan yards. That helps build capability. Then they can build from there to do more complicated functions.

So that's one area. The second is that by producing oil and gas, we generate profits and pay taxes and royalties that help the government fund its infrastructure needs, its social programs. If all is well managed, that's fine. Of course sometimes it isn't managed well. If it is not managed well, my view is that it's the role of the people that live in the country, and maybe some of the international multilateral organizations, to address those issues. Our job is to do what we do well, which is to invest, create jobs, create relationships with people who can provide support, and build capabilities.

Every now and then we have stepped outside that role. In Angola, for example, when the civil war ended, there was a unique opportunity to work collaboratively with a number of NGOs to try to address some of the pressing needs of post-war Angola. One of those needs was security of food supply. So we put together a partnership initiative in Angola, which involved collaborating with USAID and a number of NGOs who were experts in food production and education to be the catalyst for helping to jump start the country after a very difficult period. People asked, "Why did you do that?" Well, it was good for our business. We needed to help. We're not making any policy decisions for the government. We were just trying to provide an impetus to restore agricultural production to the country and provide a stimulus to some of the education needs. Also, we leveraged our contribution by multiplying it by the efforts of other organizations, and let the expert organizations manage these processes because they are the experts in them.

While traditional energy companies are changing their value propositions, so too the INOCs and their government owners are widening the scope of their engagement in resource rich companies. As Lowden described, local investors are entering with advantages of alignment with societal needs, and NOCs are becoming more capable. In aggregate, energy companies will need to bring truly exceptional competencies, such as specialized, unique technology, as well as creative partnering to the table.

Developing creative international partnerships

IOCs may approach international partnering by joining with the local private sector to form a company that retains its local identity. BP, ConocoPhillips, and Suntera each took this route in Russia, and analogous opportunities have occurred elsewhere.

Nick Butler described BP's experience:

> We tried to go into Russia as a private entity and soon found that it was impossible to work there simply as a foreign company. It was very difficult, quite painful, but very important as a learning experience.
>
> So we came back from that and decided Russia was still the place where there were resources and what we needed to do was to do it as a Russian company. That's why we established TNK-BP, and it's doing very well. The challenge is that TNK is owned by three individuals. So on one side of the table there is BP, a large institutional company, and on the other side there are three billionaires, which is an interesting phenomenon for us. And behind them there is Prime Minister Putin and the Russian government with all its complexity. But so far it's working well. It's producing more than $1 billion in cash, which we can repatriate, and we're reinvesting a lot. And we have a lot more that we've found. We applied western technology and have found another field below Samotlor, the largest and oldest field in Russia. The challenge is to manage the political environment in Russia, which will sequence the developments to their needs not to ours, and that's a continuous negotiation.
>
> The model hasn't been copied, partly because of the Russian government's actions against Yukos, which has discouraged other investors, and partly because since we did our deal, there seems to be less political appetite in Russia to allow a major foreign investment on this scale. Russian politics change. We were lucky to get in when we did. Some people say it's a great risk. It probably is. But we've now brought back quite a lot of the money we put in. The challenge is that we could probably double up our investment, because the resources are there. But that would take it from 10% to 20% of the company, and that might increase the market's perception of the risk to BP as a whole, which is not something we would want. So we feel the need to go through quite a period of bedding in and actually

showing it can work, showing we can be good citizens, building our links, being there before we grow it enormously.

The model is not transferable. I don't see any other countries that will hand over their patrimony to individuals or to international companies like BP. They might need us there, but we are going to have to work there on terms that are negotiated with governments. So will everybody else; we're not unique. That will be the case in Iraq when it settles down, and with the Saudi gas when it opens up. I can't see any country going back to the old model. And every place is different.

John Lowe of ConocoPhillips has a similar philosophy and says their deal with Lukoil started with modest expectations:

We entered discussions with the attitude: we don't need to do anything in Russia, the company is fine, we're doing great; but if there is an opportunity to do something, then we'll see if there is something we think adds value.

I think that attitude served us well with Lukoil. We weren't trying to be the big brother and tell them what to do. They actually came to us, they were interested in a transaction, and what we said was it has to be meaningful to us if we're going to partner with somebody, so the actual transaction was much bigger than the one they envisioned. We said we want access to resources. With 20/20 hindsight this is even better than we thought, just as Arco Alaska has turned out even better than it seemed at the time; it could all turn south years from now, but I don't think it will. At the time it appeared it was going to be difficult for foreign companies to get access to the best Russian reserves. Even as we negotiated, it got more and more difficult, and now it's almost impossible for foreign companies to get access to the best reserves in Russia. That might all reverse itself over time but certainly not in the near term.

We felt that a Russian company was going to have much better success at getting access to the key reserves in Russia than a foreign company would, so who is the best company to pair up with? Lukoil was clearly the best company. They were willing to do the joint venture on the YK field, which we'd been discussing for several years, which is a nice oil field, and then have us make a small equity investment in the company. Arco had done something similar several years ago, and that didn't go well for them, so Lukoil had some concerns. We said we need meaningful participation in Russia through this investment, and it can't preclude us from doing other things such as Shtokman or any other opportunities; so we want to be able to get up to 20% of the equity in the company and have a broader joint venture. We think that working together we can drive improvements; we think if you do adopt western practices, that you will gain some efficiencies, and our only way of getting a piece of that value is to own a significant piece of the company. So the 20% was negotiated, as was the broader JV. Government approval is required to go over 20%, so 20% was a logical break as well.

The government was engaged in the process, and that was another thing we did right; we asked permission instead of begging for forgiveness. We spent a lot of time meeting with government officials and getting their assurances that they wanted us. What were the success factors? I would say that our openness to find a win-win solution and our making sure that we assessed who all the stakeholders are, and addressed their issues and concerns.

Steve Lowden continued with the Suntera and Sun Energy Resources story: "Now to get downwind of those resource nationalism trends, we've partnered with the Indians. We've partnered with the Russians. The value proposition is privileged access by (a) bringing the skill sets to these entities and (b) partnering with them to help them get access to markets and to the resource." Lowden believes that his company is better positioned to offer this service than are the established IOCs:

We have created a company called Sun Energy Resources, which is a global energy investor very much looking at energy from a private equity investment perspective. So we are going to be a money manager in the business—not an operator, but a money manager. And we are going to buy into a Russian venture. And that Russian company will be our Russian tool that will do our investments and our operations in Russia and be our conduit for growth there. You can't use the word 'controlling' interest anymore, because it has negative connotations for the other partner. By definition, if I am the controlling interest, it means the other guy is not. It's a partnership, in my mind. Not a majority interest, but a partnership interest—an interest by influence.

The idea of partnering with the local private sector is also gaining traction in other countries. John Lowe talked about ChevronPhillips Chemicals initiative in Saudi Arabia:

> Clearly, the chemicals, I think, follow a similar model in that they went into Saudi Arabia 50/50 with a group of very influential Saudi businessmen, and that has worked extremely well, because this is their own money that they're investing and they're very influential in the country. They don't want to lose their money so they are looking out for our interests. It's a 50/50 venture. We got non-recourse financing; the Saudi businessmen obviously wanted the best financing they could get, so we've got some government loan support and very low interest rates from the local banks. The deals just turned out tremendously; we've even asked for a second deal with them so we could double down, and we're looking at a third deal that we could double down again. When you need local content in some less developed countries, it's very difficult to find businesses that are really legitimate and can carry their own weight; but certainly in Saudi Arabia you can find that type of person, and in Russia, Lukoil certainly represents that.

Partnering is also taking hold as IOCs seek to gain access to the fast growing markets of India and China, which are committed to opening their markets as part of the conditions of membership of the WTO, which India joined in 1995 and China in 2001. Shell and ExxonMobil have important joint venture refining and chemicals projects under development in China, but IOCs have not been successful in penetrating these markets using traditional structures.

Jeroen van der Veer of Shell emphasized partnering when describing his company's activities in China:

> In China, we have invested $3.5 billion in developing a wide range of businesses, with a range of Chinese partners. This includes completing the major Nanhai petrochemicals complex in Guangdong together with CNOOC—the largest Sino-foreign joint venture in the country—on time and on budget this year.
>
> We are using our expertise to develop the Changbei tight gas field in the Ordos Basin with PetroChina, and aim to start supplying gas to customers in Beijing and other cities next year. And we have also established a joint venture in Jilin province to pursue the potential of developing China's oil shale resources, the

world's second largest. Our retail venture with Sinopec in Jiangsu province is already serving customers in 200 service stations, with 300 more planned.

Shell is now the leading international energy company marketing lubricants in China and has third largest share of China's rapidly growing lubricants market. A joint venture with Shanghai Automotive will set up a network of fast lube service outlets modeled on our Jiffy Lube chain in North America. Our bitumen business is also growing rapidly. Together with Sinopec we are building a coal gasification plant to provide feedstock for a fertilizer plant in Hunan province, and have licensed our technology to 15 companies in China. And with other Chinese partners, we are pursuing the potential for combining our coal gasification and conversion technologies for coal-to-liquids.

We believe that "win-win" value propositions will be required in the future, and the face of the company as well as a significant proportion of the equity in many places will need to be local, not global.

Leveraging knowledge in mature regions

To paraphrase *Monty Python and the Holy Grail*, North American resources may be maturing but "are not dead yet." There are substantial opportunities in what have become known as resource plays: specifically tight gas and oil sands. Until recently, it was possible to lease large tracts of land holding unconventional resources at rock-bottom prices. More recently, that has become more difficult. In order to assure value creation, it is important to have at least sufficient and preferably superior knowledge on the resource potential and development costs. The industry is very familiar with the "winner's curse" phenomenon associated with public auctions, where the winner often is subject to regrets for having bid significantly above competitors, "leaving money on the table" and creating a challenge for future development economics. It is clearly preferable, though not always possible, to negotiate directly with the seller.

Gwyn Morgan, CEO of EnCana, tries to avoid entering into resource auctions and seeks instead to leverage superior knowledge of a play to his advantage:

The Cutbank Ridge field in BC was the biggest land sale in the history of Canada by a huge factor. It was about $400 million dollars in one day and about 80 or 90% of it was EnCana. Burlington was our partner in some of that.

What happened in that play was that EnCana is such a big landholder in Canada that we already had acreage spotted all through the play. So we were able to do our piloting completely under the radar for that play. And it gave us a lot more information than Burlington had. But because we didn't want to blow each other's brains out at the land sale, we said, "Okay, we'll come together for the land sale." Of course what happens in a land sale is that the night before the bid, sometimes the morning of the bid, the partners get together and say, "Here's what we're going to bid." And the other guys say, "Here's their numbers." And if the guy who is lower isn't prepared to raise to the upper guy, then you are on your own.

The reason that EnCana was the dominant bidder was because Burlington didn't come up to us. We would have been bidding on that one day four or five times their annual budget for land sales, probably. But we had a lot more information. We had worked the play a lot longer. And so it wasn't that they were not competent; it's just that they didn't have as much information as we did. We knew what we had. Even at today's prices, in a resource play the acreage cost is only a portion of the final cost of the play. Cutbank Ridge at 400 million bucks is probably ultimately five or 10 cents an mcf.

If you look at our history, there are very few occasions in which we've ever been involved in an open bidding process for either land or assets. It has almost always been a process where we had the knowledge advantage, or we had some other advantage. Because today in these tight gas resource plays, bidding for them when it is common knowledge with other people is too expensive. So we have never played that game.

Take the Piceance Basin for example. There's a huge resource there. We made a one-on-one deal with a company called Ballard. We thought they were only just tapping a small part of the resource. And then we kept adding land. And then we realized that Tom Brown, a large Rocky Mountain, independent had plays in the Piceance that the market gave little credit to, and where they had the lowest production and lowest booked reserves; and yet, it was where we saw most of the value of the company.

So when we made the offer, because we were operating right alongside of them, we knew more about their land than they did or anyone else. So there was no competition when we made that offer. And that's very typical of how we built the company. We very seldom will bid in an auction and almost never will win.

Thinking beyond upstream resources

Though companies continue to be focused on resource capture, the fact is that new resources such as stranded gas and heavy oil have a much higher proportion of above-ground expenditure than the prior generation of deep-water resources. These investments go to liquefy and transport natural gas by tanker; or to build huge, long pipelines such costing tens of billions of dollars to bring gas to consuming markets, as from the Caspian or Alaska; and they go to extract and convert bitumen, which at normal temperatures has the consistency of a hockey puck, into transportation fuels. Art Smith of J.S. Herold notes: "The natural love affair among energy executives at the large integrated oils for the downstream (refining marketing and transportation) has been reawakened; refining and transportation investments, in particular, are back in vogue and garnering a bigger share of the investment dollar."

This shift in emphasis, in our view, will continue over time and extend into biofuels, coal transformation, and eventually into power, requiring new business models for competing in those areas. Energy companies will be advised to integrate their strategy and organizational development processes tightly into their resource allocation process and set up the right partnerships to be sure that they have the right skills and competencies to succeed in these new business segments. Companies are facing revolutionary change in business models and must address the change holistically.

Expanding the Portfolio

Mergers and acquisitions will continue to play a part in strategy execution, even in the future environment of fully-valued assets and companies. The reason is that combinations of companies can result in a new entity that has the scale and scope to address opportunities that the pre-merger companies could address on their own. Further, assets can be worth more in an acquiring company's portfolio than in the portfolio of the company selling them. Good mergers and acquisitions are not "zero-sum games," but create positive value for both buying and selling companies. However, to be successful they must be seen as a continuous part of strategy, and must be resourced appropriately. Acquiring companies must be continuously "in the game" and should be continuously evaluating opportunities. After the acquisition, they must be attentive to the source of value in the new assets, and put resources to work on optimizing the value of the combined entity.

Establishing an acquisition program

Chesapeake's strategy has focused on balancing organic and inorganic growth, and includes as a key element continuous evaluation of a stream of smaller acquisition opportunities. Aubrey McClendon explained: "We have about 60 people working on that full time, so we are always looking at opportunities and never feel forced into making a particular deal." They also look hard at risk. "I think we're willing to think more strategically about risk to our business than a lot of people have been willing to do; or if they thought about it, they haven't come up with the answers. Natural gas prices, well I've heard a lot of CEOs say, 'Well we never hedge because we didn't know where the prices are going.' Well, I don't know either. But it doesn't mean that I can't form an opinion that is at least somewhat based on interpretation of factors that are reasonably evident to most people." Chesapeake locks in natural gas prices for the acquisitions it makes, virtually eliminating price risk from the transaction. Aubrey points out in his public presentations that, although transaction costs have been rising rapidly, the margin between the transaction cost and the value of future production has also been rising, so a well-hedged transaction can still create value.

Staying in the game

Mergers and acquisitions success is contingent on staying in the deal flow. Jim Mulva of ConocoPhillips makes a point of staying in touch with potential partners. John Lowe explained the origins of the acquisition of Arco Alaska:

When Jim Mulva became CEO of ConocoPhillips in July 1999, BP Amoco had announced the acquisition of Arco. The first thing that Jim did as CEO of Phillips was to call and arrange a meeting with John Browne, and he flew over to London to meet him in that first couple of weeks. And he told him that he didn't know what's going to happen with your acquisition of Arco, but if you need to sell any assets, refineries on the west coast, pipelines in the mid-continent, assets in Alaska, whatever you need to do, we will support you. He just kept up with the conversations, and it started out as one set of assets that changed to another set of assets, and Jim just kept assuring him that we would be there. Then it ended up being that they had to dispose of all of Arco Alaska, and we had already talked to them about what our purchase and sale agreement needed to be, so when we put our bid in all we had to say was "yeah, we agree." I think others had similar numbers to us, including Conoco. But we had the relationship, and they knew they could get to closure with us, which was most important to them.

Bobby Shackouls confirmed Mulva's attention to building relationships:

Jim made a point of staying very close; he would call me about every six months. "Let's go have breakfast. Get lunch." The very first time he called me was on September the 12th, 2001. He was stuck in Washington, and I was stuck in Omaha. And there was a rumor going around that Shell was interested in BR. And he said, "Well, I just wanted to let you know that if there was any truth to that, we certainly would have an interest in being a player and participating if you were going to go down that route.' And always the discussion that he has had with me is, 'If anything starts to happen, please let us know, because we'd like a seat at the table."

Post-merger optimization

Valero has made an art and a science out of capturing value from their acquisitions by applying their superior operating and technology know-how quickly. This started with the ability of an individual, the late John Hohnholt, to see quickly how to improve yields, and has now been institutionalized. Bill Greehey explains:

John was brilliant. Every refinery that we bought, he could figure out a way to increase yields, increase capacity, and reduce operating cost. Every single refinery. We took over Benecia, and day one, we changed the way we were running that refinery. They ran the refinery based upon proprietary ANS crude, and they had plenty of capacity in their upgrading units, but they never would go out and buy intermediates; whatever the crude unit produced, that's what they processed. And the first thing we did is, we bought different kinds of crudes, we bought intermediates, and we loaded up all the units. And we increased the capacity of that refinery.

Let's use St. Charles as an example. St. Charles was a refinery that never operated well. So we buy the refinery out of bankruptcy. During the process, the coker catches on fire, so we've got that problem to deal with. So when we buy the refinery, I said, "We are going to make sure we have no unscheduled down time." We took the best people that we had at all of our refineries and we sent them to St. Charles. They sat with the people for six weeks around the clock, watching what they were doing and explaining how Valero does it, and what best practices are. And it took six weeks before they felt like these people were all trained to where they could really do the job and not have any problems.

We then took in John Honholt's old group, all of the really smart engineers. We've got experts on cokers; we've got experts on crackers; every major unit we have, we have a headquarters staff that are experts. And so then they go into the refinery, and they study these units. And we increased the capacity of the crude unit and the coker unit; and then we took the cracker down and replaced feed nozzles, did some expansion there. Very little cost. Great rates of return. The refinery paid out in one year.

We go down to Aruba. Exactly the same thing. This time it only takes four weeks, but we baby sit with those people. We bring in a top management team—I think we brought in five people. And Aruba has gotten to the point now to where the plant manager is Aruban, which has really worked out well. We buy Premcor. Same deal. Send the people in there. Look at their operations. Sit with them. Make sure operationally we are doing best practice. Bring top people in on each refinery. And that has been the key to our success.

Our observation is that mergers and acquisitions in the future will be done to strengthen the portfolio, not to capture synergies. It will be most important to enhance value after the transaction, so retaining and motivating the acquired staff and stimulating innovation from different points of view will be critical success factors. A purely financial view will most likely end up by destroying value.

Deploying Resources

Creating opportunities leads to choices in deploying resources. Within a portfolio of opportunities, where should resources be deployed and what should be the scope of the investments in those opportunities? The energy industry is capital intensive, and one of the most important business processes is capital allocation. Because the investments are long-lived, companies benefit or are burdened by decisions taken decades before. Getting capital allocation right is essential. Capital allocation may be the most important and most strategic business process in the energy industry. Failure to allocate capital well is probably the single biggest reason for the demise of a series of energy companies. They were either unable to secure a sequence of profitable growth opportunities, or they spent too much relative to the value of the opportunities they had developed.

In the allocation process and later in project execution, access to key technologies and to good equipment and services suppliers become important in delivering capital projects on time and on budget and in operating them effectively through their economic life. Deciding which parts of these supply chains should be internal, which external, and how they should be contracted are mission-critical choices.

Making capital allocation choices

During the 1990s, capital discipline was more important than securing sustainable growth. We suspect that the reverse may be true in the next decade, and the parts of the capital allocation process that concern the origination and capture of projects will be of particular importance. The challenge for companies will move from the discipline to "screen projects out" to innovation to "screen projects in" by discovering ways to reduce costs or discover additional value. Our interviews gave some clues on how.

Establishing capital discipline. Nick Butler of BP stresses the importance of continued capital discipline:

> You can't just focus strategy on the future; you've got to pay a lot of attention to operational excellence and the efficiency of what you're doing: supply chain, costs, and the capacity to do it. The forward thinking pieces are really just an add-on to that. If you lose sight of the current business, you risk losing everything. That partly was perceived to be BP's problem in 1990. But it also brings up another part of strategy, which is not to fall in love with your own rhetoric. In 1990, there was a dependence on the price of oil staying high, and that's very dangerous. That is why we plan the business within a very tight financial framework. We still plan every project on the basis of a threshold oil price. We won't approve any project that does not recover the cost of capital and provide a return at the threshold. The whole financial discipline is that the company is balanced at the threshold and any profits above that are returned to the shareholder. So we are secure even if the price of oil falls for a long period of time.
>
> We can grow as we are growing output now within this discipline at 5–8% pa. We have a sequence of projects. The constraint on doing more is not capital, it's technical capacity, people, project management. That would take time to develop a

whole lot more. So if we were to pour another $5 billion into the upstream, which we could do, the danger is that the new projects would not be as well run because we would be using not the best people, and that we would be feeding a degree of inflation in the service sector.

We can adjust the balance somewhat, but I still think it is right to be disciplined. After all it is only two years ago that the price was about $25/Bbl. We could easily start rushing in and spending as if the price would be $60/Bbl forever. But as you know there is no economic reason why it should be $60/Bbl. It's a political decision. There's plenty of supply; it's a question of whether OPEC will bring it on-stream. Sometimes they do and sometimes they don't. They are now investing in increasing capacity, and we would actually predict that prices will fall to below $40/Bbl. So we have to plan on a reasonable, prudent basis through the cycle. We might not get it right, because we don't know what the cycle is, but you can't start planning on the top. The point is that discipline is part of strategy.

Ian Howat and Catherine Roget of Total outlined their capital allocation process, which seems to us to be exemplary in the way that it is used to catalyze a continuous conversation between the senior executives, the CEO, and mid-level managers:

In terms of strategy and in terms of investment approval, you can hardly imagine a more centralized company than Total. We think that the cocktail of risks that we're involved in, be it technical risk, geopolitical risk, market risk, just means that it's a better idea if we approve projects centrally. To give you some idea of the delegation levels, I think for the E&P side, there is a limit of 20 million Euros per project. The head of E&P can spend only up to that sum—I told you it is centralized. Exxon is the only company that could be remotely as centralized. I believe their level is 50 million, which pro-rata is about the same. In fact it wouldn't matter if we made it 20 or 50. And for R&M, I think the level is 10 million rather than 20 million Euros, so they could build a couple of service stations, but that's about it, before they need central approval. Obviously most of what we spend, about 70%, goes into the E&P division, so we're investing about 8 billion out of a total of 11 billion Euros in the E&P division. You can imagine that nearly all the 8 billion is coming before the executive committee

(Comex). Comex meets once a fortnight on a Tuesday morning, and we try to limit it to the Tuesday morning from 8:30 AM to 1:30 PM, about five hours, though sometimes it overruns. In that time we can see about five investment proposals. We try to make sure that an investment proposal does not come to the Comex unless it is mature for decision. It should not come in an embryonic state that would risk degenerating into a talking shop. Comex members come in with the idea of saying yes or no to each proposal, not to say that it's nice idea but needs this or that before a decision. There are other forums in the group for looking at projects on a more embryonic basis.

John Lowe talked about linking project approval and supply chain issues at ConocoPhillips:

Even if you have an approved project, if you have equipment orders of a certain size or other types of commitment, those decisions would also be reviewed and approved by Jim Mulva. Early on there were a lot of complaints about "this is excessive, it's not adding any value, etc." Well Jim counts how many of these approvals he does in a year, and it may be a dozen. People spend more time complaining about the controls than is warranted. What triggered this is we found ourselves in a lawsuit with a contractor in one part of the business, and a different area of the company is unknowingly about to give a big contract to the same contractor. Jim's view is we need a review if that's going to happen. At least let's talk about it; let's see if we can leverage the new business to get rid of the lawsuit—"if we give you this piece of business over here, will you drop this problem?" It's a check and balance system.

Assessing relative risk and optionality. Ian Howat further elaborated the system at Total:

There are two other things associated with the executive committee (Comex). One is called the risk committee (Corisk), and that meets generally speaking on the Thursday before Comex for the entire morning. It looks at all the investment proposals with the same agenda as Comex, and essentially looks at the risks of the projects. Its role is not to look at the strategic justification or the financial aspects of the projects, but to be sure that we have considered what could go wrong, what are the risks involved here, and have the risks been properly identified and properly expressed in the dossier. You have to understand that the dossiers are all fully written out. You're not allowed to improvise a presentation to the Comex. The Corisk meets on the Thursday, the dossiers are then distributed on the Friday to the Comex members. So they go home with five dossiers that can each run to 100–150 pages and spend every second weekend reading and analyzing the investment proposals.

As well as Corisk, the strategy department has to put a written opinion, yes or no. Obviously, we don't take any decisions, but I can bring attention through the written opinion to anything we're unhappy about, or issues that the Corisk may have with the dossier. But normally if Corisk finds a problem, it asks for the dossier to be corrected. Normally that can be done over two days, so the dossier can still be considered at the next Comex, unless there is a structural flaw. For example, we've strengthened the whole sustainable development element, and issues like what could be the impact of a carbon tax on the project and that kind of consideration sometimes gets neglected because it's new, and have to be added in for the final presentation. Occasionally there will be a legal issue that hasn't been studied such as anti-trust implications. The Corisk, as well as people from the 44th floor, has people like the head of legal, insurance, sustainable development, industrial safety, all of the people you would expect to be on it.

Some investment opportunities should be evaluated not only on their own merits, but also on the doors they open to further profitable investments. This future potential is called optionality, and is particularly important in times when traditional opportunities are becoming less plentiful. Ian Howat discussed Total's approach:

We look essentially at investments on a project by project basis within the guidelines of the long-term plan, rather than on a formal portfolio basis. The long-term plan gives us an image of the company that we're all comfortable with: that we want to grow production at a certain rate; that we have an idea what we want to do on refining in terms of investments—just health and safety and product specs, or are we actually going to make some pay-off investments. But essentially it's on a project by project basis. We'll always assume that we can find finance for it, so we don't really arbitrage between one project and another. If they're good, they'll get done. If they're not good, they won't get done. We don't do an "efficient frontier" kind of analysis. On the E&P side we're driven by issues of access to resources, so we tend to do any project that meets our return criteria.

We do look at optionality qualitatively. In fact, I used that very word last week to the E&P managers. If you go back 3–4 years, OPEC was targeting $25/Bbl crude oil in the range $22–28/Bbl and was achieving that goal very successfully. Therefore, the people in the divisions were saying to themselves, if we've got a project that stands up at $21/Bbl, it's going to fly because OPEC will keep the price higher than that anyway. We used to give guidelines on the prices to use in developing projects that needed capital. Now we're essentially saying to them that really we don't have a huge amount of idea what the price is going to be, so keep an open mind, don't kill your project because it doesn't work at $25/Bbl Brent. Bring your projects into the Comex, run the numbers over a wide range of prices, and we'll make judgments on a case-by-case basis. Of course, if you've got a project with a very long plateau, we'll look on it more kindly than one with rapid depletion. It's what we did for the Novatek deal, which was the deal in Russia for 25% of the number two gas producer, but unfortunately didn't work out.

That was a billion dollars, but a billion dollars is a relatively small sum for us; it was just a question that if things in Russia worked out okay, the value would be terrific. It was putting $1 billion to work with the knowledge that, if the worst came to worst, we'd lose it, all of it. But the optionality there, if things went politically right in Russia, at some stage we would have been allowed to export some of the production, perhaps do some LNG. It was a small amount of money with all this optionality behind it. Deer Creek in the Athabaska tar sands again was the same kind of thing. We're saying to people we should look at deals like that where you can get your hands on some reserves, and in the next 20 years if technology evolves the way we hope it will, the reserves will be worth some money. And if it doesn't, we'll write off some of the money we spent, and it won't be a disaster.

When I say we're giving fuzzier guidelines, I mean for completely new stuff, the big ideas like the tar sands, because we're trying to encourage creativity at a time when it's difficult to establish guidelines. Because we don't know what the oil price will be, and if it remains high, we don't know what the fiscal terms will be, and we don't know what the cost structure will be. We have to recognize that we have a much more uncertain economic environment than we had five years ago, and we've got to encourage people not to kill their projects. If they've a gut feel that it's a good project, then they should bring it along to the Comex.

Flexible resource architecture

In chapter 6, we introduced Michael Porter's Five Forces that determine competitive intensity, one of which is the bargaining power of suppliers. As the relative bargaining power changes, it forces a reappraisal of where to establish the boundaries of the corporation. This issue is also related to the investor value proposition. A value proposition focused on acquisitions and synergy-capture through consolidation may favor extensive outsourcing. A value proposition focused on organic growth may benefit from performing more functions internally.

In the 1990s, the pressure to reduce costs was intense, and companies seized opportunities to outsource activities that they deemed "non-core." The logic was that the suppliers had lower cost structures and by specializing could build competencies that were superior to those in the energy companies. The trend was accelerated because many of the service companies working for energy companies had a lot of idle capacity and would discount their services in order to increase capacity utilization. The trend progressed to the point that one of the authors coined the term "virtual oil company" in 1995 to describe the possible end-point. Indeed, some smaller companies were close to virtual, retaining employees only in the areas where they added most value, such as prospect evaluation. Suppliers also benefited from downsizing in energy companies and were able to enhance their technical capabilities by hiring their customer's cast-offs. The classic example was in information technology and occurred across all industries. First, the data centers were outsourced to companies like EDS, IBM, and Accenture; then network and telecommunications administration went out the door, until most IT professionals serving energy companies were in fact employed by someone else.

Despite all these efforts, the energy industry is still less collaborative than most other industries. In a 2006 survey of CEOs, IBM Global Business Services uncovered that oil and gas industry CEOs were only half as likely as CEOs from all industries to recognize collaboration as "of critical importance" and only a quarter as likely to claim a "very large extent of collaboration" between their companies and partners (fig. 8.4).

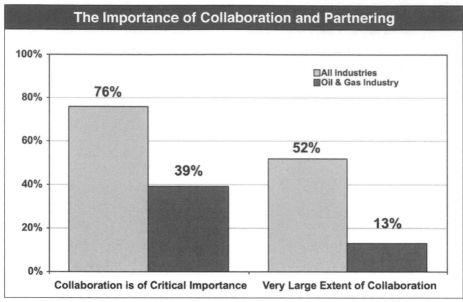

Figure 8.4: The importance of collaboration and partnering. *Source: IBM Global CEO Survey 2006*

Changing power of suppliers

As the next up cycle gains momentum and a phase change in the global energy complex begins to take shape, the relative bargaining power of suppliers and customers has changed. The need for companies committed to growth to control their own resources has started to trump the desire to hold costs down through competitive bidding. In some cases, outsourced activities are being brought inside to reduce costs; but the primary motivation is to assure the ability to deliver on the complex growth projects that are being developed. Oilfield service and equipment suppliers to the energy industry are at full capacity, and are seeking higher margins. Companies are responding by insourcing some of the activities that were routinely outsourced in the 1990s.

Aubrey McClendon talked about drilling at Chesapeake:

> You can't find gas without rigs, and so we said let's go build a bunch of rigs. Worst case, they are a good hedge against rising service costs and at best maybe they provide us some operational flexibility that we wouldn't otherwise have and other companies wouldn't have. Maybe it would give us an advantage in acquisitions because we can take an opportunity set that might take Company

"A" 10 years to drill up because they only have access to two rigs, and we can throw five rigs, 10 rigs at it and shorten the value-capture time down to two to three years and capture a lot of NPV as a result of it. What's more remarkable is when you take a guy's hat, and you peel off the drilling company name, and you put Chesapeake on it, miraculous things happen. All of a sudden your interests are aligned, it's not us versus them, things don't get dropped down the hole. Trips go faster, and they understand that their job isn't to kind of maximize the number of days on location; instead it's to minimize the number of days on location. It's to make holes to get gas to surface as fast as possible.

Honestly, it's not much more incentive than saying "if you come to work for us, if you're good you'll always have a job, whereas if you go to work with XYZ drilling company, you know that every three to five years they fire half of their whole staff. They have to, because the business is cyclical. Well you know: guess what, in 2006 we're probably going to be running 85 rigs. Forty of them will be our own. The first 45 we lay over are not going to be ours." So guys quit working for other people. They see that job as an up and down job; they come to us and get the same high wages that you would expect in an up and down industry and say, "You know what: I don't have to live life like a vagrant, I can build a brick house in this part of Oklahoma because this company is going to drill wells here literally forever." Tiny subtleties like that make an enormous difference.

Buyer's need for competition

BP was a leader in outsourcing in the 1990s. Nick Butler explained their current position:

There is an issue of where the boundary of the company lies, what do we need in house and what can we outsource. We have been innovative and have changed that in the past, but if service companies moved in certain directions, we would undoubtedly move it again. At the moment it works, but nothing is static forever, and some suppliers may choose to go in a different direction and become competitors. That's fine. We'll always need a service sector and not bring it all in house, but the boundary line is not rigid.

We do have global alliances and partnerships in some areas—I think in the IT part of the business. This isn't my field. My

instinct is that we feel happy with the transactional base, but in
the knowledge that there are a limited number of very high-quality
service companies with which we do business. I think we'll always
want there to be some competition, but we have very good long-
term relationships with all of them. We have so many projects that
we can work with all of them. To only have one would expose you
to a lack of competition and possibly a lack of innovation.

Gwyn Morgan likes to keep knowledge inside the corporation, but also likes to
contract out as much work as possible:

Another thing, we try to keep our life simple. And we try to
make sure that every service that can be done competitively is done
competitively. We keep the brain trust inside, the guys that actually
understand these plays and design the work program. But as much
as possible, we have the whole work done by contractors.

What we do is design the rig. And then we go out to three or
four contractors. We get bids from them. And in the case of the
Piceance Basin, we've got two different contractors who are going to
be competitively working on our rig design. And we always like to
have people that are side-by-side and see how they are each doing,
because we just really believe in a competitive model. There are no
worse performers than monopolists. So if you have one, there are
going to be all kinds of reasons why they wouldn't perform as well.
It's human nature. If somebody is competing beside you, you are
going to perform a little better. Like, when you go for a run, you run
a little faster when somebody is beside you.

Balancing costs and competencies

Total attributes its position on outsourcing to its French heritage. Ian Howat
told us, "One thing we're very conscious of, too, is that we've never outsourced our
core competencies to the extent that some companies like BP have done. That is very
clear, that over the past 10 years BP has seemed to do a lot of outsourcing in areas
that we would not have considered. It's not really in the French culture. Although we
have good relations with service companies, we're very wary of outsourcing anything
to do with reservoir engineering. In order to have healthy relationships with service
companies and sub-contract appropriately, you have to preserve your knowledge."

Rick George, CEO of Suncor, has brought more of the engineering and
construction management inside:

The way I would describe that is, we did a really major project for the company in 1998 called Project Millennium. We determined at that time that we were not big enough to manage that EPC work on our own. And we relied heavily on outside contractors.

What we found was a totally unsatisfactory result. I don't really like to use the term "insource" or "outsource." We determined at the end of Millennium that we're going to be into construction for a very long period of time. And so, Suncor today is half about running an operating oil company and about half about running a construction company. I didn't want to be vulnerable in dealing with the EPC firms to the extent that we were at that time. We still are very actively using engineering firms. They're a big part of us, but in each one of these projects there is leadership control, and control in the field. It's handled by Suncor employees who move from project to project and who have a learning curve.

John Wilder of TXU found that outsourcing can lead to excessive costs and complacency: "We had outsourced all of our legal function. There was no limitation around whether or when one of our managers could pick up the phone to call any one of those lawyers and seek legal advice or assistance. So we insourced legal and really put some cost controls and internal oversight around it. Fifty million dollars a year of legal fees is now down to $20 million."

Concern over supplier capabilities

Rick George is now concerned about supplier capabilities as a result of the prolonged down cycle in the industry:

Just one quick war story: When I graduated from engineering school, I went to work with an EPC unit in Texaco. All we did was run big refinery, petrochemical, and other projects around the world out of Houston. I was one of 500 engineers. The industry was then going through cycles. And what would happen is, you'd go down to the low of a cycle; people would pull budgets back in, and you'd shut down projects, even if you're in the middle sometimes. And so eventually the oil companies said, "Well, we can't deal with the cycles. We'll rely on Fluor, Bechtel, and the contractors of the world."

For the first couple of cycles, there were enough engineers coming out of the oil business that understood the business; they went into the EPC firms and did a really good job. Each progressive cycle after that, it got weaker and weaker, because every time the cycle came back, the EPC firms would move a good oil and gas guy to some business that was long haul such as power that required engineering manpower. And then they'd ask, "Oh, do you want to go back to work in the oil patch again?" And the answer would be "No, that's too cyclical." And so what happened was that we got less and less talent in the EPC pool.

And it's industry's fault. I'm not blaming the EPC contractors. It's the cyclic nature and the fact that we weren't willing to invest in the low part of the cycle to keep human talent rising in this industry. So the net result is that we're short of manpower everywhere. We're short of manpower in engineering firms. We're short of manufacturing capacity. We're short of offshore drilling rigs. We're short of drilling rigs onshore.

This has been a long cycle. The cycle started in 2002, maybe a little bit earlier, depending on how you measure it. It's going to last a long while. But I think you've got to look at the preceding 20 years of ups and downs to realize how we got ourselves in this position.

Now we're integrating Suncor people with the EPC companies; so we're not going back and forth. We don't do a lot of designing, but we've formed this engineering group where we have 300-some people here. It's really around management of projects, not real engineering design. You're making the decisions around design, and you're looking at supply chain management issues.

Technology deployment

Most companies have a process to align their technology spend with business goals; but more challenging is to define a technology strategy that drives resource allocation and spending to achieve genuine differentiation from competitors in some key areas, and assure quality in all areas. The key issues are: assuring that the right expertise is developed internally, standardizing where possible in order to maximize the yield from technology investments, and focusing on the specific technologies that add most value to key businesses. Finally, technology migrates rapidly through the energy industries, as new approaches are increasingly developed by suppliers rather than customers. However, there are ways to keep a strong grip on the intellectual property embedded in technical know-how, which includes not just the physical equipment and processes but also the ways in which they are deployed in business processes. Our interviews shed light on all these topics.

Technology positioning. There are three broad options in defining strategic positioning for a particular technology:

- *Leader* implies a commitment to be the first to develop and deploy a new technology, and to define the "dominant design" in the industry.
- *Fast follower* implies a commitment to adopt and deploy a new technology as quickly as possible.
- *Value-driven* implies timely application of good quality technology at the lowest cost.

Within each option there is a choice between internal and external sourcing, with important consequences for the deployment of the company's limited technology expert resources. "Leader" strategies imply radical commitment of money and expert resource: typically they are 10 times as resource-intensive as a "fast follower" strategy. "Fast follower" strategies involve a significant amount of creative effort in learning and customization. "Value-driven" strategies require skills and processes either for internal provision or to be effective as an "informed buyer."

Within the energy industries, technology leadership has largely passed from the companies to their suppliers. Only ExxonMobil has retained a substantial basic research and development capability. Several companies aim to be *fast followers*, applying new technologies quickly, especially in challenging environments such as deep water. In many cases, smaller independents are more inclined to experiment with new technologies from oilfield service companies than are larger companies. Valero appears to have a *value-driven* approach to technology. Rick Marcogliese describes the philosophy: "We are all about technology application. And we look at the marketplace for technology, whether it's technology companies like UOP, or Shell Global Solutions, or Chevron; EPC contractors like Fluor or Foster Wheeler; Catalyst vendors. We apply what's out there: we make an evaluation; we make the best technology application." He contrasted Valero's positioning to that of Exxon, where they aspire to a leadership positioning in technology: "If you contrast that to an Exxon situation, they develop their own proprietary technologies. But from an operating organization point of view, there is also kind of an internal bias to using the technologies that your technology arm and engineering arm has to offer. And that can be a limitation on using the best of what is out there." So there are pros and cons in aspiring to be a *technology leader*.

The issue in a strategy of sourcing technologies from the outside is often that of sustaining the ability to be an intelligent buyer. Marcogliese continued: "We've done a great job of hiring senior professionals from the industry who have 20-plus years in their field. It started out small; but as we've grown: we have now two cat-cracking experts; we have two hydrotreating experts; we have expertise in each process technology, and then we have expertise now in the mechanical disciplines such as metallurgy. We've got a metallurgist and a senior professional in machinery as part of the staff. So we have internal staff across the spectrum of all technologies, but the emphasis is on application, not on technology development. Our niche is: 'Let's have somebody else develop the technology, and then we'll go ahead and apply it.'"

Gwyn Morgan explained that EnCana also focuses on technology application rather than pure research:

> We don't have an R&D function. It's interesting. Over the years both AEC and Pan-Canadian companies tried the R&D function, and we found that it just never seemed to add value. However, we do participate in industry R&D: we've got one of the only commercial CO_2 injection projects in North America here in Canada, and it has been very successful.
>
> We work very closely with the drilling contractors, but one of the most critical areas is completion techniques. We are one of the largest, most important customers of all of the major service companies. BJ, Halliburton, Schlumberger, you name it, because we are so big. And we work together. We've never regarded technology as being in any way proprietary.
>
> The application is a different thing. You can't stop people from knowing each other's stuff. But it's the way it is embedded into your processes that's important.

Technology timing. A crucial element in all of these choices is the time taken to technology deployment. Time to deployment is often ignored because it is harder to assess than cost and value impact. Ultimately, however, most technically-driven differentiation comes not from ownership but from embedding technologies quickly and effectively into operations, and technology strategy must recognize that.

Managing the ways in which technologies are deployed can be a source of value in itself. A pro-active strategy to transfer technology to external providers can free up scarce internal resources to focus on new developments and applications. Since most energy sector technologies migrate quickly, this process of commercialization usually makes strategic sense.

Don Paul described Chevron's approach of leveraging its in-house research capability with external sources and cautioned that the full cycle of developing a new technology can necessarily be long, particularly for downstream process facilities:

> Chevron sees partnership arrangements with suppliers, governments and universities as essential mechanisms to develop new technologies. Since virtually every new technology component

must be integrated into large, complex systems, Chevron sees a primary role of its internal technology subsidiaries to integrate and deploy "technology solutions" matched to its core operating businesses and new business developments. Key proprietary platform technologies such as reservoir stimulation and catalysis, are an essential underpinning of major technology solutions to support oilfield management and heavy oil conversion. Within Chevron, internal management processes, integrating both business and technology perspectives, ensure ongoing alignment and proper direction of R&D programs and partnerships.

An important aspect of processing technology which many people do not realize, is that the time from lab research to new production facilities is very long. One of the things about putting in refinery-like applications, whether its either in a refinery or in an oil field, you are not going deploy at scale a multi-billion dollar facility lasting decades until you are fairly confident it is going to work. The complete development process involves laboratory bench-top research, leading to a controlled pilot plant, followed by a large scale experimental facility, then finally to full-scale production plant. So you go from a tenth of a barrel to two barrels a day pilot, to 2,000 barrels a day semi-works, to 100,000 barrel a day production. A semi-works is the first deployment at scale in a manufacturing site. Once that works successfully, a full-production facility can be implemented based on the whole range of business factors. It's a long process, one that can easily take a decade.

I think that the big change that is going to happen in the product business is that the feedstock is going to diversify once you've decided that you have to create synthetic fuels. The feedstock range is potentially much wider. Today, many companies are moving in this direction. So I think that's the shift. For synthetic fuels such as GTL and biofuels, the processing facility is located at the feedstock source. It's a distributed manufacturing model at different scales. For example, the biggest bio-processing facilities in the world are more at the scale of large breweries. And they are small compared to refineries. A giant one is 20,000 barrels a day. A typical bio plant is more in the range of 20 to 30 million gallons a year: 2,000 barrels a day. So I think that biofuels, if it plays out, is going to have to drive a distributed manufacturing system. You're simply not going to have a 200,000 barrel a day facility. Will that make some refineries redundant? That's an interesting question.

Protecting proprietary knowledge. Notwithstanding the rapid migration of most technology, a coherent strategy towards intellectual property is valuable. It is sometimes possible to protect innovations by simply limiting external access to them: subsurface characterization and modeling and also process integration are key examples. It is rarely possible to keep E&P technologies proprietary through mechanisms such as patents, and the value of royalties is immaterial in the context of E&P cash flows. But patents can be a useful tool in managing the commercialization process with service companies.

Gwyn Morgan illustrated by referencing a particular resource play how EnCana tries to build and protect its proprietary technology knowledge:

> Like all these resource plays, the first ones are not commercial. Too high a cost, and so on. We realized, based upon our experience historically, that when we really got going, we would be able to continue to learn enough to make this work. And so we quietly assembled a bunch of land. We bought little companies that were pursuing other plays entirely, and we got some Crown sales. And we kept on learning more, and then we got more and more confident. And eventually we gathered up about 1,200 square miles of land. We controlled the whole trend, which is the way EnCana operates. We like to control the whole resource play because once we start applying the technology, we are giving all that information to everybody else.
>
> And, you know, the play is producing around 350 million a day; but it's got a long ways to go. And the cost of drilling has gone way down, and these horizontal drilling rigs now are purpose-built for us; and the crews all understand exactly—because we always have the same crews on the same rigs—and they now have turned this into a very big commercial success.
>
> The reason I give you that example is, that with the technology that we are using, we were pushing the ability of the service companies to provide what we needed. But now they have developed more and more sophisticated approaches. We were able to put the crews in place, the rigs in place, all the equipment, exactly what we want; and we just kept driving down cost and learning more and more. And because you own the resource, everything that you do and learn incrementally can be applied to a lot more acreage

in terms of value creation. And so we truly are an unconventional company in how we think, what we do, our resource base. And you can probably see why we felt focusing exclusively on this would make us even better. No distractions.

Organizing Effectively

Organizations should be structured according to a classic principle: "form follows function." This is applied in two senses: form reflects or mirrors function, and form can only be created after function has been determined. It is important to establish the direction and the strategies required to fulfill the shareholder value proposition, decide on the corporate boundaries and which aspects of supply chains should be performed internally, and then to organize in a way that will promote the desired results. This will vary by company and by business segment. Inevitably, the issues of organization revolve around the advantages of accountable business units, the importance of functional excellence, the role of the center, and the value of integration. Within these issues, there is a need to address regional and local stakeholder needs.

The energy industry has used two main organizational approaches to improve returns on capital employed. The first has become known as *atomization* in which companies strip out supervisory layers in the organization and give considerable autonomy to business units. Business units may comprise a single processing plant, an oil or gas field or basin, and a marketing region. Each business unit commits to financial results, is responsible for delivering them, and has considerable decision-making authority to deliver on promises by enhancing revenues or reducing costs, subject to corporate standards on ethics, safety and environmental performance. BP was innovative in strengthening the atomized model through "peer groups" in which business units with similar value drivers would be responsible for the collective results of the group as well as for the local results of each unit. Thus, mature oil fields from around the world could be aggregated together in one group, inland refineries in another, and growth projects in another.

A more conventional approach is to create a strong functional organization and impose best practices on all the regional operations. ExxonMobil has been successful in driving down costs, deploying technologies throughout its operations, and developing superior project execution capabilities through a strongly centralized organization model. While the motivation in the past has been primarily cost reduction, we believe the motivation in the future will focus more on delivering growth with the least possible resources. The need for organizational effectiveness is shared, but a number of mental models and mind-sets will be sharply different. Several of our interviews threw light on these issues.

Accountable business units

Gwyn Morgan believes in accountable business units:

> We are, by the way, a decentralized business unit organization, so it is anything but hierarchal. I think this is another very crucial thing about the management and the success of the company. I've always believed the top-down hierarchical organizations are not going anywhere. So we created a company with business unit structure; business units and teams have their jobs and they tell us what they can accomplish.

> Based upon that, and a competition for capital against all the projects, they get their allocation and their budget. And then they are measured against what they said they could do. They have to come back every year and say, "Here's what we said we could do, and here's what we did."

Functional focus

Rick George, on the other hand, is moving away from autonomous business units at Suncor:

> As recently as a year or two ago we were a very highly decentralized company, so the refining and marketing operation in Ontario ran very differently from oil sands, very differently from the natural gas business, even though we had an integrated strategy model. So what that meant is you have four different supply chain management systems. You have six different maintenance systems. And you're going through such rapid growth that every once in a while you've got to stop and fix the process. So we're right now in the middle of the implementation of the ERP SAP (corporate financial and operational information system) program.

> So we are functionalizing the company, which is something I understand Exxon did back when they merged with Mobil. Suncor is a very late follower in that. And there is good reason for that, in the sense that we were growing so rapidly, and decentralization worked really well for us for a long period of time. People felt real ownership

for their piece of this strategy. They knew their businesses and were very focused on returns, focused on how they could bring about growth in this model. The trouble is, when you get to a certain size, you lose efficiency, and that's really one of the things that drives us. Now, I wouldn't be surprised to see us five years, 10 years, down the road moving back the other way, but I think that what we've got to do right now is drive efficiency and effectiveness in these functional areas.

In our view, both George and Morgan are right. George has an integrated strategy of extracting, transforming and delivering molecules derived from oil sands, primarily to motorists. Autonomous business units hinder rather than help this strategy. Morgan is developing a series of very different resource plays based on a core competence of reservoir understanding. North American natural gas is a fungible commodity that can be sold and delivered at hubs to a variety of marketers. It is not necessary to forward integrate. The technology platform is shared, but performance in the different resource basins is dependant on the nature of the rock and subject to local permitting processes. Independent business units make most sense for his business model.

The theme of centralized strategy and decentralized operations was echoed by Rich Marcogliese of Valero:

We do have a centralized process for strategic planning, with the refineries participating. I think that has worked in John Roach's group very well, that he has technology people and development people in the same group. And so we have a centralized review process, and centralized planning process; but it's not one in which we say, "All right, refinery manager, we'll tell you what..." If a refinery says, "Well, that technology is not going to work for these reasons," there is going to be this exchange: "Well, what will work?" We went through our own experiment: "Should we have a very prescriptive corporate organization which says how everything ought to get done, or one that allows for decentralized decision making?"

And what we found was, when we became too prescriptive, the plants started to say, "Well, look, I'm not in control of my own resources any more. You guys decide. We're not involved. Vendors, contractors, they're not accountable to us any more. They all want to go tell San Antonio a good story." And this happened at a time that we started to see some deterioration in our operations, and we started to ask ourselves why. And that's the feedback that we got.

So we discovered we could go down a bad trail if we don't give the refineries some sense that they have local decision-making. So we try to avoid doing one size fits all. Everybody has got to become VPP Star qualified (top performance on ASHA's Voluntary Protection Program) over time. We want everybody working on reliability programs. But we're not going to tell you what contractor you've got to use, or kick out an incumbent that's doing a good job. And we try not to issue a lot of corporate mandates, because it may not match up with local priorities. And that is a little bit of a difference from large companies.

Thus the role of the center is to guide rather than direct, and to persuade local operations that they will find it easier to meet their performance obligations if they take advantage of certain corporate initiatives. However, if the center becomes too coercive, the damage in terms of local sense of responsibility for outcomes will exceed the potential benefits of standardization.

Operational integration

Rick George distinguishes between strategic integration and operational integration. He maintains that Suncor for many years has had strategic integration between upstream and downstream business units:

We haven't had too much trouble with alignment. We've had some people leave, but not many. On getting alignment, one of the things we've always determined is that we were not going to divide this company between an upstream and a downstream company. The goal is taking those heavy bitumen molecules and moving them to the market.

So most of the conventional players, even today—the Exxons and whomever—they're divided upstream and downstream. You're either an upstream or you're a downstream person. At the top of one of those pyramids, you might have a 50% shot at getting the top job. It's usually the upstream, but every once in a while the downstream wins the top job. The two groups don't mix a lot. What we said is, "Now, listen, we're going to be in this oil sand business for a hundred years." It's all about that molecule, that bitumen molecule, at the mine face or SAGD and saying, "How do you get it to the marketplace?" We try to move people around various groups because, again, it's all one business. It doesn't matter where you're located.

Increasingly George is moving away from independent business units to capture more of the potential operational integration. In our view, unconventional resources demand both strategic and operational integration, and we believe that the specialist companies such as Suncor and BG have an advantage over the so-called "integrated companies" that normally separate upstream from downstream activities. This separation works well when there are liquid and transparent markets for the produced commodities and a market-based transfer price, but integration is necessary for resources for which those preconditions cannot be met.

Addressing Stakeholder Concerns

The drive for cost-effectiveness has led to organizational innovation and higher returns on capital employed. However, there has been some collateral damage other than decay in the industry's talent base. Another casualty has been the regional leader with real decision-making authority. This function has been superseded by central control in functional organizations, or dispersed decision-making among business units in atomized organizations. In both situations, the reporting relationships go back to corporate headquarters. The regional leadership role in many IOCs has devolved into that of a diplomat, conveying messages between company headquarters and host country leaders. This presents a challenge in properly addressing stakeholder concerns at the country level.

We have seen that access to opportunities depends at least in part on the development of value-adding relationships with stakeholders in resource-rich countries. Good relationships with outside stakeholders also underpin an energy company's license to operate, and ability to fulfill its shareholder value proposition. If a company lets its reputation slip, the repercussions can be felt globally, as noted in chapter 4. The industry in aggregate has been derelict in its failure to educate the public, and is reaping the unfortunate consequences. Companies make excuses that public education is a long-term investment with uncertain returns, and that a single company cannot change the industry's reputation. They also argue that companies should not be involved in local politics. But as many have said one way and another, change begins with the individual, whether a person or an energy company.

Decades of inadequate investment in public relations have contributed to the current low reputation of the industry. This in turn has made the industry an unwelcome neighbor in communities, subject to acute NIMBY (Not In My Back Yard) opposition when it tries to site new facilities, adverse rulings on MTBE and ethanol, restricted access to public lands, drilling moratoria offshore, and endless hearings on oil prices. Had the industry acted in a collaborative way to educate the public honestly and without condescension on the basics of the supply chain, the sophistication of the technologies being deployed, the realities of geopolitics, and the true costs and value of environmental rules, the authors believe that the returns on the investment would have been tremendous. This is another example of "the tragedy of the commons"[8] whereby each company, pursuing its narrow interests, fails to safeguard the health of "the commons," in this case the image of the industry among the general public.

Developing public support

BP has adopted a "progressive" stance relative to most energy companies with respect to the environment and public interest groups as mentioned in chapter 4. BP and Shell come under significant societal pressure in Europe to adopt a "progressive" stance, and the pressure has hitherto been less intense in the U.S. Pat Yarrington, vice president of policy, government and public affairs at Chevron, is an experienced energy executive with keen insights into the industry. Yarrington sees the need for a shift in U.S. policy with respect to energy, and blames the industry for failing to educate the public and its elected representatives on the issues. She is particularly concerned by the politicization of international energy sourcing:

Political realities

Fundamentally there is an issue of growing bilateral relationships between new consumer countries and old producer countries. That is potentially problematic for the old consuming countries because they need to view energy as a strategic commodity, and not all of the consuming nations really have that view.

So from a policy standpoint—I am thinking particularly of the U.S.—there needs to be a real shift in strategic thinking around energy policy, and to link it with foreign policy and environmental policy. And have those tradeoffs very explicitly identified for people, so that when they are making the choice between environmental or foreign or energy, they understand what the consequences of those policy choices are.

It puts U.S. companies at a strategic disadvantage. If you have foreign governments with producing companies—INOCs—for whom they look out for strategic partnerships, bilateral relationships with producing countries, and the U.S. does not, then I think it is problematic for US companies. We need to continue to raise the relevance of energy and the strategic importance of it in the U.S. governmental dialog.

Industry inaction

She blames the industry for failing to present itself properly over the past two decades:

> The industry, in a way, abdicated a very important role of communicating to people the value proposition of energy: the efficiency of the industry, the size of the industry, the reliability of the industry, the relative affordability of our products over time, and the values of the industry. We've also not communicated sufficiently about the enhancements in the industry, whether it be through technology, or whether it's through the environmental footprint. A lot of the progress that's been made in the industry in the last 20 years we haven't communicated in a sustained, comprehensive, impactful way at all. The industry hasn't been a sustained issue for the public because we've had relatively moderate energy prices. We've had surplus capacity relative to demand. We did have a period right after the spike in the '70s when we had a huge personal consumption pattern change, associated with the higher energy prices. And so demand fell off substantially, and it has really taken quite a long period of time for demand to get back up to production capacity limits. For the American consumer it has been a wonderful period of time.

> But we lost those 20 years of trying to tell people about the industry, and what the industry stands for, and what it has done, and just the complexities and technologies, etc., associated with it.

> And why haven't people seen the need? I think we have a whole generation of managers in senior positions in these companies who have spent their entire careers in the '80s and '90s. It's hard for people to completely change from their entire managerial experience to be effective in a completely different environment. When you've learned you don't control the revenue streams, and when you've learned that your revenue stream can be cut in half in a very short period of time, that long term 30-year outlook—"Let's

spend money on things that will not pay off for years"—just is too difficult for people to see the need for and to sustain effort around.

Public awareness

As a result, the vacuum left by industry inaction has either been filled with poor science, ill thought out concepts, or has been left unfilled. Immediately after the hurricanes of 2005, Chevron executives were trying to explain to the press the damage to facilities in the Gulf of Mexico, and found an astounding lack of basic knowledge on the energy system. One of the editorial writers at one point exclaimed in surprise: "You mean you've got pipelines out there, too?" In fact, damage to pipelines and gas processing plants has been the cause of much of the reduced production from the Gulf of Mexico following the hurricanes.

Another example is the decision to forbid drilling of the Destin Dome prospect in the Gulf of Mexico:

I mean, literally, we've flown Congressional delegations out to it. We've got a platform in the eastern part of the Gulf at the edge of Florida waters; producing from it, you could almost drill diagonally over to Destin Dome. But it's in the federal offshore Florida waters. And so, instead of Destin Dome gas coming in and serving Florida, we're bringing Angolan LNG.

U.S. consumers don't want refineries in their backyards. They don't want LNG plants in their backyards. They don't want pipelines. They don't want anything having to do with energy in and around where they live. And yet they continue to consume an increasing amount. So we really haven't done a good job at all of informing them about the industry; telling them that there are choices that have to be made.

Yarrington has started to work on this problem through advertising and a new Web site sponsored by Chevron, www.willyoujoinus.com, to encourage public engagement and awareness:

We call our ad campaign the "real issues" campaign. It was really designed to try to raise awareness about energy issues. We felt that there were myths out there about hydrogen being the next savior. Or you had Matt Simmons' book implying we're running out of oil. Or we had these unrealistic political solutions. For example everybody thought a few years back that fuel cells were going to yield the be all, end all. There is just a lot of misinformation out there about the energy industry, on the demand side, the supply side, on the alternative energy sources side.

So we just decided "we are going to start educating." We are raising awareness for people, so that they begin to have a few facts in their arsenal when they are thinking about policy choices; or thinking about what this means to them, how they want to live their lives, and the choices that they want to make. We also felt there isn't one easy solution to it. They are very complex issues. And, long-term, it is going to be collaboration between the producers and the consumers, between bilateral governmental entities. It is our manufacturers and the energy efficiency with which they make their products. It is a whole system that has to become more energy efficient, more energy knowledgeable. Otherwise we run the risk of higher prices, and curbed economic growth. And that's not good for the stability of the globe.

So we created an online forum—we have constructed it as a pro and a con, point-counterpoint set of arguments. We come up with the topics that we think need to be debated. We go out and search out writers, those who have written on it elsewhere and put together a pro and a con. Then we just encourage folks to express themselves. It has been amazing. We have thousands of registered users—people participating from all over the globe; people who want to continue to hear the next debate.

After all the blogs come in we have the Aspen Institute provide an executive summary. And that gets sent back out to all the people who participated if they chose to be contacted; again by e-mail. And about every two to three months, we look to change the debate, summarize that last debate and send it out again to the people to keep a dialog going. There are some very good opinions expressed, with some very thoughtful, provocative dialog going on between people of different opinions. And I think it is very healthy.

The reality is that when you go out and you talk to people on the street, and you ask them about energy, or oil, or natural gas, or anything like that, they are not well informed. And they would like to be well informed. They just don't know how to get that information and not have it be a biased view. They are skeptical of oil industry people parlaying the information. We don't have as much credibility as we would like to have. They are skeptical of our trade association, the American Petroleum Institute (API), at least in the U.S., as a credible source, and they really don't know where to go to get valid information. They don't believe necessarily all the hype of the hybrids or the hydrogen or all that stuff. They are very practical, down to earth people; and they just want to have an ability to get a little bit more informed. And that really is the point behind the issues campaign and "willyoujoinus."

Grade school education

There's not any initiative in elementary schools of any substantial nature. I chaired an effort with the API committee a couple of years ago. It was called the "Envision the Future" initiative. It was really geared towards an approach of "How can we be seen by the policy makers to be a solutions provider, not a *put your heels in the ground and dig in* every time something comes forward? How can we get to the point that they trust the industry; that they trust the statistics that we are putting out?"

Fundamentally, this detraction of the industry and lack of knowledge of the industry, and, frankly, the taking over of young minds in schools by the environmentalist groups, unanswered, has been occurring for 20 years.

And kids aren't looking at the industry as being technologically driven. In the U.S.—they are not going into math and science. They are not going into the fields that energy really needs. We have dug this hole ourselves. And we are going to have to, as an industry, dig out of it. But it's a 20-year dig.

Establishing community relations

Rick Marcogliese of Valero talked about his expectations from refinery managers on community relations:

We do expect the plant managers to be involved in the community. We modify that based upon the situation. If we've got an operation that just is in the ditch and can't get out, we tell the plant manager, "Look, we don't want you to spend your time at the Rotary Club. You need to spend your time on your coker, on your compressor platform, until this situation gets squared away." Where you have a properly functioning operation, we expect the plant managers to be involved. We've got things like formal volunteer councils at every refinery. We have charitable budgets. That is part of the culture.

Usually there is a local budget, and there is also a corporate match that goes along with that. And there is some discussion with the corporate group and Mary Rose Brown's group on best practices for community relations. It is based on what we have learned from Corpus Christi.

Corpus Christi was Valero's first refinery where they developed much of their operating philosophy. For example, their Corpus Christi refinery developed a close relationship with the local community college to assure that there was a steady availability of qualified operating staff, and Marcogliese confirmed that this practice had been adopted in other refinery locations.

Rick George is strongly committed to contributing to the well-being of the communities where Suncor operates:

We set up a foundation. So you take part of your money and you're investing back in the communities. The way it still works in some companies today is, everybody comes in to see the CEO and asks for money. And that, to me, is not very sustainable long-term. What we did about five years ago or more was, we formed a foundation and we funded that foundation at the rate of about $8 million a year. They're building up a reserve, and then they invest back in the community on a very professional, long-term basis.

We're not here for the short-haul. So, you're working with a farmer, just think about it. You're going to have to work with that farmer and his descendants for a hundred years. So start treating him like that. You may have certain rights to drill a well on the guy's land. You also have a moral obligation to treat him with respect and try to work with him and try to understand what he's telling you.

You're not going to make everybody happy. Leadership is never about that. But you can show people respect. You can have a dialog. You can try to work things out. You can kind of bend over a little bit once in a while.

Proactive incident management

The energy industry, and particularly the petroleum industry, deals with hazardous materials and operations, often in hostile environments. It relies heavily on a broad variety of technologies to perform and control its operations. The industry is subject to dramatic accidents that can result in injuries and fatalities as well as damage to property and the environment. Such accidents inevitably draw media attention. We describe below three examples of incidents, the challenges they created, the way in which the involved companies responded, and the lessons that were learned.

Europe. Ian Howat of Total reflected on his company's experiences with two major accidents, and his comments are characteristic of the sober and thoughtful attitudes found in most energy companies:

> On the CEO role in environmental incidents, Total had two major incidents that have taught us an awful lot: the break-up of the Erika off Normandy in 1999 that spread a lot of fuel oil on French beaches, and the explosion at the AZF fertilizer plant near Toulouse in the south of France in 2001 that killed about 30 people, but affected some 30,000 others.
>
> The industry has not been very comfortable with the emotional aspect of management. For example, obviously, the proposed solution for Brent Spar was in engineering terms very well thought out. Unfortunately, Shell just hadn't realized that the emotional response to the idea of using the sea as a dumping ground would be so violent. Also Greenpeace in its communications multiplied the amount of oil that was in Brent Spar by 100, then took two years to apologize for the error, and by then everyone had forgotten about it.
>
> I think we're beginning to realize that we're a very high profile industry, often on the front page of the newspapers, and we're going to stay on the front page for as long as oil prices stay high. I think that the CEO has an increasing role now to explain, not to a technical audience, but to the general public what the company is doing, how it is behaving itself, and what its role in society is. I think probably Exxon does not spend enough time on it. In our opinion, Exxon does a lot of work on things like global warming; it just doesn't bother to communicate it. Actually, they don't do much financial communication either: they don't talk much to analysts.
>
> But having said that, I think we're all climbing up a learning curve and a long way from an optimum position, because you're never going to be loved, and anything you say is sure to be used

against you. There's a huge amount of disinformation. The Internet is a marvelous machine for spreading disinformation and is very unmanageable. So the whole communications thing has become very complex to manage. I'm not sure we're being successful on the whole as an industry. The scale of the industry and the scale of its profits means that *a priori* people are against you, so we're always fighting an uphill battle. The question is can we communicate more, and we certainly learned a lot from our two incidents.

If I look at it bluntly, the Erika disaster in legal terms was like if I was moving furniture and the van crashed. That was the limit of our formal liability, because the ship had all the certifications. Total has always disclaimed responsibility, but we agreed to spend 1 billion francs to clean up beaches and reprocess the sand. But the French public has never accepted that anyone other than Total was responsible. The fact is that we had the deep pockets and were the only French company that could be held accountable. So that was a bit of a communications disaster, because we handled the situation really from a technical point of view and explained what our legal position was, but that wasn't enough.

I don't know if we'll ever know the cause of the AZF disaster; we don't after three years. And from a public relations side, this sort of thing is very difficult. We are investigating from our own side what may have happened, but there is also a penal commission and a civil commission at work. But the French people are suspicious of our inquiry, because they think we're just trying to cover our tracks and conceal the evidence. All these things are very difficult to handle— it's a big company against individual people, so it's a question of size, but also of power. We think we settled the AZF claims in an exemplary fashion. We even twisted our insurer's arm into going as quickly as possible. This is recognized in political quarters but not by the general public. There's always some disgruntled party who thinks they should have got more than they got.

Since these incidents, we have instituted training programs and have put together a better incident response process that can assemble teams quickly, and a crisis group; so we're better prepared now.

As we pointed out in chapter 4, the key is to move beyond the pure technical response to the situation and address at a personal level the concerns and fears of the communities affected, recognizing that the public has raised its expectations to include needs from the ego and self-actualization levels of the Maslow hierarchy.

Gulf of Mexico hurricanes. We talked with Rick Marcogliese about Valero's response to hurricanes Katrina and Rita as an example of incident response: "For the hurricanes, we activated an incident response center here in San Antonio." There were numerous stories of support for employees and communities affected, including dispatching barbecue on a helicopter to provide food for the Port Arthur refinery after Rita:

> I had my own air force for while! It was quite a deal because, especially during Rita, we had four plants down at the same time. I just never would have imagined we would have had something on that scale all at once. But we had tremendous resources at our disposal. And the other thing is, our plant managers responded in outstanding fashion to support their communities, and they did it exactly as we would have wanted them to. So, yes, we do expect the plant managers to be incident managers.

> Bill Greehey would say that refinery managers know they don't have to come in and ask for permission on certain things. The small example was emergency services in the parish where one refinery sits. They were running out of fuel. They couldn't get motor fuel anywhere. Well, the plant manager just opened up the refinery gas pumps. "Take all you need, just come in": police cars, ambulances, whatever it is. And so they all came to us. Refinery Managers didn't have to call and ask for permission.

> We also gave it out to employees who couldn't get fuel: "Just go ahead and fill up." Obviously, that was necessary so they could come into work, but the help went further: "Look, you need to live in your house, but there's no electricity. Well, here's a portable generator." And "You've got a tree in your driveway. Well, here's a chain saw."

Valero's strong culture of community responsibility is central to former CEO Bill Greehey's value proposition and results in positive behaviors at the site of the incident, without the need for central direction.

Canada. Rick George recalled a recent major accident in Suncor's oil sands plant:

> When I think about that, how you do in the bad times is actually more important to me because anybody can run one of these companies when it's operating well. And January 4, 2005,

was a good example for us—a day that'll go down in history here. We suffered an extremely dangerous explosion and fire in the oil sands plant. It would have been very easy to get into a blame game and to say, "that happened because that particular unit wasn't built right or they didn't operate it right or they didn't inspect it like they should."

How this company did react to that was everybody pulled together and said, "Okay, listen. What do we need to do?" So the refinery guys called. "Can we send somebody up to help? What do you need?" The major projects guys went into emergency session about what was required. They're pulling drawings out within an hour of the incident. Thank God, we didn't have a single injury.

I was up there the next day. You've got to be careful not to show up so quickly that they have to worry about taking care of you. But I was up the next morning. I walked the site, talked to the people who experienced it. It's more about the time than it is about what you can actually do. You can't really do a lot.

I also was on the phone the next day to shareholders. I didn't have the answers. I just said, "Listen. Here's what we know about what happened." The thing about emergencies is you don't want to jump, because you don't really get the true picture of what happened for about a month. That's why as a manager you never want to panic. One of the things that I'm proud of is that our management team didn't panic. And we led the investment community through it. The stock price did not suffer the next day tremendously. We went down a buck or something like that. I said, "You guys give me three or four weeks to figure out where we are." Came back in three or four weeks: "Okay, here's where we are. Here's what we think it will take. Here's what it will cost. This is what the insurance coverage looks like. This is the outlook." And we did the same with our employees. Again, it's just kind of that open leadership style. And, you know what? The shareholders stuck with us.

Conclusions

Energy companies are just beginning to understand the magnitude of the challenges they are facing. They will need to change their entire operating models to address the different resources and types of partnership that will be required by the markets of this next half century. The high prices of today are not just part of a normal investment cycle, but signal a phase change in the global energy complex. Responding to this change will require different strategic directions than those of the past, and strategy execution will require different ways of thinking. We believe that there is much to learn from our interviews, but that the learnings will need to be adapted and extended to the newly emerging reality.

In executing strategies, leading energy companies will still be obliged to secure commitment from their employees and support from their shareholders. They will need a paradigm shift in their value propositions to governments, national oil companies, and citizens of resource rich countries if they wish to gain access to resources and markets; they will need to rethink their approach to partnering. They will have to further improve their ability to deploy resources by inverting their capital allocation process from screening projects out to screening projects in. They will further need to reappraise their approaches to outsourcing and technology development and be sure that they have access to the scarce technical resources that they need to implement their strategies.

This will end up with a shifting of the corporate boundaries and require some rethinking of how the extended enterprise will be managed. Although some companies have stepped back from collaboration and outsourcing, the authors are persuaded that the industry has much to learn from other industries on the importance and the value of effective and more extensive collaboration with customers, suppliers, and external stakeholders. In many cases, energy companies will need to realign their organizations with the true value drivers of their different business lines and move away from a functional emphasis on efficiency towards a focus on effectiveness. They will continue to require accountability from their businesses to deliver on promises, but they will need to do that within a more integrated approach of bringing more difficult resources to market. Finally, they must work harder at going beyond merely securing a license to operate to become an integral part of the solution in satisfying the needs of the communities where they do business, recognizing that different stakeholders are at different places on the Maslow hierarchy of needs.

We have described how the industry has allowed itself to be characterized as an unresponsive, polluting, and venal enemy of society. To create a more positive and accurate image would have required collaboration among the tribal chieftains of super-majors, independents, refiners, gas utilities, and power companies, as well as their suppliers, and funding of joint education programs. With public opinion on rock bottom, new collaborative efforts are underway. The jury is still out on whether the industry can reposition its image.

The authors believe that the route to *terra incognita* will be difficult, will require expert navigation and, most of all, will require partners. With partners, "positive sum games" can be designed that create new wealth that can be distributed equitably among the partners, and a diversity of talents will be available to address implementation issues as they arise. Without partners, no company can hope to address the needs of all the stakeholders with an interest in the global energy complex. Partners will include other private and national energy companies, suppliers and technology developers, international agencies and NGOs, and local and regional communities and political leaders. Partnership will require sensitivity to the hierarchy of needs of others, and recognition that finding the right technical solution is only the beginning of the journey.

1 Adam M. Brandenburger and Barry J. Nalebuff, 1996. *Co-opetition.* New York: Doubleday, p. 8.

2 Harvard Business Review January-February 1995, Letters to the Editor, p. 142.

3 John P. Kotter, 1996. *Leading Change.* Boston, MA: Harvard Business School Press, p. 21.

4 Edwin C. Nevis, Joan Lancourt, and Helen C. Vassallo. 1996. *Intentional Revolutions.* New York: Jossey-Bass

5 Kotter, p. 108.

6 Larry Bossidy, Ram Charan, and Charles Burck, 2002. *Execution: The Discipline of Getting Things Done.* New York: Crown Business.

7 Quoted in a *Houston Chronicle* article on the 2006 G8 conference by Julie Mason.

8 In the original example, individual herdsmen could graze their herd on common land at no cost. Their incentive was to expand their flocks, but in aggregate this led to over-grazing and the loss of sustainable grazing land for all. Regulations were required to save the herdsmen from themselves.

9 Leading in Turbulent Times

Leadership has many dimensions. Leaders are the captains setting the company's direction, but they also have to have on board the first mate and boatswain to propel this vision into action, which is particularly important in turbulent times. In this chapter, we will review what our interviewees told us about leading large energy organizations. There are many thoughtful writings on leadership, and we have drawn from several of them as well as from our interviews in proposing a framework that seems to work for the energy industry, particularly for the turbulent times that lie ahead.

Leadership Models

We learned from energy company leaders that they convey a sense of purpose to the organization, they deliberately shape the culture and values, they create an environment in which decisions are made, not deferred, they demand and achieve high performance from everyone in the organization (and their suppliers), and they pay a great deal of attention to their people (fig. 9.1).

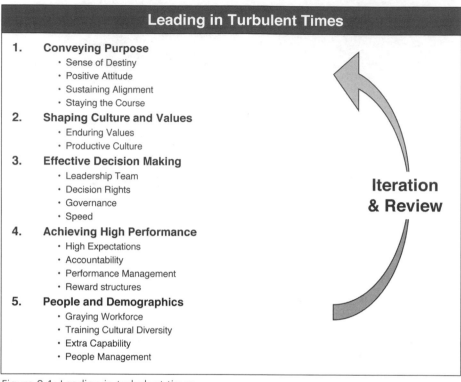

Figure 9.1: Leading in turbulent times

Is this the right leadership model? As we will discuss in the next chapter, the concept of leadership evolves over time. The important point here is to define an archetype that covers the key leadership dimensions in a thoughtful and useful way. In order to give the reader some comfort, our first objective is to correlate this leadership framework with the perspectives of several leading thinkers on the subject. Specifically, we will highlight the thinking of Warren Bennis, Howard Gardner, John Zenger, and Joseph Folkman.

Conveying purpose

In conveying purpose, the energy leaders seem to embody a sense of destiny with a long-term view of the company and articulate the corporate narrative consistently and persuasively. Warren Bennis calls this "managing the dream."[1] They maintain an aura of personal optimism, which others in the organization come to rely on, and they keep the organization aligned and on target, while encouraging debate.

Bennis refers to guiding vision and passion, and creating the trust that will get people on your side. Gardner, after analyzing leaders from politics and the world of ideas as well as business, is struck by the story that leaders tell and how it resonates with the sense of individual and group identity for those to whom it is addressed. Zenger and Folkman assert that "The highest expression of leadership involves change, and the highest order of change is guiding an organization through a new strategic direction, changing its culture, or changing the fundamental business model."[2]

Shaping culture and values

Energy leaders recognize their responsibility to shape the culture and values of the organization. They insist that the leadership teams "walk the talk" and lead by example, fostering a progressive environment. They also are prepared to remove people who may be highly motivated and technically competent, if they do not live the values.

Bennis, Zenger, and Folkman believe that character and integrity are central to leadership. Gardner talks about the leader being the embodiment of the ideas he or she is championing. People are not perfect, and leaders are people. Everyone has flaws. Leaders with serious flaws in character and integrity place the organization at high risk, as the Enron scandal exemplifies. Actions speak much louder than words. "Walking the talk" is essential to shape an energy company's culture and values.

Effective decision making

These energy leaders create an environment in which decisions are made thoughtfully and in a timely manner. They select a team with whom they are comfortable. They make it clear where the decision authority lies. And, they spontaneously (and in advance of Sarbanes-Oxley) establish a governance system that assures stakeholders that the company operates ethically.

Setting the context in which decisions take place as a matter of course is not easy, especially in turbulent times. Bennis discusses leadership attributes of curiosity and daring, and we would agree that leaders should demonstrate and encourage curiosity to uncover new opportunities and daring to take risks. John Browne, CEO of BP, and Jim Mulva, CEO of ConocoPhillips, have shown both of those attributes in their Russian ventures. Gardner sees the organization as the vehicle for realizing the story, but notes that leaders in the domain of ideas found it very difficult to organize those they wished to influence.

Achieving high performance

All of these leaders demand high performance of themselves and of the people in their organizations, and most have set up reward systems and cultivated a motivating environment that drives high achievement through accountability. Zenger and Folkman have determined that a focus on results is an essential competence for leaders, and we find their formulation to be representative of how our interviewees talk about achieving high performance. Gardner points out that this focus on results can be achieved by direct or indirect leadership, and cites Jean Monnet as an example of a leader who worked through others to achieve his vision of a European community, effectively causing them to achieve the high performance he encouraged.

People and demographics

Finally, we believe that the people, demographic, and diversity issues faced by the industry represent another aspect of the looming crisis that leaders will have to address personally. To do that with credibility, they will also have to take a more proactive position in creating public awareness and improving the industry image as described in chapter 8 in order not only to retain their license to operate, but also to attract and retain the talented people that they need.

There is no leadership without productive followers. Turbulent times, in particular, require different points of view that come from a diverse workforce. Developing followers manifests itself as *interpersonal skills* for Zenger and Folkman and as the *audience* for Gardner. We see this as a broad array of conversations with feedback loops, rather than simply telling the story to the audience. In this way, the story becomes more and more refined, and the sense of purpose is reinforced. Strong interpersonal skills are required to encourage these conversation to be productive rather than lapsing into a Tower of Babel.

In the energy industry, the imperative for what Gardner calls "the issue of expertise" almost goes without saying. All the energy leaders we met have spent their lifetimes in the business, have mostly a technical background, and started their careers in the field. ConocoPhillips is an exception in having a preponderance of financial executives at the top. Most energy companies are dominated by engineers, with an occasional scientist to add zest, such as BP's John Browne, who earned a bachelors degree in physics. But nearly all can debate seismic maps with their geoscientists and discuss catalyst selectivity with their refiners, as well as having personal experience of negotiating with the highest levels of government. There is little chance of an individual lacking in personal capability rising to the top of these organizations. By contrast, business skills such as negotiating with the highest levels of government and dealing with outside stakeholders are learned largely on the job. Lack of business education can be a major competence weakness, and the authors contend strong energy leaders need a blend of both business and technical skill training.

In summary, Table 9.1 maps the leadership framework that emerged from the interviews to models of several respected authorities on leadership. The relationship and linkages are useful validations to proceed forward with this energy industry leadership framework.

Table 9.1: A Map of Leadership Models

Energy Industry	Bennis[3]	Gardner[4]	Zenger & Folkman[5]
Conveying purpose	Guiding vision passion	The story	Leading Organizational change
Shaping culture and values	Integrity Trust	The embodiment	Character
Effective decision making	Curiosity Daring	The organization	
Achieving high performance		Direct and indirect leadership	Focus on results
People and demographics		The audience	Interpersonal skills
Expertise taken as a given		The issue of expertise	Personal capability

Now, we turn to the thoughts of our energy leaders from the interview process. Our theme is that leadership can be learned more effectively from listening to the stories and commentary of real leaders than reading generic leadership texts.

How Leaders Convey Purpose

The fundamental questions for any energy employee considering following their leader are: "Why should I join you on this voyage? Where are we going? And what is in this for me if I come along?"

Sense of destiny

We found that the leaders who had built their companies from the ground up tended to have a strong sense of destiny. This was less pronounced in the leaders who had been appointed from the outside, who were highly focused on their duty to shareholders. The leaders of the super-majors differed in being very conscious of their place in a continuum of first-class leaders committed to move the institution forward.

Dave O'Reilly, CEO of Chevron, with characteristic modesty, summarized the super-major perspective: "I think every generation of leaders wants to leave the company in better shape for the next crowd of managers. To me that's very important."

The long view

Rick George, CEO of Suncor, articulates his sense of destiny in terms of the longevity of the resource that he is developing:

> I think this comes from strength around this vision of Suncor— we're going to be here producing oil sands for 100 years. And you just put that in your brain and think about that. So where a typical oil company will be in a certain country, they know they've got about a 20-year run on their leases with the ability to sell it when they get to a certain point. But in Suncor we're going to be producing oil sands for a hundred years. If you think about how much our views on the environment have changed over the last 10 or twenty years, then think about the next 100 years and what the demands are going to be on us. So, my view is that you've got to stay way ahead of the game.
>
> Why does Suncor invest in wind power? Is it because it gives us a high rate of return? No. It's because it's another 100-year view item. My own view of the wind power business is that it is probably the lowest return business we invest in, but I like the technology. The turbines we get in the next five to 10 years may be the ones we put in for the next 100 years. So when you take that 100-year view, the nice thing about those windmill sites is we will get to keep them for a hundred years. I do like the fact the whole course is renewable.

We discussed the fact that George had started very young as CEO, and has been in the position for 15 years: "I would just say that probably there is even a little bit more weighing of that obligation, of thinking long term, as opposed to I'm here as a shepherd, I'll watch the flock for five years and then head off to the sunset." Steve Lowden, CEO of Suntera, has a sense of destiny for his embryonic firm: "We look out 10 years, and by this time Sun Energy and Suntera are multi-multi-billion dollar companies. And we're going to look back, and we're going to find the IOCs will be a quarter of the number they are today. And the NOCs will be bigger than the IOCs and they'll be international. And I hope that I'll be playing with most of them." And he thinks it all comes back to the attributes espoused by Bennis: "I think executing the strategy is as much about will, desire, passion, and ability to choose the right people in the team. I think mankind has shown that you can do whatever you want to do if you set your mind to it, and you know who you need to go to for help. Then I think we can achieve whatever we wish. Of course, your business model has got to work as well."

Overcoming adversity

Aubrey McClendon, CEO of Chesapeake, exemplifies the determination that comes from successfully overcoming adversity. He talks openly about the rocky road to success.

> Every month I meet with our new employees, 30 to 35 people a month; and I sit down with them as part of an all-day orientation. I spend about three or four hours with them, and we go around the room and everybody has to tell me who they are and where they worked.
>
> Then I take them through the history of the company. I say we started in '89 with no money and a few people; went public in '93 in the worst IPO in America: no one liked us for our strategy. From '94 to '96 we were the best stock in America, up 8000%; '97 to '99 were one of the 10 worst stocks in America, down 99%. From 1999 to 2005 we're one of the best stocks in America.

By telling this story, he is creating a sense of adventure, commitment and resilience that binds employees together.

This sense of overcoming adversity to inspire change also propelled John Wilder of TXU after he became CEO:

> Well, the way I approached it wasn't anything ingenious. It was kind of a standard model. You create some sense of urgency, a sense that things have to be different. That was pretty easy. It didn't have to be manufactured here. We were taking 300 seconds to answer the phone for a customer. We had $14 billion of debt, but we generated only a billion and a half of EBITDA. We had a business worth $4 billion go bankrupt in Europe. We had billions of dollars of lawsuits against us. We had $2 billion of underwater commodity contracts.

It took a week to make a list so long that if someone would just listen—which our employees did; hopefully our customers did; and our board did—we have a problem and we've got to change.

Positive attitude

It is hard to follow someone who always has a negative point of view. Bill Greehey, Chairman and former CEO of Valero, emphasized the importance of a positive attitude:

If you don't have a positive attitude in the energy business, and a "can do" attitude, you're lost. Years and years ago, when I was first running LoVaca, under quasi-receivership by the court, we had all these bad things happening with the Railroad Commission. They came out with two doomsday orders. If the orders had been carried out, LoVaca would have gone bankrupt and the company wouldn't survive.

When the first one came out, I told the board, "I've been talking to the Railroad Commission. They are not going to put this in effect; they are not going to do it." And about three days later the Railroad Commission came out with another order rolling back the price to the original contract. Our average cost was 60 cents, and they were going to revert to the contract, 22 cents, which means we are bankrupt day one. And so I said again to the board, "I talked to them and this is just to get Coastal to work on the settlement and get this company spun off. And I promise you that these orders will not go into effect." This pastor that we had on the board—a Presbyterian minister—said, "Greehey, I worry about you." I said, "Why?" And he said, "You are so optimistic that you are not realistic." And I said, "You know what? You are a man of the cloth, and I have more faith than you do."

But I think you've got to have confidence in the people. No matter how bad things got in the company—this is leadership. People always knew that we would find a solution and we were going to be fine. Nobody worried.

Sustaining alignment

Employee alignment requires significant "navigational leadership," particularly given the diversity of cultures and mindsets around the world. Nick Butler, group vice president of strategy for BP, addresses the increasing challenge of alignment for super-majors straddling the globe in their operations and for their designated leaders of large organizations in foreign lands:

> The nature of what people are doing is different. You can't manage by command and control. We can't instruct on a day-to-day basis what that person should be doing in Russia. They have to have that leadership capability. That's a high level of person. It's an unusual skill, and it's questionable how far you can train it. You can make mistakes in some of these appointments, and we have. You have to set boundaries, but within quite broad boundaries they have to have trust and clear responsibilities to be able deal with events and circumstances.
>
> This is one of the biggest things for a global company: how you manage the reach and spread. I would say that one or another global company is going to fall down on this. There has to be an underpinning of shared beliefs, but that's quite tough when you have 100,000 Russians who have never been here. We create quite a large sense of network for the top 50, the top 300, and so on. But there is a large number of staff beyond that who are local and not international. Imbuing them with a set of common values is quite difficult, because you're just doing it on the Web site or on paper. So all the management do a lot of town halls and get around and talk—try to disseminate messages.
>
> Schlumberger has a unique model for this. But it is easier to do that at the professional level. We have a large number of people who are not professionals; the majority of our employees are basic workforce, upstream and downstream. Schlumberger can move their people around. Most of our people don't move across countries. We can make it work with the professionals, but it is more difficult with the rest of the work force.

Gwyn Morgan, CEO of EnCana, recognizes the difficulty of achieving alignment: "If I summarize my views on leadership generally, it would be that most organizations in the world are misaligned. Almost all government organizations are terribly misaligned. Union-based organizations are misaligned. And a lot of private sector companies are misaligned. In other words, there is a misalignment between what you're trying to create and what you want to accomplish, the behaviors you are going to get, and the rewards that people will get for doing those behaviors."

Alignment can be powerful when it exists in an organization. Morgan gave a specific example from the divestment of EnCana's North Sea and Gulf of Mexico assets:

> Some people say it was a bold decision. To us, it couldn't have been an easier decision. The management team was very much on board. One of the big things about an oil company is that you can often fail to make decisions. Decisions get pushed around, or the facts and the true picture get pushed around by people worrying about what it means to them. We've managed to avoid that.
>
> In fact, our leaders in the North Sea were saying to us, "We've looked at what you can do in North America. We can't match that. We can't ask you for more money for the next play because we can't compete with the other opportunities." And they have done well. They discovered Buzzard, and they were well taken care of. The Gulf of Mexico guys also told us we should sell the Gulf of Mexico. And they knew they'd be well taken care of too, and that we'd handle it well. We had that sort of understanding—just a sense that the company has a track record of making sure people are recognized for what they do. And if they work hard and do real well in the sale, that would be a good thing for them.
>
> The other way around, as well: there are penalties for falling out of line. If you can align including the ethics and values, and if you can put in place great people, just sit back in awe at what happens. I've been awestruck by what this company has accomplished—more than my wildest dreams, simply on the basis of a few principles, a foundation, and alignment.

Staying the course

Sustaining alignment is closely related to another key aspect of leadership purpose—knowing when to plot a new course or when to stay the course. Dave O'Reilly notes the constant distractions of modern times: "The other thing that is different today is the transparency of the business—the communications, the instantaneous flow of information from all over the world, the way the markets function, the role of the financial markets. The fundamentals still drive the business in my view, but the day-to-day and week-to-week volatility that we see is to some extent a product of the immediate information and immediate flow of money, sometimes on rumor, every now and then on facts. That is a change. The leadership challenge is to not get over-preoccupied with this, to stay steady and not be obsessed by looking at the screen. After all, our business is not the blip on the screen. For most of us, the business is still about supplying energy: finding it, developing it, transforming it into whatever shape it needs to be, and getting it to the market."

We also asked Jeroen van der Veer, CEO of Royal Dutch Shell, what will be distinctive about Shell in the future, recognizing that it has been previously renowned for its "long view." Van der Veer said, "These strengths are as important to Shell today as they were yesterday, and we continue to build on them through our long experience in scenario planning and our experience of operating in more than 140 countries. My vision for Shell in 2015 is that we still be primarily an oil and gas company, together with base chemicals, coal technologies, and a growing renewables business."

Leaders Shape Values and Culture

In a business context, values are the standards and principles that govern a company's activities, while culture comprises the ways that an organization expresses its values in day-to-day work. Both are manifested in the actions and behaviors that an organization requires or condones. Most companies claim a similar set of values centered around integrity and performance; but cultures are quite distinctive to individual organizations, reflecting how the values or principles are embodied in policies and actions. Business cultures may be formal or causal, authoritarian or egalitarian, risk-avoidant or risk-taking, disciplined or permissive, conservative or flamboyant, collaborative or internally competitive, bureaucratic or entrepreneurial, process-oriented or results-oriented, and so forth. In practice, culture is always more complex than these simple dichotomies suggest.

All the leaders we spoke to were very conscious of their responsibility to be good citizens, and we were again impressed by the fundamental integrity of the energy industry. We do not know whether this industry is different from other industries in this regard. However, success in energy is so dependent on the trust of partners that misbehavior tends to be heavily punished if ever a company is viewed as untrustworthy. Companies can get away with being tough and difficult if they bring finance, technology, and execution capabilities to the table; but they cannot get away with failure to deliver on promises or any sniff of dishonesty. We believe that the prevalence of partnerships, particularly in the upstream part of the industry, creates an internal policing that demands ethical conduct.

Enduring values

Dave O'Reilly of Chevron started our conversation by observing:

The thing that has become the most apparent to me has been the importance of the value system of the company—how critical that is in an organization that is big and diverse with multiple cultures, multiple geographies and multiple lines of business. You can't possibly micromanage everything. One of the most important issues is communicating the value system in a manner that is fairly simple, and then checking on it periodically to make sure that the other leaders of the organization live up to it. As CEO, it becomes very clear that that is one of the major roles, to be the keeper of the value system.

That's not something that develops in one chief executive's career. It's something that gets passed along from predecessor to successor, because you don't just create a value system in a few years. It's something that builds with time. There are probably strengths and weaknesses in any company's value system. So, I don't take credit for the strengths because I think that they are in many respects the result of the efforts of my predecessors. Making sure that I personally see that we build on that and hand it over in better shape to the next generation is very important to me.

The value system has to include learnings from the acquisitions we made. It would be very simple to say that is only a Chevron value system, but I think it has to include elements of the other companies, because we are an amalgamation of multiple companies. If you go back far enough, you can even think of Standard of Kentucky and all the earlier moves. When we brought Chevron and Texaco together, I personally worked very hard to ensure that we agreed early at the leadership level what the value set is for the company. We worked hard to get everyone on the same page—not just understanding it on paper, but operationalizing it—understanding what does it really mean.

Helge Lund, CEO of Statoil worked hard on the values of his organization:

The first year in Statoil I spent a lot of time working on the value statements and the leadership principles just to make sure that people really understood them. It was not a campaign—it was really the heart and soul of leadership.

This is the most important part as there are people in many organizations that have values only considered to be nice words, "Just another fad." So you have to change the systems; you have to change leadership development programs, you have to change reward schemes, the succession schemes, and so on and so forth. People must see that there are subsequent changes. After we introduced the new values, we also introduced a new 360-degree evaluation scheme so, for instance, members of my executive committee are rating me on multiple dimensions directly related to the values. Also the employee surveys are connected to the value statements.

Internally I illustrate leadership in a matrix, and we are building a performance system based on these metrics. There are two dimensions of the performance system. One is what to deliver, the other is how. Of course all of us would like to be in the northeast quadrant but sometimes we are northwest or southeast. There are also a few in southwest and if they are unable to move out of this corner, they should not have leadership responsibility in Statoil.

The philosophy behind this is that complicated organizations need to work with other parts of the organization. Even though you delivered in your territory, if you do it in a way that destroyed people or company values elsewhere, you destroy more value than you create in your own unit. Over time you always have to remove this kind of behavior or you have to bring them to match the values. Then this has to impact your governing systems and your operating procedures.

If people see over time that those leaders that are delivering on both "how" and "what" are promoted they will grow confidence and start living the values. I also believe that they have to see the commitment from the CEO and from the top management team. I invested a lot of time in the process of developing it, and I continue to spend a lot of time on promoting it internally in Statoil.

Gwyn Morgan of EnCana has devoted considerable personal time to articulating the values he believes in and has prepared a corporate constitution for EnCana:

> I come at this from the point of view of a leader that has been building a company for three decades. You start with principles. And you never move from those principles. You can change your tactics. You can change your strategy. But you never change your fundamental principles.
>
> So part of those principles is that you create an ethical foundation in the company. Everybody knows how you are going to do business, what is expected, and what is beyond the pale. You can't live in EnCana without living the constitution. Every one of our major policies, whether it is ethics and integrity and so on, starts with the principle in the constitution that applies. And you can't work for EnCana as a contractor without knowing what the constitution says. And the contract says you have to understand it and live up to it.

The EnCana Constitution declares four "shared principles":

- *Strong character:* We understand that sustained shareholder value can only be delivered by people of strong character. We lift one another up to greater success, we are determined, dynamic, and disciplined, and we can be counted on.
- *Ethical behavior:* We function on the basis of trust, integrity and respect. We are committed to benchmark practices in safety and environmental stewardship, ethical business conduct, and community responsibility. Our success is measured through both our behavior and our bottom line.
- *High performance:* We focus where we passionately believe we can be the best. We are accountable for delivering high-quality work that's continually enriched by open, dynamic lookbacks and learning.
- *Great expectations:* We have great expectations of one another. Living up to them will enable us to experience the thrill and fulfillment of being part of a successful team, and the pride of building a great company.

Gwyn closed by saying: "There is a tremendous pride about being the kind of corporation that you say you are going to be. There is a huge problem if somebody steps out of line, including the CEO. It doesn't matter who it is. People know."

Company first mentality

John Lowe, with ConocoPhillips, talks about how the values of their CEO, Jim Mulva, permeate the organization:

> I guess it all starts with leadership. People who do well in ConocoPhillips put the company first and themselves second; self-promoters don't do very well. I think that comes from Jim; he's definitely the opposite of a self-promoter. He's relentless and tireless in his pursuit of making ConocoPhillips a better company. He's given up his life to the benefit of the company. He spends every day, 365 days a year, committed to making ConocoPhillips the best company he can make it.
>
> The other thing is to put his personal interest out of any consideration of any transaction. He talks about how you have to use your head instead of your heart. It would have been very easy for us to say, "Why shouldn't it be PhillipsConoco?" Phillips was 56% of the total. "Why shouldn't it be headquartered in Bartlesville?" People on the street in Bartlesville said, "Geez— there are only two things that were important in the transaction, where are we headquartered and what's the name. You lost both of them; how bad a negotiator are you?" So if a condition to do the deal with no premium, to do a merger of equals, is to say: "It's ConocoPhillips, and its home is at Conoco's headquarters here in Houston"—well then that's fine. So it's the ability to put aside emotion and make decisions in a non-emotional way that adds value to the shareholders.

In a similar fashion, Jeroen van der Veer of Royal Dutch Shell said, "I value teamwork very highly and have sought to develop a culture of 'Enterprise First' in Shell. This means putting the overall good of the company ahead of personal ambition and departmental priorities, and it is based on the three pillars of leadership, accountability, and teamwork."

Employees matter most

Bill Greehey of Valero elaborated on how employees matter most:

> People just forget that management is simple type stuff. Do the right things for the right reasons. Put your employees first. Treat everybody with respect, whether it's an employee or customer. The

principles are very, very simple. Carrying those principles out is the challenge.

We got rid of three top executives in the last four or five years. They were good at their jobs. They worked hard. They loved the company. They loved the culture. But they were never part of the culture. They always were talking down to their people. They were never developing their people. Their people were unhappy. And they weren't operating as a team. And you can't have that in an organization. If your culture is treating all people the same with equal respect, not talking down to people, then you've got to carry out that philosophy. If somebody is not doing it, you need to get rid of them.

At every management meeting, I preach about that. And when somebody gets promoted, I say, "If you want to succeed, care more about your people that work for you than you care about yourself." These are simple principles. They are just not followed in business. And we got rid of three really outstanding people that could never get developed to where they could bring their people along and treat them as equals. And it'll show up in reliability and safety at a refinery. If you have a safety and reliability problem, you've got a people problem.

John Wilder of TXU very much agreed with Bill Greehey on really caring for your people:

TXU had always stressed safety and had a good longtime record. But I wanted us to be tops because you're talking about people's lives. I'd hear about a safety incident and then I would ask about it repeatedly: "How is George doing? Has anyone seen his family? Who has been to the hospital? Give me the name of the hospital. I want to go visit him." You know, just trying to send signals out to people; when an employee gets hurt, it matters. And I started to use really personal stuff with management to remind them that they were responsible for making sure their employees worked safely, for insisting on safe work practices. I made sure they knew I was holding them accountable.

And we had an incident where an employee was killed. Electrocuted! And I was meeting with the business unit that had the fatality. The business unit had a great year. So we are talking about their bonuses, and they're bragging about their great year. Their safety statistics were fantastic from a recordable and a minor injury standpoint. So I said, "Listen. I really want you guys to tell me. Do you truly believe management needs to be paid at these levels when one of our employees has died?" I said, "I don't buy it. I think from me, on down to all of you, we need to take a hit in our bonus." I asked, "Are you satisfied with what we have done here? Here we are talking about this big bonus we are going to give ourselves and one of our employees is dead, for goodness' sakes." I made the point that as senior management, we don't escape the accountability—we can't just push it to the middle management.

Productive culture

A productive culture is one in which, as a North African client and soccer player once expressed it, "players focus on kicking the ball rather than kicking other players." A productive culture not only enables high performance by encouraging objective assessment of improvement opportunities; it also makes companies into desirable business partners, and opens up new opportunities.

Culture pays

For Bill Greehey, the culture that he deliberately put in place in Valero was an important competitive advantage in executing his strategy of growth by acquisition:

We acquired Benicia, which was our third acquisition. We did the Basis deal first, which we bought at like 10 cents on the dollar. We hit a home run. It put us in the big leagues. What's interesting about Benecia is that it was not going to be sold to the highest bidder. It was going to be sold to the best bid.

This guy, Gene Renna, the senior VP of ExxonMobil in charge of refining, said "We're going to take a look at all the different parts of your business. We are going to look at your involvement with your employees, what you do with your employees, your benefits, your salaries, your policies. We are going to take a look at the community, what you give back to the community. We're going to look at your safety record. We are going to look at your environmental record." And when they compared us with UDS and Tesoro, which were the other bidders, we were the low bidder, but yet we had the best bid.

And when Gene Renna called me up, he said, "I've got some good news and I've got some bad news." He said, "What do you want first?" And I said, "Give me the bad news." And he said, "Of the three companies, you've got absolutely the lowest bid." And I said, "Well, Gene, what the heck is good about that?" And he said, "The good news is, you've got the prettiest face." This means that we had the best bid. And he said, "The only thing is, you're going to have to raise your bid." He said, "And even when you raise your bid, you're still going to be the lowest bidder. But you've got to get up to $900 million." And I was kidding with him. And I said, "Gene, I would never buy anything with a nine in front of it.' And he said, 'How about $895?' That's how we bought Benecia.

And I'll tell you, an important part of the culture is that when we started this company, spun off from Coastal in 1980, we said in our mission statement, "We're going to be good corporate citizens. We're not just going to take from the community. We're going to give back. We're going to make it better. We're going to give our time, and we're going to give our money." Our employees take so much pride in what the company does in giving back to the community, and their involvement in the community. We have volunteer councils. We do all these special things. And when people get excited about that, they get excited about the company and their jobs. You'll never find anybody happier than our employees at Valero.

As Chesapeake expands geographically, Aubrey McClendon stresses how they are trying to retain their distinctive culture and encourage knowledge sharing:

We have asset teams with probably 25 people that report up to three different regions all which are headquartered in Oklahoma City. We have almost 25 field offices, but they are only gatherers of information and implementers of decisions. They are not individual fiefdoms run by guys who are competing against this division or that division.

Instead, we get everybody where we can look them in the eye. We believe that we've got a very distinctive company culture. We've got a very distinctive company campus. We want everybody to have exposure to that, including relatively open and frequent interaction with the guys at the top. We are highly visible on campus. We have no high rises. We have a whole series of four- and five-story buildings; it's a hard driving place, but also a very human and a very organic place. One of the things I have always said is that a company's levels of hierarchy rise to meet its levels of floors in the headquarters building. With that as a truism I want to be in a five-story building; actually, I'm in a two-story building. I think that employees really respect that. So that culture for us is so unique that we wanted it cooked and eaten in Oklahoma City by everybody.

Customer focus

John Wilder of TXU determined that reorienting his employees to a strong focus on the customer would explain why he was making tough decisions to energize the significant internal change process:

What I did a little differently here than at Entergy is really zeroing in on the customer. And I had learned, was more experienced, and more mature as a manager. I tried to make every decision in the context of our customers who now have the freedom to choose. They're in control. They're making the decisions here.

Our back office was woefully inefficient. You could just tell it by walking around. And you could see it all in the numbers. So one thing I wanted to do very early was shift execution risk to higher-quality management. Just give it to the pros. And, as you would expect, the employees in the back office, they just couldn't stand me. They thought I was a traitor. "Now how could you come in and take away my lifetime job."

But here's how I tied it into customers. "This decision was for the customers. It is taking us 300 seconds to answer the phones for the customers, and that's unacceptable. They don't have a good platform to interact with us on the Web. They don't have the ability to pay our bill with credit cards." I could go on and on, but we had this kind of very vocal exchange. And it was a good exchange. It was not one that our culture was used to. Some of the people wanted to take me on. I said, "Let's have a good debate. This isn't about you. Let's be clear. This is not about you. This is for the customer." It was a defining event for us as a company. This is the kind of business we're in. It's about serving the customer better. And we took our average speed to answer from 300 seconds to 15 seconds in 60 days.

Building professionalism into the culture

CEO Abdallah S. Jum'ah of Saudi Aramco talks about the importance of a professional code of conduct:

> I think the most important aspect is not really what we do, but it's how we do things. I think the respect that Saudi Aramco has among its suppliers, its customers, and its partners is that they like to deal with us because of our corporate culture. We will do whatever it takes to satisfy our partners, to satisfy our joint ventures, to satisfy our employees. There is a corporate culture here that is built, foremost, on reliability. People like this. We are professional. We are efficient. Our partners like to work with us because we are above board. And if you sum up the reaction of any employee in Saudi Aramco, what we feel is this pride on the part of the employees about being connected with an organization of this sort.
>
> We have 52,000 employees. I go and tell our young people in the plants, particularly the people in the remote areas, "Don't think parochially, you are not to think on this platform that your dos and don'ts on this platform do not shake the world. They do shake the world. You are a link in the chain that takes us to the market." And I tell them, "NYMEX is shaken every day up and down by a pump going down in the refinery on our East Coast. So if you don't do your job well, the world is going to be different. And you are only one in this organization." So I think that people feel that they are part of the bigger world, that they affect it. This is the culture we are trying to impart to the people.

We asked Abdallah S. Jum'ah if Saudi Aramco would be the Standard Oil of the 21st century. He said:

> I think our culture would even strengthen our position. I give credit for this to our American connection. There is no question about it. Because unlike any other national oil company, we continue to be proud of our connection with our American ancestors. Yesterday I was talking with the former chairman and CEO of Saudi Aramco, Frank Jungers. He said, "You know, I'm very happy that the work ethics and so on that prevailed during our time is continuing."
>
> We had a good base, number one. And we built on that base. Seven years ago my management team and I sat together and asked,

"What made Saudi Aramco what it is today? Let's identify the elements that make us different." And we articulated our corporate values. And we distilled all of that talk into 10 corporate values. These corporate values are our concern for human resources, fairness to our employees, safety of our employees and facilities, our work ethic, and so on. We sent these to all our employees and we reminded them, "Look, what is making us successful today are these values. So we have to keep them." Today any young man coming into the company, first we drill these ideas and work ethics.

I think one other aspect is that the CEO and the lowest level employee are subject to the same dos and don'ts. It is very important. For instance, take our work ethics and conflict of interest policies. I sign it every year. And every other employee signs it every year.

And it's clear, and people know it. Employees and officers can have their own businesses; but they shouldn't deal in any way with the company. I think that the business ethics and conflicts of interest policies of the company are creating a sense among the employees that, "You operate for the company; you don't operate for yourself." We would fire anybody who violates them. There are no sacred cows, executives or non-executives, if they don't abide by these rules, we just fire them. And we have done it.

Leaders' behavior matters

How leaders behave personally, their role modeling, has a tremendous influence on the culture of the firm. Ian Howat explains how Total's CEO, Thierry Desmarest, shapes a productive culture:

The group business review is held for E&P and for R&M once a month. The agenda is totally under the control of the heads of E&P or R&M. We debate what we think about them. I've seen many times Thierry Desmarest say, "Well I like this aspect, but I don't like this other aspect. It's got too much risk here and too little reward there. Go away and work those things and see whether you can come up with a better deal."

It's not obligatory to bring along embryonic projects, but the head of E&P uses it for that, because E&P is essentially an investment business; it's nothing more than the sum of all the projects. So the head of E&P likes to come along and say, "Let's look at this project; it's nowhere near ready for approval yet, but what do you think about it?" There have been some projects that we've actually seen five or six times on the way to approval, and Thierry Desmarest doesn't mind at all. He tests their ideas. He shows by the way he asks questions what we're looking for and indirectly imposes a quality standard. He makes it clear when he thinks people haven't done their work, but he doesn't do it disagreeably. He doesn't tear people to shreds. He's very polite and really almost kind to people, which some people might find a little strange.

By encouraging an open, non-judgmental dialog, Demarest shapes the culture in favor of creativity and objectivity. Through his questions, he conveys what's important to those developing investment opportunities. In this way, investment opportunities that would be killed in less open cultures are rethought and enhanced in Total.

What Leads to Good Decision-Making?

Good decision-making is not just the matter of choosing this or that, yes or no, approved or rejected, decreeing, or taking a vote. It is rather the preparation process in an organization that leads to the judgment, determination, or conclusion. Conducting this process properly is the role of energy leaders because good decision-making drives superior performance.

The right leadership team

Having a high-quality, aligned, well-formed leadership team is recognized as a precondition for successful decisions. John Lowe noted that ConocoPhillips' Jim Mulva, took some time to find the right leadership team: "Jim is a big believer of having the right people on the bus; and you have to have the right people, people who believe in what you believe in and are committed to what you're committed to. That is why you saw so many changes in the first couple of years. We found there were people who weren't committed, and that made for some rough days. Jim's view was that he was being held accountable by our board, so he needed to be able to hold everyone else accountable and the board supported him on this."

Bill Greehey of Valero also feels it is important to get the right "people on the bus." He recounts how he secured Rich Marcogliese from ExxonMobil to strengthen his leadership team. He urged us to "talk to Rich. He's a good guy to talk to because he's in charge of all of our refining operations and our Canadian operations. Here's a guy that was one of the stars in Exxon. Exxon would not let me interview him. They said, 'He's one that we are going to keep from Benecia.' And I kept telling them, 'That's not fair. We are making dual offers to all of the top management, and we're not making a dual offer to him. I don't think it's fair to him.' And finally they said, 'Well, he would never come to work for Valero.' And I said, 'Why? Let's see.' And I spent a lot of time at the refinery; so he and I got to know each other really, really well. He understood the culture. He understood the strategy. He jumped at the opportunity to work for Valero."

Aubrey McClendon thinks the Chesapeake leadership team is a clear source of competitive advantage. He describes the leadership structure and its continuity as follows:

> If Chesapeake has been at all successful, it is because of my partnership with Tom Ward. I think we have a very unique setup; I don't know of another public company where you have two guys at the top. You can't have two CEOs and two COOs, but we've got the closest thing to that, with me being CEO and Tom being COO. Tom is a co-founder of the company, and we've been partners since 1983. He runs the operations side of the company, which any COO could do. The difference is Tom owns just as much stock as I do, and I've got somebody who's really a soul mate. I can really sit down and talk to somebody that has as much skin in the game as I have.
>
> Whereas at the end of the day in a traditional organization, where you have a CEO alone at the top with a lot of his own thoughts, but never truly sure that all the thoughts he gets from down below are honest thoughts, spoken in complete comfort and confidence that there are no negative ramifications to saying, "That is a bad decision." Those conversations don't happen very often in other companies, whereas they can happen pretty openly and freely in our shop. I think the sense of mutual trust and respect that Tom and I have for each other permeates the whole organization. Remember, this is an organization that 10 years ago had 10 people. Today I've got 13 top people, senior VP's and above, and 11 of those have been with us 10 years or more. It really is the original group of people that built the company, so they know the values.

Steve Lowden, CEO of Suntera, affirms the importance of the leadership team relationships in creating an efficient organization: "I think organizational fluidity in communication is extremely important. And I think that the biggest element inside an organization is trust. Trust makes an organization very efficient. Lack of trust makes an organization hopelessly inefficient. And I think for trust to work, people have to know each other. And they probably have to know each other socially as well as business-wise."

Helge Lund notes that it is not just about picking leaders, it is also about forming them:

First of all, I think the key parameter in leadership development is to give people assignments that are of a higher complexity than they think they can solve. I got mine in 2001 when I took over AkerKvaerner. I accepted the responsibility as CEO for a company that had more than 40,000 employees, faced serious financial challenges and served clients all over the world. Of course then you learn everything the hard way! For any person—regardless of being a leader or not—I believe being given "stretch assignments" is the most important way of learning.

Secondly, I think you have to have people around you that care about you in a day to day situation. I have never used a coach, but I like to talk to many people that I trust and give me precise feedback. I'm not so sure that I buy the idea of the CEO being the loneliest person in the world. I believe if you develop good teams around you, you should be able to sit down and also share with them your concerns or your thinking as part of a process. You should also have them quite close to you otherwise I don't think you have succeeded in being a leader.

Then I believe there is a formal element to it as well. We are changing the leadership and competence development in Statoil by introducing more leaders training leaders. Previously in Statoil, there have been too many courses and seminars and programs that have been driven entirely by external consultants. I am not discounting that—they can have a value and they can hold certain parts of the seminars. But I think the best way of developing a culture and establishing leadership standards and principles is by having the corporate management involved in the training of younger leaders.

We are changing that now, first by developing a clear profile and clear aspirations for our leadership. Second, by having all our executives taking on the responsibility as teachers in the leadership training. This is much more effective and efficient than having people outside the company designing our programs. All together, leadership development includes several factors; but if I should pick one, it will be to give people stretch assignments and follow them up closely.

So to build a distinct leadership profile, a part of that is being able to simplify and being able to see the bigger picture, and being able to renew the picture. Also to be able to dive into and understand not all the details, but the important details, is incredibly important. In many organizations there is less and less space for taking on responsibility and being accountable. In Statoil I want leaders who are hands on, who take responsibility and who "live and die" with the quality of our decisions.

Of course we should take due consideration of all the regulations, rules, governing procedures, but eventually we—as leaders—have to make up our mind and propose and decide something. This is what I mean when I say that perhaps in an industry like oil and gas sometimes the systems overtake good leadership.

Decision rights

For good decision-making to occur, an organization needs clarity about who has the responsibility and authority to make the decision. Bill Greehey was clear on where responsibility lies:

I had a chairman's committee; and everybody that worked for me had a voice in everything that was going on. But I made it very clear to them that, "If we can't reach a consensus on anything, then I'm the one making the decisions." And I made a lot of decisions that weren't very popular at the time. I'm sure a lot of people disagreed with what I was doing.

The no-layoff policy has been one that a lot of people fought me on. They just thought it was ridiculous to buy UDS and not cut people. But, if you ask people about my leadership, people would say that, "He always puts me first." All the employees have tremendous loyalty because they know I have put them first. They are going to be well taken care of. I respect them. I treat everybody equally.

Greehey's comments are echoed by Abdallah S. Jum'ah of Saudi Aramco:

If you come to our management committee, the way I handle a meeting is telling them, "Look, let's not be at each other's throats." When we have a presentation, I give every one of our employees a chance to say what he wants. I often tell them, "You have two minutes, or three minutes. Just come to the point. No speeches. Make your comments, and make your recommendations." I don't speak at the very beginning because I don't want to direct the discussion. And then, we debate. It's not a democracy. At the end of the day, there is one who is going to make that decision, and I'll make it. But still, we debate.

The other thing is that people today in Saudi Aramco are encouraged to speak up. We call it an innovation campaign that we put on a web site. "Look. Anybody in this company who thinks that we should do better, or has an idea, we promise you we will look at it." And lo and behold, now we are getting 5,000 to 6,000 hits every month on ideas. Some of these ideas have been patented. Some are very simple. And some are very complex. But we look at each and every idea. So there is a sort of a democracy—anyone can help. Anyone can do it.

The corporate strategy sets the direction of the company; so the decision-making process that establishes strategy is vital to the company's future and probably the most important area on decision rights. Total sees strategy as a conversation rather than a destination. Ian Howat explains:

Putting a project in the long term plan or budget doesn't imply it will ultimately be approved. Obviously there's a debate about the project at the long term plan meeting but we may say "Okay put it in the plan to see what it looks like." But the real strategy development is done in a very implicit way during meetings of the executive committee and the business reviews as well. You're following the company in real time. It's just a constant dialog with the CEO and the heads of the main functional businesses and main business divisions going on the whole time. It feels like a very small company; everyone is talking to each other the entire time. You mustn't get the idea of the strategy coming from some central decision and then communicated in a formal way. People are engaged in a continuous conversation on: does that fit with what we're doing or should we be chasing that project or not?

If you're talking about a big project in a risky country, it actually speeds up the project development to have the decision-making centralized. Because the country manager is not going to be comfortable investing $2 billion in a risky country, he's going to keep putting off the decision. If you do it centrally, he can bring the project to the executive committee and say, "Look, I know we've got a problem with our current production. But we have another project which is a bit different." Then he can see the way top management is thinking about it, and the dialog helps shape the project. That's probably how we would define strategy—as a permanent ongoing conversation between the CEO, top management and operating people on which projects to select and which to deselect, with a clear direction in mind.

The decision making process also permeates deep down in Total's organization. Ian Howatt explains:

The other thing you should take into account is that the people who come to present at the executive committee can be several levels down in the organization who are actually involved in developing the project. It's not the head of E&P who presents, or even the head of the Americas division; it's someone who has been involved in the work. And that is fantastically useful. It allows a lot of people at different levels in the hierarchy to see how the executive committee functions, to see the kind of things the executive committee is interested in and the kinds of questions they ask. Thierry Desmarest's questions are part of the communications effort; he's there for the whole debate. People can understand the reasons for his decision. It allows the executive committee to see the people coming up through the organization and to judge the quality. They can judge very quickly in that kind of discussion.

Governance

Hitt, Ireland, and Hoskisson do a good job of describing corporate governance in their book *Strategic Management: Competitiveness and Globalization.* "Corporate governance represents the relationship among stakeholders that is used to determine and control the strategic direction and performance of organizations. At its core, corporate governance is concerned with identifying ways to ensure that strategic decisions are made effectively. In addition, governance can be thought of as a means used by corporations to establish order between parties (the firm's owners and its top-level managers) whose interests may be in conflict. Thus, corporate governance reflects and enforces the company's values."[6]

Governance has come under close scrutiny in the United States as a result of several huge corporate scandals, with Enron being the noteworthy villain in the energy sector. In response to the public outcry, Congress enacted the Sarbanes-Oxley Act in 2002 placing new, more stringent rules and procedures on public corporations and setting out criminal penalties for top management where problems arose. These criminal penalties are based on the premise that "management should have known," which was a reaction to the Enron leaders' testimony that "I didn't know."

Energy leaders have mixed opinions on Sarbanes-Oxley. They recognize that it arose from real abuse, but find that it has added costs and complexity that will not produce real benefits. Most worrying to the authors is that it makes organizations look inward and upward rather than outward, at a time when just the opposite is called for by the energy trends. Societal expectations everywhere are that those closest to the "mine face" should have responsibility and authority to deal with issues that arise. Sarbanes-Oxley pushes organizations into procedural steps that dilute this authority, and put pressure on the CEO to know every detail of all the company's diverse operations—a quite impossible task in the large companies. Thus, many leaders have turned to overly extensive bureaucratic controls to manage the company in order to avoid the risk of criminal liability. Of course, this is how societies work in overreacting to major perturbations and the authors expect in due course there will be a cycle back to a balanced normality. Nonetheless, the interviewees' discussion around Sarbanes-Oxley gives good insight into the governance process.

Bill Greehey of Valero said:

> Let me tell you what Sarbanes-Oxley did. When Sarbanes-Oxley came out, if you looked at everything Valero was doing as a board, we were doing everything under Sarbanes-Oxley rules. You must make sure that all the board members are independent. Making sure the compensation committee is independent, audit committee is led by a strong financial guy. Making sure you have a lead director. I was the only inside board member. Every single thing under Sarbanes-Oxley, we were doing. We were doing environmental audits as part of the audit committee 20 years ago. We were the leaders in everything.

When Sarbanes-Oxley came out, all these articles get published. Board members read these articles, and it's like they are looking for something that you are doing wrong. And this company has never done anything wrong. As a management, as a board, our reputation is our word is gold—it's the values. That's the way we run this company. The board is completely informed of everything.

But Sarbanes-Oxley put the fear of God into board members. The minute it came out, board members started looking for problem areas. There was an article on airplanes. And then there was an article on hunting leases. All of a sudden they want audits of airplanes and hunting leases. We keep a log of the airplanes that anybody can look at. It's never used for personal use. I have never used the airplane for personal use. Nobody can use it for personal use. Hunting lodge is used for entertaining, very inexpensive. It's just crazy now what they're doing, checking all these internal controls. We haven't come up with one discrepancy in all the years.

The accountants and the lawyers, they realized that this is really a bonanza for them. And everybody is writing articles just trying to scare the hell out of board members on Sarbanes-Oxley. But you know what? The rules for a board member have never changed. It's the business judgment rule. And the board's biggest responsibility is to make sure they've got an honest CEO that they can trust, where everything is brought to the board.

But Sarbanes-Oxley was the best thing that happened because companies weren't doing the governance part that they should have, and they are still not. For example, I think if you're the CEO of one company, and I'm the CEO of another company, you shouldn't serve on my board. I certainly shouldn't serve on your board. I think that is absolutely a no-no. That's still permitted under Sarbanes-Oxley. And I wrote all kinds of letters to the New York Stock Exchange and said, "That's a conflict, and that's something that ought to be addressed."

Bobby Shackouls, former CEO of Burlington Resources, thinks that Sarbanes-Oxley has seriously dampened his enthusiasm for running a public company, but he does not include that as a factor in the sale of Burlington Resources to ConocoPhillips. Shackouls expanded on his views of Sarbanes-Oxley:

It's just a lot of overkill that didn't really accomplish anything. You can't legislate morality. And I think that's what that was an attempt to do. But Congress felt like they had to do something. And in the process, what they created was a bunch of busy work that really doesn't accomplish much. All the things that Sarbanes-Oxley causes you to do and document today, we were already doing. We were having disclosure control committee meetings, but we just didn't call it that as an example.

And then the other thing that it does is to set up traps. So if something does go wrong, it makes it easier for them to get somebody that they can tar and feather. All this was set up to prevent the defense of Skilling and Lay, which is, "I didn't know." Because with all the stuff that I'm putting in the file right now, I could never say, "I don't know." But I could have never said that anyway. That goes back to what I said earlier. You can't legislate morality. If you write a set of rules and you've got somebody that doesn't have the moral compass to follow those rules, he's going to sit down and try to figure out a way to get around them. No matter what they are. And smart people have figured it out.

Speed

The pace of change is accelerating to the extent that even companies in long-cycle industries like energy must consider the speed of their decision-making to gain competitive advantage. Frank Chapman, CEO of British Gas, notes the value of speed, and the difficulty of being agile and big at the same time:

The difference between British Gas and Shell? Size is very important. We are the size where we can participate in any world-scale project that Shell or anyone else can do. But we are much smaller. Therefore all of these things are much easier to oversee. They are all material.

So we have a philosophy that we deliberately do not participate in things where we have a small equity stake. Some U.S. power companies have a sort of philosophy of investing broad and wide, low equity stakes in many projects, and thereby spreading their risk. In fact, it's an approach that magnifies risk because you become highly leveraged to other people's execution capability. And so we are very small in number of assets because we tend to focus a lot of energy in single large projects.

This scale thing, don't underestimate it. If I take Exxon and Shell, they are companies of a size that requires they be run differently; they actually require different levels of process, different complexities in their processes. And as a result of that, they will move more slowly, and for good reason. BG is actually able to move very much faster without any diminution in the quality of the process simply because we are running fewer of these things.

Helge Lund, CEO of Statoil, takes a systemic approach to assuring speed and effectiveness in decision-making:

I think certainly that speed could be improved dramatically in this industry in terms of the decision-making. I came from the contracting industry. In that industry, long-term was next Friday. So you had a need for rapid decision-making. The value of leadership being bold was much more valued. Statoil has an ambition of speeding up our decision-making process. But when I look at some of our competitors, I think we are rather quick on our feet!

That is part of value creation. It is so important to work with a corporate culture that will be able to pitch together strong quality processes, but without jeopardizing the value of being decisive and being able to move quickly. I highly regard leaders that can implement quickly.

Most of my thinking starts with the culture and the values in the company. People need to deeply understand what the company is based on, because that creates some certainty of what they can do and not. People running around in the organization, uncertain of that, are wasting a lot of time.

The next layer is strategy and ambition, clearly articulated in terms of directions. In our case, they are clear. Those four elements: rejuvenate NCS, build a few strong international growth positions, grow the gas business, and improve profitability downstream. Very clear to people what they should do in terms of long-term strategy.

And then I have implemented a new operating model in Statoil that is very clear. When I meet the different business areas, the topics and agendas for those meetings are clear. They have a good transparency at what point they need to come up with ideas, proposals and projects. These elements give people some certainty so they can act with speed.

I should put one more topic to this, that I feel it is extremely important. And that is the profile of Statoil's leaders. What should a Statoil leader look like? We are in the process of defining that. In many big companies, systems can become more important than leaders. But I cannot go to my supervisory board and say I didn't deliver on my targets because the systems in Statoil didn't allow me to do that. Part of the reason for a distinct leadership profile is to be able to cut through the bureaucracy—being able to see the bigger picture, being able to innovate the picture. Also, being able to dive down and understand the important details is incredibly important.

Corporate staffs and all the functions are sort of taking over management. There is less and less space for taking responsibility and being accountable. I would like to raise the profile of Statoil leaders. Then they have to live and die with the quality of their decisions. Too often I see people coming to me saying they read up what the agency people have said, what the controllers have said, what this and that quality assurance process, instead of going forward themselves. Of course they take due consideration of all the regulations, rules, governing procedures. But eventually they have to make up their mind and propose something. I think maybe in an industry like oil and gas that sometimes the systems overtake good leadership.

How to Achieve High Performance

There must be a focus on results to achieve high performance. An energy company has to get things done and must hold its employees accountable for delivering on promises, measure their performance individually and in groups, and compensate the high performers commensurately with their contribution. All the leaders we spoke to consider that sustaining continuous pressure to perform is an essential part of their jobs.

Results can only be evaluated in relationship to goals and expectations. The leaders of high-performing organizations set goals that surpass previous accomplishments and challenge existing capabilities. They then make it clear that they expect to achieve those goals and will not accept less than full commitment and full effort from every member of the organization.

High performance is also about delivering on promises. Companies that fail to live up to the commitments they make become undesirable partners, and that can be the kiss of death in the energy business. High performance includes operational reliability, which is essential for safety and avoidance of environmental contamination. And ultimately, high performance is about continuous learning and application of the lessons learned to further improve performance. We found that companies that take continuous improvement seriously have formalized a "look-back" process to be sure they understand what worked and what didn't work from past decisions, so they can improve the results of future decisions.

High expectations

Mediocre goals do not produce high performance. The only chance for an energy company to achieve superior performance is to have high expectations. Bill Greehey of Valero sets high expectations for his people:

> I don't scream and holler at people. But everybody knows that I have a high standard for performance. And no one wants to disappoint me. And as a result, people get here at 6:30 AM and they leave at 7:00 at night. They are here on weekends. Everybody is working together as a team because they are happy. And it's the right philosophy. I told Bill Klesse, "The one thing I hope you do as CEO—I hope you have fun." Because if it's not fun, being a CEO is a killer of a job.

John Wilder, Chairman and CEO of TXU, stressed high performance in turning around TXU and using the past to help anchor the future.

> I started trying to get this performance dialog going. But no one wants to say someone is a bad performer. We had this huge cost problem—two billion bucks of SG&A. From just rough calculations, we probably needed to be at about $700 million or so. I went back in history; in the mid-'90s we ran at $700 million SG&A levels. So I put it in the context of performance. Then I could start using terms inside like, "Now, why can't we run the company with the kind of performance we used to? Those managers ran this thing with $700 million. We've done it before. Why can't we now? Come on, guys; you've been around. What went wrong? Why did it go from seven hundred to two billion?" So even things like cost, I tried to put in a context of high performance. Not in terms of, "Let's go

slash a bunch of heads," but that industrial companies run at low cost levels. That's your badge of courage, a badge of honor, not to run at high cost.

Wilder went on to confirm the importance of having people at the top of the organization who embody the high-performance mentality:

We are trying to change the selection, the development, and the de-selection processes. A lot of the officers we had when I started are gone. It was a kind of cultural house-cleaning that really energized the employees because underperforming executives were a big complaint.

I told the general counsel, "I'm going to call you with some names. And you are going to call them and work out a separation agreement." Now that's not a sustainable organizational model, but I wanted to start with the senior people. That sends the signals to everyone that there are no sacred cows. The neat thing about it is that we started getting self-selection. Literally, I was like one day away from giving the General Counsel one guy's name, and he goes to him, and says, 'You know, I don't really think this is the kind of place where I want to work.' And the General Counsel said, "Okay, let's work it out."

Accountability

Decision making improves when people know they will be held accountable. BP is widely recognized for the change in culture that they accomplished in the 1990s from a rather bureaucratic organization to a culture built strongly around accountability. Nick Butler of BP elaborated: "Accountability is included in setting what is acceptable and what is not acceptable. It's setting up the system through which people are accountable both for delivering against objectives and for the way they do it. John Browne has progressively adopted a system so that people's rewards and chances of promotion and at the extreme, whether they stay with the company or not, are determined by delivery and by behavior. They are both judged. It cascades down from the top, with each level assessing the level below. It's sequenced well. There is a 360-degree assessment. Of course it's subjective. For some people there are numeric measures, but for most people it's someone's judgment, a qualitative analysis; but it's reasonably well accepted."

John Lowe says that accountability is central at ConocoPhillips: "When you think Jim Mulva, you think accountability; he's very high on accountability. It's critical;

don't try to hide anything; be completely transparent. It may get us in to trouble sometimes, we're so transparent; we don't try to sugar-coat anything. We're just very honest, very consistent."

While Lowe thinks that the cultures of Conoco and Phillips were generally a good fit, he notes, "There were very clear differences between the cultures of the two companies on accountability; Jim is very results oriented. Each company had an operating plan, and each of us would update our operating plan. With Jim, you are held accountable to your original plan. What did you say you were going to do and what have you done? Conoco, as far as I could tell, really didn't hold people to any plan; they would keep updating and keep updating. So there really was no accountability tool in there. Jim is obsessive about a plan and consistency and accountability."

The practice of understanding in depth the reasons for good and bad performance and learning from them is fundamental in improving the decision-making process. Gwyn Morgan, CEO of EnCana, stated: "One of our key philosophies and cultures is called, "Look Back and Learning."

Duane Radtke, CEO of Dominion Exploration and Production, thinks similarly: "We focus very much on a look back process. And let me tell you, our people hated it when we started it. It was tremendous work to go back and sort out all the data. We now have a four-year database. We look at what we forecast. How much it was going to cost. What rate it was going to come on. How long it was before it was going to come on. What reserves we predicted we were going to book. And we use that as a learning tool in our forecasting going forward. The look back process helps retain objectivity in capital allocation." A strong look-back process also provides a set of priorities to fix. Because that is how an organization learns to do better. Most of all, it imposes a fact-based culture, where decisions are made on rational grounds, not on the amount of noise that a manager generates. It creates the foundation for a high performance organization by encouraging continuous improvement.

Performance management

Accountability and performance management go hand in hand. Rich Marcogliese, vice president of refining for Valero, helped us understand the different approaches of ExxonMobil and Valero:

> Exxon had a philosophy when I was there that within a certain category of jobs people perform at a high level and they perform at a low level. They enforce a very strict performance ranking system. But that creates an environment where there are very high performers, and then there are low. I have a view that there are

certain exceptional performers that drive the business more so than others. But then there are a lot of people that just do a very good job, day in and day out. It's very difficult to differentiate them from one another.

So in Valero's system, if I've got 50 people in my department, I don't have to tell five, "You're the last five in the group; and you're really not doing very well." There is not a structure that forces that to happen. In our system it is more that we're going to recognize the exceptional people. If there are people that really aren't pulling their weight, usually they don't stick around very long. But within those two extremities, there are a bunch of people that love the company. They are highly motivated. They work hard. We're not going to create a strong differentiation and put labels on people.

In the former Exxon operation that we had in Benecia, it made such a difference with people to get unburdened by that. As a result, I think we get strong teamwork. And, I want to say that people in Exxon are professional enough that I wouldn't say that that's a big problem. But the nature of that system, I've got to tell 20% of the people that, "you're below average." It is very de-motivating to people. So we don't do any of that. We give the plant managers pretty broad discretion in how they allocate merit treatment. And what they do is work the extremities. But then, within the extremities, there are just a plain bunch of good, hardworking people who all get rewarded pretty similarly.

The business cases are another difference between the two companies. We are not trying to grow an executive of the chemicals operation 20 years from now. We are not trying to groom somebody who is going to be the country executive in wherever. Our pyramid isn't as tall. We are trying to build expertise in refineries. We have movement of management spurred by growth, but we don't have this development system that is constantly trying to identify the very top people so we can move them around. If anything, we have kind of a little counterbalance to that; we want to maintain expertise in the refineries. And our best refinery operation is where we've had the plant managers there four or five years plus.

What I like so much about Valero is the philosophy that you get better business results from motivated organizations; and we're going to establish an employee culture that supports it. But it's not only the employee culture; it is the rewards system that goes right along with it. It's the orientation towards growth in this business and success. It all comes together, and as a result, you have people at Valero that work very, very hard. And they're not working hard because they think they are going to be 10 points up in the ratings.

Some companies have location-specific bonuses based upon operational metrics. We don't do that. We have the same, all-boats-float-on-a-rising-tide approach; and so even if a refinery has not had a good year for whatever reason, we won't trim a refinery's bonus because of that. Let's face it—the guy who is an operator on the wastewater unit, or a guy who is a pipefitter, did not have a lot to do with the fact that something went wrong and the refinery's metrics weren't all that good. Would you refrain from rewarding people over something that they may not have had any control over? The refinery manager may be a different matter, but even there, Valero emphasizes qualitative more than quantitative results.

Rick George, CEO of Suncor, has installed a comprehensive performance management system for setting goals and measuring progress:

The annual incentive program is one that's based upon an annual set of goals, and that's a very detailed roll-up. So those employee goals roll up to the management goals, and those goals roll up to the CEO's goals.

Well, let's take safety for example. It's evolved over time, but we have very rigid targets around lost time incidents and recordables frequency numbers. But, on the other hand, we started a program two years ago called Journey to Zero. And we'd actually try to measure how each business was performing against a targeted implementation of the Journey to Zero.

George elaborated on the continuous improvement of the performance management goal system:

Leadership goals have evolved over time. Where initially it was to try to identify high potential people, now, in each business we're measuring things like your engagement with employees. We do an employee survey every other year. Two or three years from now, that may change what those targets are.

Environmentally, we measure how many contraventions of regulations you have in your business through the year. We're going at that on a much more proactive basis, targeting reduction of emissions, energy intensity productions, and how are you doing against water usage. You get more sophisticated. When you've accomplished certain things, then you go on and try to accomplish others. It kind of continues to build.

It is important to recognize that achieving world-class performance is an ongoing task. Abdallah S. Jum'ah of Saudi Aramco describes where his organization is in the performance management process:

> We toiled with the issue of KPIs [key performance indicators]. But today, this is part of what we call our corporate imperatives. We continue to think we are doing a good job, but how do you measure it? How do you make people accountable for what they do? Yes, a plant is running, but are people accountable only if it shuts down? But it doesn't shut down. It is running. So what's the fuss?
>
> Today, we have driven this performance spirit down to the lowest operator. When you go to Ras Tanura or Jeddah refineries, you see competition among these young people on, "How can I produce the same thing with less energy? How can I save on power? How can I save on fuel? How can I reduce my giveaway in the refinery?" So we encourage people.
>
> But more and more we are in the process of developing our KPIs all over the company using a balanced score card. Every organization is now toiling with this. At the end of the day, I think we are moving, but we are not there yet.

Reward structures

Performance management systems are interlaced with reward structures. There are many different components of a reward structure beyond compensation and promotions, but these two are certainly key aspects. Over the years there has been much debate about whether compensation is a motivator or a hygiene factor. Those arguing a hygiene factor contend that when compensation does not fit the market place, it will be a demotivator; however, pay is not the primary driver of employee performance. Others think compensation can be a big driver if structured properly.

Whenever we brought up the issue of compensation, the response, as expected, favored relating it to performance. Companies differed in the extent to which employees were exposed to compensation risk and the balance between short-term and long-term packages. Total has moved in the direction of performance-related pay in recent years, as Ian Howat explains:

> Let me tell you a bit about how the company is actually managed. Total is very decentralized in terms of day-to-day management. There are objectives given to field managers, and they are held accountable for those. There is a whole remuneration package that includes basic salary, plus bonus, plus share options. The judgment on bonus and share options is based on three factors: the performance of the individual, the performance of the group he works in, and the performance of the company as a whole. So there is a mixture. We want people to look at their own performance and their business unit performance, but we also want them to look at the performance of the group as a whole.

> We consider ourselves to be a typical European company on the balance between fixed and variable compensation—more like Shell. BP has become an American company along with Exxon and Chevron in terms of compensation packages. In Total you don't have the kind of extremes of compensation that you get in the U.S. companies. Also, the stock option attribution is very different. We've started handing out free shares, which is something the French government has now allowed us to do.

> The whole idea of the remuneration system that we have was to take Total from where it was in 1990, a civil service ethos, to some notion of being a quoted company—to get people to start thinking about creating shareholder value and so on. So we have focused much more on individual responsibility through its linkage to bonus and stock options, and I think it has been very successful. But we didn't want to go to the other extreme, to the American extreme. We're still very conscious that we are a quoted company; we do a lot of work on financial communication. We also believe, like Shell, that there are a lot of other stakeholders that we are also responsible to, and we don't want to diminish that. I think we have a nice balance now in the way our remuneration package is framed, to send a correct signal to individuals. In my opinion, one of the problems with Enron was that in their business development the developers had a big chunk of their own money in the project, for example with Dabhol. We would not dream of going that far. We think it was excessive.

Bill Greehey of Valero was typically forthright:

> I think compensation should be primarily related to performance. The bonus is performance-related. We have three measures. We look at where the stock was at the beginning of the year, and where it was the end of the year. And so we evaluate that, and part of your bonus is determined by that. Then part of it is determined just strictly on the budget that's approved by the Board. Then the third one is an industry-wide comparison on how we do on return on investment.
>
> That's our bonus criteria. And then the compensation committee has a 25% up or down. Last year we had 140% return to shareholders. The year before, we had 108%. So everybody got 200% bonuses. In our company, everybody normally gets two weeks' bonus. Everybody got a month last year and the year before.
>
> Then we have performance shares which are based upon a three-year running average. One year drops out and another year gets added. We compare ourselves to the peer group on how we stand. If in that three-year period you are number one, then the number of performance shares you've got can double. If you're number two or three, then they can be 150%. If you're average, then they are normal. And if they are below average, then they go down to where you can get nothing.
>
> Obviously, if the company doesn't do well, you don't make any money off options. It's all performance related. Since our performance has been so great, everybody has been well compensated, especially on options. And we've got a lot of millionaires with options.

Greehey emphasized the philosophy as a moderate base salary with considerable performance-related upside: "Fifty percentile average on base salaries—we probably don't even average that in the industry. But when you take a look at our bonus, at the stock options, and restricted stock, we are way ahead of our peer group. But, it's performance related."

Rich Marcogliese compared Valero's approach to his previous employer, ExxonMobil:

> The system that I remember in Exxon—the base pay rates are very high. Success sharing or bonus payment only originates at higher levels in the organization. At Valero that originates right down with the hourly paid employee. We maintain if Bill Greehey

and Bill Klesse get a bonus, then the hourly-paid employees are going to get a bonus, too.

Exxon's value proposition is centered on a fulfilling career, high base pay, and generous retirement benefits. That's right. What I would say is, doing it our way supports that notion of excitement and dedication for the company. We are trying to grow a business. If we have a spectacular year, everybody is going to benefit. It creates a greater connection with hourly-paid employees that we are all working towards that objective.

EnCana has a similar approach. Gwyn Morgan explained:

Everybody in the company has what we call a high-performance contract. Everyone!—including the field operator who is turning valves. They agree annually with their supervisor. And that's tied back to their bonus. This time of year, everybody is going over their bonus and their HPR (high performance contract) with their supervisors. If you are a lower level, an operator in the company, it could be 10% of salary. In the higher levels, it could be a much higher level. It could be 100%. Then we have something called President's Awards. That's for accomplishments beyond and above anything that would have been expected. That can be up to 100% of salary as well and can be applied anywhere in the company. And it is. It can be a large amount. It is a big deal in the company.

The high performance contract goes to what you are going to accomplish that year and what your objectives are. But also there is a whole section on living up to the corporate constitution.

Raising performance expectations

John Wilder used rewards to raise the performance expectations at TXU:

I think the second day I was here I called in an employee who was working on one of our most important business initiatives. It involved our customers. Called her in and said, "You know, our performance here is totally unacceptable. You're going to fix it, and fast. And when you achieve it, I'm going to hand you a big bonus."

So this is like a three-minute discussion. And the employee jumped into it, and took a real leadership role on the project. The work generated real value for the company. The news about the bonus raced through the whole organization, and it was a good sense of symbolism that customers matter, and that performance matters because performance standards that seem impossible really aren't.

But, the really great thing about this story? This person decided TXU wasn't the place for them, and I agreed. She didn't think we were going to pay the money. The person said, "John, I don't know how to ask this. I feel like we did agree on that bonus thing, and, you know, we are going to hit it. I am leaving, and I would like to have some of it." And I said, "Some of it? You're getting all of it. I don't care that you're leaving. You made it! You get the money. It's fantastic."

In a start-up company, Steve Lowden of Suntera has a slightly different point of view since he is recruiting an experienced group of people:

Performance metrics and compensation can help or hinder trust. I guess I start from the perspective that we are going to recruit people that have done it before. They know what to do in their sphere. They are looking for the pinnacle of their career to be this. It's not a route to something else. This is not a stepping-stone. This is the stone. And also, they are very self-confident and probably reasonably affluent.

If I mix those all up, that would suggest to me that remuneration isn't the be all and end all of this drive. It is just useful. It's more of a reward tool that we are going to use than an incentive. Recruiting is incentivized by the business model. The trusting and friendly relationships, the excitement of the business, the winning business, and the rate of growth have been the main drivers. When it is successful, we'll get paid a lot of money. They will get paid through the success of the firm. It's total sharing. It's going to be more in long-term compensation than short-term—which is why it is very important for this to be the pinnacle of their career. They know each other, and they trust each other, and they are excited.

There's just not going to be any short-term payout. Basically, the employees get paid when the shareholders get paid. And the shareholders are very close to the management. That model wouldn't work in the public environment. Because in the public environment, there are too many outside factors that affect share price. In the private environment, it's really easy to see how much money you are making.

In a big company, you can't differentiate between who is carrying their weight and who is not. And it's a problem in big companies. In this environment it is going to be extremely easy. And also, we are not in an environment where we need to do a G.E.—to let 25% of the company go every year. We're just bringing in star players.

It is important that all the people that participate in this thing want to get richer than they are, or want to get rich enough to look back and say, "I created something that I am very pleased with." But it's more important for us to look at the legacy we are going to create. That's a bigger driver than remuneration—what I want to create.

People and Demographics

Ultimately, people are the key ingredient to the success of any company and, in fact, any industry. So, how does the energy industry get the people it needs to meet the challenges of *terra incognita*? Unfortunately, the energy industry has a problem. It hasn't shaped its image in the marketplace; so others have shaped it as low technology, polluting, and politically expedient. This caricature is unfair, but as Pat Yarrington of Chevron said: "I think the industry in a way abdicated a very important role of communicating to people about the value proposition of energy. As a result, the industry is no longer a favored career choice for new talent. Further, the reputation as a declining smokestack industry has been reinforced by the substantial job losses of the 1980s and 1990s. The industry must invest in transforming this message and present itself as a dynamic, growth-oriented, high-technology center of rewarding and fulfilling career opportunities with solutions to environmental issues and exciting travel and cultural experiences in interesting places." This is the industry that the authors joined in the last growth cycle, and it is a more accurate description of what the industry should be today.

Graying workforce

Dave O'Reilly takes note of the demographic issue facing the industry:

The supply of talented people, certainly in North America and Europe, is declining. I think there's an inflection point coming up; a lot of the people who came into the business when we did will, over the next decade, reach the mature stage of life that we're entering. Though we're still quite young by the way! The question is who is going to replace them and where are they going to come from. I think that coming change is a lot more apparent now than it was six to seven years ago.

Our philosophy is that there are four things you have to do well [operational excellence, cost reduction, capital stewardship, profitable growth], and then you need to build the capability to do them. Operational excellence is the start—if you can't execute, forget strategy! Cost management is still important. Capital stewardship in our business is absolutely essential. And then getting that balance between performance and growth right is the fourth one. We have for a long time been very strong on organizational capability. Organizational capability is how the organization works effectively together to accomplish what we need to do in those areas. But at the root of that, we must have the raw materials, which are skilled people. You have to have the right people with the required skills. Then you have to get those people to work together.

This is an area that is now more of a challenge. Demographically, you can see it coming. One thing about the world is that demographics tell you a lot—you know to a fairly high degree of certainty what the population and its composition is going to be 10, 15, 20 years from now. So we're spending a lot more time on the hiring and developing of people, not just in the conventional areas of the past, but also from other areas. We're sourcing talent from parts of the world now that we weren't doing even 10 years ago. It's all about understanding where the talent is, accessing it, bringing it on board, and developing it. We're beginning to see the consequences at the leadership level. The issue for me is that it is not an option. We have to go that way if we are going to be successful for the long term, because we're not graduating enough scientists and engineers in our traditional recruiting grounds to meet our needs.

It's a valid point that we have been able to reduce costs through flat organizations enabled by an experienced workforce. It may be necessary to design different organization structures with a less experienced workforce. It's tempting to look back and think you know everything when you don't! If you look at the history of the industry, particularly from a U.S./UK viewpoint, there was an influx of people into the business in the 1940s from the military services in both countries. That generation retired at the time we came in. We're in the midst of another sea change over the next decade. We're probably at a peak experience level at this moment for most companies. We've had the benefit of this enormous group of people that are extremely talented, and very experienced, and are at their peak performance—with a median age of late 40s, perhaps. Over the next 10 years, we're going to go through a period of less aggregate experience. So we're going to have to compensate for it either by better development programs, better training programs, or changed organization structures. Those are all issues that we face right now.

Clarence Cazalot of Marathon is also concerned about future human resources:

What keeps me awake at night is access to resource and qualified people to carryout the work we have begun. Companies like ours have built some pretty attractive retirement plans, so in today's environment, people can elect to take a lump-sum payment, walk across the street and get a sign-on bonus and maybe the same or higher salary. So the challenge of attracting and retaining people and incenting them is ever greater. We're going to have to work differently from how we have in the past. We're going to have to let some of our senior people work three days a week if that's what they want, rather than have them retire from us so we lose them. Possibly work remotely. We're going to have to think differently about how we get the job done so we can keep the folks we've got longer while we have a chance hopefully to fill in with younger people coming along. We're all out there today trying to promote more and more students into the sciences and engineering. I think that's working and we're finally convincing kids that this isn't a low tech business, it's a high tech business and it's not a sunset industry, it's an industry that the world depends on greatly and really has tremendous growth opportunities, but we need smarter people to tackle bigger and bigger challenges. Part of our solution has to be sourcing internationally. Yesterday we had IBM, SAIC, and EDS in, because we have outsourced many elements of our IT to those three companies, and we talked about people. We've all got the same challenges, though they need IT professionals not engineers.

They have operations in India and other parts of Asia and Latin America and they are addressing their staffing needs, at least in part, that way.

I came into the business in 1972 and have moved many times and have done everything the company has ever asked me to do. Today, people coming in have a very different view. Typically, they've got a spouse who works, they don't want to go international, they're not prepared to move around very much; it's a very different mentality. The challenge is that the business isn't just going to be in South Texas and Louisiana, it's going to be global. But it's also not about having expats out of the U.S. and UK, it's about developing a global work force. Our model for doing that is Schlumberger. Schlumberger has done a fantastic job of building a global workforce. Developing the capacity of nationals from the countries in which we do business is a major challenge and an opportunity. In fact, given the changing nature of the relationship that exists between the IOCs and the NOCs or host countries, having the ability to develop the talents and skills of the local workforce can provide a competitive advantage. The host governments and NOCs are looking for more that just technical skills and investments, and workforce integration is one of the elements of a mutual gain approach.

Schlumberger Consulting Services has determined that there will, in fact, be an adequate total supply of technical graduates for the industry. The problem is in regional imbalances. North America and the Middle East have huge projected deficits in technical talent, while China, India, Indonesia, and Venezuela have substantial excess supply. Russia is in slight deficit; it develops huge numbers of technical graduates who are in high demand due to the level of activity in the country. However, taken together with the Caspian region, the deficit disappears and a slight surplus of Russian-speaking graduates emerges. The challenge, therefore, in OECD companies is to accelerate learning for new entrants, leverage technology, strengthen organizational dynamics, and make it attractive for mature employees to continue working. Even then, there will probably be a need to tap into the talent pool of the countries with excess supply, possibly by setting up new technical centers in those countries.

Training cultural diversity

As Cazalot implies, the world has become a global village, but not all the villagers have common backgrounds. Nick Butler explains what BP is doing:

We're doing two things. The natural growth of the industry is the growth of demand (say 2-3% p/a) plus replacement of production decline. Also, our business is shifting from OECD to non-OECD. So three to four years ago it was 80% OECD. By 2010 it will be 70+% non-OECD, because that's where the growth is and that's where the resources are. Some of the skills are the same and some are different. In terms of the core skills of engineering, geology, commercial skills, we have to make sure we are attractive as a company to the best graduates. We're trying to do that, partly with pay, partly by changing the image of the industry, showing that it is not dying and not dirty. We do quite well in attracting the graduates we want in the UK and the U.S., the core markets for these skills.

The challenge is that there aren't enough engineers. John Browne is President of the Royal Academy of Engineering this year, and they are going to look at what they can do worldwide to increase the pool. There are quite a lot of people being trained in places like India and China, but they're not all being trained to Western standards. So we will have to have a conscious, active program to do that. It's partly making sure we are chosen by the best graduates, and partly it's expanding the pool of really qualified people that we choose from.

The other part that we haven't got right yet is making sure that the people that we do appoint know how to work in these complex environments. We have some experience of people who have only worked in the Anglo-American world going to complex places and not understanding the cultural differences. So we have a leadership education program on culture, diplomacy, and these operational issues to broaden the managers' range of capabilities when they are asked to be head of Colombia, Azerbaijan, Vietnam, or Russia. It's working, but it will take time, which is why the pace can't be rushed. We have a lot of number two or number three people in places like Trinidad, Angola, Baku, Siberia, who look pretty good, but you can't push them out to a really complex operation just like that.

Steve Lowden of Suntera has similar thoughts:

> There is a huge demographic bust in the Western business. The age bulge is moving into the 50s next year, I believe, the peak of the age bulge. And that means most companies have got most of their senior people that are five to 10 years from retirement. Well, people don't do wonderful things when they are five to 10 years from retirement. You may, and I may, but most people don't! But when you've been in a big company for a long time, I'm sure the mind-set is to be much more risk-averse than it would have been if you were 25. And another thing is the willingness of people to travel to difficult places, which this is all about. It's much less when you are 51, and you've got two kids at a university in the U.S. than it is otherwise.
>
> I think the IOCs have got less demographic flexibility than they have ever had before. Whereas, in India, you have the opposite; in China, you have the opposite. It's the complete opposite, where the peak bulge is 20. And they are looking for adventure. Russia is kind of in the middle. Russia doesn't have the declining population. It has a growing population if you adjust for immigration, but you do have a lot of very smart, very well educated Russian oilmen who want to go and seek their fortunes. So we can plug into that resource base. We provide the experience; we've done it before. We provide the framework and the work for them to go off and get it done.

Extra capability

Ian Howat believes Total has lost less capability through post-merger synergy capture than the Anglo-Saxon firms. In this regard, Total resembles natural systems that preserve some deliberate redundancy in order to be able to adapt to changing conditions:

> There was an analyst from UBS who came over to our senior management conference last week, and obviously he didn't say Total was the best in everything we did; but one thing he did say was that Total was the big winner from these mergers. We were the company that did deliver all the synergies we promised and clearly nobody else did. In fact 50/50 may have helped in synergy capture because we got to know both sides better and people were not afraid of telling where there were opportunities. Sometimes when

I address investor meetings, I'll be asked whether I think Total may be overstaffed or understaffed. And I say, don't look at the 25,000 people who work for Hutchinson, for example, our specialty chemicals division; that's not overstaffed because it's a manpower business, and would be dead in the water if it were overstaffed.

Where we may be overstaffed is in the 15,000 people who work in the E&P Division. If you compare our E&P division with Exxon, for example, you'll probably find we are relatively overstaffed because we think you have to sustain your knowledge. And we chase more growth opportunities than they do. We have consistently outgrown all the other super-majors in organic terms, and we think you don't grow unless you have some fat in the system. You have to be prepared to chase all these projects and have a 90% attrition rate. Other companies may also get more "auto-censure" than we do. It is very difficult to measure attrition, because you don't know what would have happened to the project you don't do. But our process is very rigorous. It's the sins of omission that may kill you in the end. The sins of commission: you can see what they are and correct them; but the sins of omission are swept under the carpet and forgotten.

People management

There are three aspects to the people issue: 1) to gain access to a stream of highly qualified candidates, 2) to persuade the best to join, and 3) to make them want to stay. Rick George of Suncor understands this well:

Listen, I think it boils back down to really good people who want to work together and who have enjoyed success. Success breeds success. I think people here feel very comfortable. They really want to work hard. They understand the strategy and know it, and they want to deliver their piece. If you can get a group of people to work in that kind of a fashion, as opposed to sitting around and blaming each other for why things don't get done, it's amazing what you can do.

I'll give you one analogy. When I started working for an oil company, they seemed to run their control systems to manage the 2% of people who were going to mess around full time. I think you should try to run a company for the 70%–80%–90% of the people who really want to succeed. And by succeed, meaning do a good job, have a lot of pride in what they're doing. It doesn't matter to me whether you're a truck driver in this company or you're the senior vice president, people want to take pride in what they do. So run your company for that. Then have them help you take care of the 2% of people that don't want to work.

From half a world away, Saudi Aramco takes a family view of the workplace, as Abdallah S. Jum'ah describes:

We take care of all employees. We have a medical organization that takes care of 400,000 people. And these 400,000 people are the employee, his immediate family, his wife, his children, and his father, and his mother. That medical is really unlimited. For former employees, it covers the former employee and his wife. We provide interest-free loans to our employees to build homes, so that we know they are comfortable. We provide schooling, and help them in schooling. Our training and development is, of course, examples.

Today, we are the preferred employer in the kingdom. Every year high-schoolers come and compete. We have 10,000 high school students, high performers, come knocking at our doors. We choose only 200 to 300. And we run them through undergraduate and graduate school, and so on. A lot of the people today in the company were trained by the company.

We are very demanding, but we also make them comfortable. I think, at the end of the day, that they want to raise children. They want to have healthy children, and so on. If we can reassure them that we'll take care of them and so on, they work very hard. That's why you see young people staying long hours, and not complaining. And I heard recently something very interesting. Somebody was telling me, "I have never heard anybody in Saudi Arabia who left, who retired, who does not bless day and night the organization."

Conclusions

Leaders must learn the lessons from the past to develop the wisdom and foresight to get the best performance today while plotting the course for the future. These interviews are lessons from the past, much of it the very recent past, and thoughts about tomorrow to help navigate the enterprise today. Therein lies the leadership challenge: getting the maximum achievement from the organization's capability today while having the foresight and courage to lead the organization into a brave new world when the trade winds foreshadow a shift in direction. Through their stories, these energy leaders provide useful lessons in the subtle art of leadership. Mature leaders teach the leaders of tomorrow.

Unfortunately, people outside the energy industry often denigrate energy leaders, casting them in the stereotype of J. R. Ewing—conniving, insensitive, and simply out for the almighty dollar. The public's view has been shaped to see them more on the "dark side of the force." Certainly, every barrel of apples can have a bad apple or two. This is true of energy leaders, politicians, media, and every other occupation. But in hearing our many leaders talk, one gets a sense of their strength of character and capabilities. They are real people with their own sets of strengths and weaknesses too. Each leader we interviewed brought different perspectives, but also many commonalities, such as a sense of purpose, imbuing a positive attitude to assure alignment, and ability to stay the course during set backs. They shape culture and values rather than letting them rest in the hands of fate. They create a context in which good decisions are made naturally because the leadership team trusts each other, decision rights are understood, and governance is supportive. All the leaders we spoke to are committed to maintaining high performance through well-designed accountability systems, proper performance management, and long-term and short-term incentive plans. Finally, an essential part of leadership today is creating an attractive place to work, particularly recognizing the industry will lose a large proportion of its experienced staff in the next decade. Overall, their stories and views provide powerful insights into how to run an energy company in turbulent times.

1 Warren Bennis, 1994. *On Becoming a Leader*. Addison Wesley, p. 192.

2 John H. Zenger and Joseph Folkman, 2002. *The Extraordinary Leader*. McGraw-Hill, New York: pp. 54-70.

3 Bennis, pp. 39-41.

4 Howard Gardner, 1995. *Leading Minds*. Basic Books, pp. 290-295.

5 Zenger and Folkman.

6 Hitt, Ireland, and Hoskisson, 2001. *Strategic Management: Competitiveness and Globalization*, Southwestern Press, pp. 402-403.

10 Next Generation Energy Leaders

As with most human competencies, leadership is a combination of nature and nurture. Peter Drucker makes a pragmatic and wise observation: "There may be 'born leaders,' but there surely are too few to depend on them. Leadership must be learned and can be learned."[1] Business leadership continues to evolve, although some attributes, such as sound character, transcend time. Different situations require different leadership styles and capabilities. Leadership is particularly important in times of change.

Leadership can be learned; however, it also makes great demands on those who become leaders. "It takes strong vision, conviction, and energy to create and animate an organization; it takes great judgment, wisdom, and skill in pulling large groups of people together to institutionalize processes on a global scale with a population that varies widely geographically and in age. And it takes learning ability and personal flexibility to evolve and change organizations. . . . The one thing that is becoming clearer and clearer is that the institutions of the past may be obsolete and that new forms of governance and leadership will have to be learned."[2]

Leadership is an evolutionary process with each new generation building on the former. In this chapter, we will begin by taking you through the different eras of business leadership starting with the Industrial Revolution. This will set the stage for the next-generation energy leaders who make the transformation from oilmen or utility administrators to energy executives. We will integrate the learning from throughout the book as a mentoring aid for future navigators in sailing through these turbulent times in the energy industry. At the end of the day, it is the role of energy leaders to chart this future. Next-generation leaders must move beyond simply reacting to the times toward shaping and driving the future, as the original empire builders did when they ordered the chaos of the early petroleum industry into a more productive course.

Business Leadership Evolution

Business leadership has evolved through several dominant models since the middle of the 19th century. At each stage of this evolution, new ideas altered the existing model in response to changing economic, demographic, and societal challenges. The energy industry has sometimes led the evolutionary process and sometimes lagged as its leaders strove to control the sharp swings and disruptive cycles and improve productivity in an intensely competitive environment. At every step of the way, however, it has adapted current leadership concepts to the extraordinary scale and complexity of supplying the energy to power the world economy.

Empire builders

We saw in chapter 1 that some of the early leaders had similar ideas and traits. Gowen, Frick, Rockefeller, and Deterding were all trying to control what they viewed as wild aberrations in the market place and built their dominant empires to do that. While they had somewhat different personalities, their styles were similar. These early leaders were strategists and visionaries. To realize their vision, they ruled their companies with an autocratic iron hand. Rockefeller, perhaps more than any other man, symbolized the strengths and weaknesses of these early titans.

Some called Rockefeller "the most important man in the world" during his time.[3] He was a great philanthropist in his senior years and yet he used unmerciful tactics in destroying his competition during his prime. He started from sparse beginnings but developed an incredible fortune. Rockefeller saved money to make money. "This, then, was the man in 1860—frugal, calculating, money-bent—cautious in trade yet daring, quick to seize yet ready to wait, and withal 'good'—that is a steady attendant at church and Sunday-school, serious—that is eschewing all amusements which might be called frivolous, the theater, cards, and dance."[4]

Rockefeller entered the refining business with a passion and searched to find good bargains in the up and down cycles of the industry. He was driven by the desire to be rich and the need for control. "It is quite probable that Mr. Rockefeller, natural trader that he was, learned early in his career that unless one has some special and exclusive advantage over rivals in business, native ability, thrift, energy, however great they may be, are never sufficient to put an end to competition."[5] He thought perhaps transportation was that advantage. He was cunning and secretive in gaining railroad rebates that created his advantage and showed no remorse for this behavior that society eventually condemned. The apparent paradoxes of brilliant insight, ruthless response to competitors, and deep commitment to charitable causes mirror striking similarities with Bill Gates of Microsoft.

Rockefeller's strengths were not only in his business savvy but also in his vision of the scope and scale that would make the oil business highly efficient and orderly. It could be readily argued that such stability, scale, and efficiency served the public and the economy well in this early stage of industrialization. Rockefeller indeed felt that consolidating nearly all oil refining into a single, giant entity was required to save the business from excess capacity and devastating price fluctuations.

He was clearly interested in power but held it cloaked and hidden from public view. When Ida Tarbell exposed this mystery, the great empire was broken up after embarrassing public scrutiny through the anti-trust trial. Rockefeller's leadership clearly had fatal flaws that ultimately undid his great vision in its inability to address the public's requirement for fair play. He also appears to have underestimated the public's dislike of excessively large and powerful organizations and the government's willingness to assert political control over economic activity. But his legacy for shrewd business decisions and operational excellence lives on in ExxonMobil, which has been the most successful IOC in the 20th century, as does his discomfort with public and government relations.

Pioneering management principles

The next leadership model began to evolve around the turn of the 20th century. These leaders were still builders, but they also began to define more clearly the principles of management. Henry Ford was most symbolic of this next era with his engineering mindset. His Model T Ford introduced in 1908 not only transformed personal transportation but also set in motion the phenomenal growth of the oil industry to supply the Model T with gasoline. To meet the growing demand, Ford employed precision manufacturing, standardized and interchangeable parts, and division of labor. By 1913, he developed a continuously-moving assembly line, which significantly lowered costs.

Ford was part of a generation of leaders who followed what Frederick Taylor described in his 1911 book, *The Principles of Scientific Management.*[6] These ideas had been evolving since Adam Smith advocated the benefits of division of labor in saving time through specialized tools and skills. Taylor saw the dawning of a new era of leadership, "The old fashioned dictator does not exist under scientific management. The man at the head of the business under scientific management is governed by rules and laws which have been developed through hundreds of experiments just as much as the workman is, and the standards developed are equitable."[7] The drive was towards maximum efficiency based on careful specification and measurement of all organizational tasks. Even President Theodore Roosevelt got on this bandwagon when, in an address to the governors at the White House, he proclaimed, "The conservation of our national resources is only preliminary to the larger question of national efficiency."

Taylor's management principles set a framework around clear delineation of authority and responsibility, separation of planning from operations, incentive schemes for workers, task specialization, and the idea of management by exception. The model focused on the worker level and optimal work design against a backdrop of building economies of scale supporting to the underlying strategy of cost leadership. For Ford, scientific management brought terrific success: by 1918, half of all the cars in America were Model Ts.[8] Fordism had profound impacts on society in fostering the middle class and stabilizing economic policies through mass production and paying relative high wages.

Ford, however, got caught in this mindset of management by exception, which became his strategic error. The concept simply posited that you could not have workers on an assembly line off doing their own thing; the assembly line would crash. Management intervened when there was no choice because an exception had occurred to the plan and program. The very essence of management was planning, organizing, directing, and controlling. This *modus operandi* ran entirely counter to systemic change, in that change was by exception. Ford's success had been so outstanding—60% market share by 1925—why would one want to change such a successful business model? So Ford got stuck in a rut, unable or unwilling to change the concept of his car. Companies that have been highly successful in a prior era often find it difficult to make the transition into a new era.

Management principles were expanded by Alfred P. Sloan, who was a consummate leader. He led General Motors to implement change in styling every year, pursuing planned obsolescence. He differentiated the various lines of GM cars on price, from Chevrolet, Pontiac, and Oldsmobile to Buick and Cadillac, to appeal to every buyer segment. He had something for everyone. The growing middle class signaled that society was climbing up Maslow's hierarchy of needs with a broader array of tastes and income levels. In order to run this market-segmented organization, Sloan decentralized the business into GM divisions. Max Weber, building on Taylor, described this bureaucratic theory of dividing organizations into hierarchies, yet establishing strong lines of authority and control to the corporate center.[9] This approach reached its apogee in the 1960s with growth of conglomerates such as ITT, which under Harold Janeen accumulated a wide range of disparate businesses managed through financial controls from the center. Big oil companies took on these principles to run their expanding empires.

Its superior understanding of the changing times propelled General Motors rather quickly into a sales leadership position in the early 1930s ahead of Ford Motor Company; Ford has not been able to regain its market leadership ever since. But Sloan brought much more to leadership, becoming the first celebrity CEO. He was the only corporate leader to be included in Howard Gardner's book on *Leading Minds* examining the greatest leaders of the 20th century.[10] Bill Gates in his book *Business @ the Speed of Thought* wrote, "I think Alfred Sloan's *My Years with General Motors* is probably the best book to read if you want to read only one book about business. The issues [Sloan] dealt with in organizing and measuring, in keeping [other executives] happy, dealing with risk, understanding model years and the effect of used vehicles, and modeling his competition all in a very rational, positive way is inspiring."[11] David Farber notes that "Sloan's tenure at GM provides a useful reminder that great corporate leadership is mainly honest and usually brilliant."

The availability of rigorous financial and statistical tools drove Sloan to measure the performance and profitability of the business. At the same time, GM was the nation's largest advertiser. He kept risk taking alive in spite of a hierarchical, rule-bound, decentralized organization. And he managed largely by consensus. But one of his other great strengths was his engagement with the public. Sloan felt strongly about the role of modern business corporations in society and pursued institutional and personal campaigns to affect public policy. It is probably worth recognizing that automobiles are on a far shorter marketing cycle than the energy industry, with styling changes each year. The need for change in energy companies is, thus, less frequent, and this reduces the need for communicating with customers and the public. Nevertheless, it is regrettable that the energy industry has produced few leaders capable of engaging with society on the vital role that the industry plays.

From discipline to performance

The struggle for higher wages and improved working conditions has been led by unions played since the second half of the 19th century. Unions such as OCAW (Oil, Chemical, and Atomic Workers) still bargain today for pay, benefits, and workplace rules. The growth of unions served to balance the negotiating position of worker with increasingly large corporations. Perhaps more than that, unionization was a response to the dehumanizing aspects of scientific management, which often treated workers as robotic machines monitored by time and motion studies. Around the 1930s, management began to pay more attention to the uniqueness of the individual as researchers applied the behavioral sciences to the workplace.

The ability to manage people through understanding their needs became another important dimension as society climbed up Maslow's hierarchy in the 1950s and 1960s. This was a period dominated not by corporate leaders, but rather by management theorists like Elton Mayo, Douglas McGregor, Chris Argyris, Frederick Herzberg, and David McCleland. The paradigm shift was expressed by McGregor in his concept of Theory X and Theory Y. In Theory X, the underlying assumption was that people had to be coerced into work, whereas Theory Y envisioned people with natural inclinations toward self-development and productivity. The manager in Theory X was more like a disciplining parent, while in Theory Y the manager served as a mentoring adult. Herzberg recognized that higher-level needs were motivators, whereas lower-level needs were considered as basic hygiene and did not act as motivators, but instead required consistent performance to prevent them from becoming de-motivators. Higher-level needs involved the work content; the lower-level needs focused on working conditions.

Renis Likert studied management styles through behavioral methods, examining organizational performance differences among autocratic and more consultative and participatory styles. Higher performance tended to be associated with managers who engaged their people through consultation or participation rather than just telling them what to do.

In parallel with the introduction of behavioral science, another sequence of innovations stemming from W. Edwards Deming's thinking on Total Quality Management evolved into concepts of Continuous Improvement and later to Six Sigma approaches developed by Motorola and GE which became corporate themes in the 1980s and early 1990s. In total, these innovations transformed how leaders viewed systems of people and machines by empowering individual workers to improve the manufacturing systems in which they worked. No longer did all ideas and insights flow from the top down; now the full experience and intellectual capability of the workforce could be brought to bear on a series of small improvements, which when added up became a major source of competitive advantage for companies such as Toyota. Leadership under this new concept evolved from giving orders to providing direction and encouraging collaborative, objective analysis leading to sustainable performance improvement.

The energy industry makes such huge investments in the hardware side of the business that, in comparison to many other industries, the workforce to run the facilities constitutes a much smaller portion of total costs. The energy industry pays well, however, because it does need enough of the right-skilled people to ensure that plants and equipment are always operating. Notwithstanding the relatively low contribution of labor costs, energy companies have been aggressively downsizing from the mid-1980s as they strive for ever greater cost-efficiency.

The decision-making side of the energy business is primarily run by highly technical people, primarily engineers, trained in the physical sciences rather than the behavioral sciences. Often behavioral theories are considered soft management issues, lacking the hard objectivity found in the engineering world. People management skills have not been the greatest strength of energy leaders, though there has been significant progress in recent years as companies recognized that effective teamwork was essential for execution of complex projects.

However, people management influences worker productivity. And in the cost leadership drive, improved marginal performance can make a significant difference in a commodity industry. Bill Greehey, recently retired CEO of Valero, is an example of a strong people-leader who reinforced this point in our interview with him: "Why do we always beat our competition, our peer group companies? We beat them every single year. And I don't think it's strictly on strategy. I think it's on culture, because we attract the best people. They are motivated. We don't have any turnover. They are happy people. We are always in the '100 Best Companies to Work For.'"

In many respects, behavioral management skills were the next leadership step in moving up the proverbial S-shaped growth curve of industrialization built on a massive scale of people and machines. Running big corporations required not only the management skills of planning, organizing, directing, and controlling, but also people skills to motivate employees towards personal and corporate goals. On Maslow's hierarchy of needs, the socialization process of belonging and feeling needed was demanding more attention. People were beginning to look toward their ego needs, which as Herzberg pointed out involved job content. People are motivated by interesting and challenging job assignments, which in turn drive performance.

From hierarchies to networks

The information revolution had been developing for several decades before it began to show a real impact on productivity in the 1990s. In many respects, it was initially just the next stage of the industrial revolution, with machines doing the work of humans, but in this case in backroom and technical activities instead of operations. Many clerical tasks were automated through programs running on mainframe computers, beginning in the mid-1960s with IBM 360s. The dramatic increases in processing speed and capacity to handle enormous databases enabled computers to generate complex mathematical representations of physical systems, including seismic analysis and linear programming. The impact on leadership, however, was rather insignificant in this early phase.

A change began to occur in the mid-1980s and early 1990s as data grew into information, computer processing approached real time, information was accessed through workstations, and enterprise-wide systems evolved. The time from data capture to action shrank. Knowledge became more widely available. The way in which work was done could be redesigned from individual tasks to integrated jobs. Information technology increased at such a rapid pace that fast-cycle industries were born.

Lessons from fast-cycle industries. Computer chip manufacturing was one of those fast cycle industries, and Intel's CEO Andy Grove stood out as a next-generation leader of this information era. In fast-cycle industries, leaders have to face the brutal facts quickly to bring out improved products continuously, and truth-telling becomes all-important. "Grove escaped natural selection by doing the evolving himself. Forcibly adapting himself to a succession of new realities, he has left a trail of discarded assumptions in his wake. When reality has changed, he has found the will to let go and embrace the new."[13]

In such a rapid-fire process, debacles can occur, such as the Pentium chip flaw. Intel was late in responding to replacing hundreds of thousands of chips. But Grove and his team learned a new lesson—that the rules by which companies operate had changed: "Intel would have to court its public, be sensitive to the needs of vast numbers of customers, and build equity of goodwill."[14] As discussed in chapter 4, this is a clear lesson for energy leaders: the importance of building goodwill equity with the public, so when a crisis does occur, which it will, the public shows some tolerance.

Grove was a transformational leader in a business that was always transforming. He could sense the "strategic inflection points" and develop a new path with his team. Grove was always teaching, another sign of the new leadership era.

The information revolution developed for several decades with an intense focus on hardware, which provided the incredible processing speed that improved at the rate of Moore's Law. It then shifted to a driving focus on software that involved decision-making and stimulation of thought processes. On top of that, the internet set in motion the ability for that knowledge to travel at the speed of light around the globe, creating what Thomas Friedman describes as a "flat world." This real-time flow of information provided the opportunity to redesign administrative and technical organizations that had been built on the linear, sequential format akin to an assembly line, reflecting Frederick Taylor's scientific management principles.

Reengineering organizations. Michael Hammer and James Champy promoted the idea of radically redesigning and reengineering the enterprise, with information technology as the enabler, to achieve lower costs and improve the quality of services.[15] Enterprise Resource Planning (ERP) systems, such as SAP programs, allow companies to enter information one time, and through embedded relational databases, consolidate and report it according to the requirements of management. These systems eliminated the need for multiple data entry and removed the associated errors and requirement for reconciliation of competing databases; they also required that business processes be aligned with the capabilities of the systems in order to realize the promised benefits. As a result, organizational layers collapsed as business processes were redesigned. Decision-making decentralized to "whole jobs," but also centralized as information flowed quickly to headquarters from around the world. Virtual teams became a new way of working, bringing in the most relevant expertise wherever it resided.

Evolution to super-leadership. The collapse of layers required that mid-level staff be empowered to deal directly, without supervision, with customers, suppliers, and peer companies: "The importance of those in the ranks is enhanced because they stand at the point of intersection between an organization and its customers and clients. So an organization cannot be truly responsive to the needs of those it is configured to serve unless its front-line people are given the autonomy and support. This is the true reason that the top-down hierarchical style of leadership is widely perceived as doomed to failure, even by those who are not precisely sure why this should be so."[16] Now many people in the organization are becoming self-directed, with a wider scope for making decisions, setting priorities, and taking action. In this information era, skilled leadership requires the ability to ignite these self-directed individuals through what Charles Manz and Henry Sims refer to as Super-Leadership.[17] So front-line staff now needs to be trained and coached in leadership, not of subordinates, but of relationships with third parties, and the training must also equip them with the capabilities required to succeed: "Repeatedly, we have been told by employees that the capability to do their respective jobs is the most important determinant of their satisfaction."[18]

Change is the constant. Transitions have occurred everywhere. Few people recall that "cc" stands for carbon copy. Whatever happened to the makers of carbon paper, to the telex purveyors, to punched card data entry, to mimeographed duplication, to the real cut and paste, to typing pools, to telephone operators? Supply chain management systems have reduced inventories to "just in time." In essence, the information revolution has altered the way people think about change. Change was managed as the exception in the industrial revolution; now change must be managed as the constant having an ever-increasing pace in the information revolution. No industry remains untouched by this relentless sea of transition.

With such constant change, industries no longer promise the "job for life" that had seemed to be the birthright of prior generations. As a result, the nature of a career has changed: "A career is now not so much a ladder of roles, but a growing reputation for making things happens. Influence, not authority, is what drives the political organization today in all organizations."[19] A new infrastructure is now in place to facilitate employee mobility, with portable defined contribution plans, 401(k) retirement accounts and quicker vesting. Compensation packages have come to include significant differentiation based on performance. Almost everyone is exposed to variations in their employer's stock price. E-mail and cellular phones have revolutionized communications so people are on the job 24/7. Increasingly people are working at least partly from home. Airfares are a fraction of what they were in real terms, and business travel affects most people rather than a few leaders. Leadership now requires developing entirely new ways and new cultures to retain and attract the workforce necessary for an energy company to progress. This is not a static transformation, but requires an ongoing change process that Arie DeGeus has described as the Learning Organization.[20]

Global networks. The information revolution has also fostered the "global village" where, as we learned in chapter 9, demographics and diversity are key challenges for energy leaders. Steve Lowden, CEO of Suntera, highlights the demographics conundrum: "The average age of technical staff in most IOCs exceeds 50 years. This is not true in INOCs or NOCs, and they will provide much of the future talent for IOCs as well as for their original companies." This increased diversity will have significant impact on energy leadership. "They will have to understand the dynamics of diversity in the workplace, including the basics of cultural differences in worldview, motivation, problem-solving techniques, work styles, and communication styles."[21] So energy leaders will need the skills to navigate through both global commonalities and local diversity.

A different energy world. During this information revolution over the last four decades, the world's population has approximately doubled, energy demand has grown mightily, and the oil market has moved from the dominance of the Seven Sisters, through nationalization and on to emerging competition from INOCs. Seismic technology has progressed from very rudimentary print-outs that geophysicists colored with crayons to three-dimensional visualizations that allow teams to "walk" through formations. Drilling has moved from vertical to steerable. Tankers have doubled in size, refineries convert the entire residual fraction into transportation fuels instead of sending it to power stations, and you buy Twinkies rather than oil changes at gas stations. Natural gas is fast becoming a global commodity. Japan has advanced from exporting a few cheap, low-quality cars to the United States to become home to the premier global automobile companies. And now in the 21st century, China and India have become the new energy demand pacesetters as they merge into the developed world.

In the narrow space of our working lives, we have witnessed major changes in leadership models, in organizational constructs, in compensation programs, and in careers. These changes have been driven by advances in technology, shifts in social norms, and new business models. Famous companies have disappeared and new companies have become famous. So the next-generation leaders will need to review carefully the trends and changes embedded in the strategic assessment we have described in chapters 1–4, review their shareholder value propositions and strategic choices (chapter 5 and 6), and stand on the shoulders of their predecessors (chapters 7–9) as they adapt to the whirlwind of change ahead.

Leadership Challenges

The early oil industry titans were proud to be part of the society of oilmen who knew they were helping transform the world from an agrarian to an industrial economy. As we have learned, the oil industry in its early years produced a set of iconoclastic leaders. Over time, each modern petroleum company has developed a culture that reflects its particular legacy. It is easy to see the Rockefeller lineage in ExxonMobil; and William Knox D'Arcy's bold prospecting expedition to Persia, which created Anglo-Iranian, lives on in BP's current Russian venture. BP's governmental heritage as an NOC undoubtedly made it easier for John Browne to understand the political realities of climate change and the machinations of Kremlin politics; ExxonMobil's legacy of discomfort for government and the press, however, may make it more difficult for company leaders to see climate change as a legitimate societal as well as a scientific issue or to venture deeper into a seemingly paradoxical Russia.

Companies originating in the power business developed their cultures operating in regulated markets, where they became expert at managing the complex relationship between company and regulator. In this world, growth in profits requires growth in the rate base, and cost reductions must be passed back to the consumer. Compared to free markets, the regulated world may seem oddly inverted and distorted, but in fact it is much easier for many consumers to understand. The regulated legacy of these companies has provided them with formidable competencies in influencing government policy, but it has also left large gaps that need to be filled if they are to be successful long-term in competitive markets.

Leaders of companies with legacies in regulated markets and those with legacies in the international oil business will find themselves increasingly in competition with each other, and the winners will be the companies that are best able to evolve in a changing business climate.

Building on cultural legacies

These cultural legacies have become embedded in the leadership styles of the different companies. BP and Total tend to value and, therefore, produce "heroic individuals." ExxonMobil and Chevron are process-driven cultures that subordinate individual initiative to a disciplined adherence to carefully constructed, honed, and trusted ways of doing things (fig. 10.1).

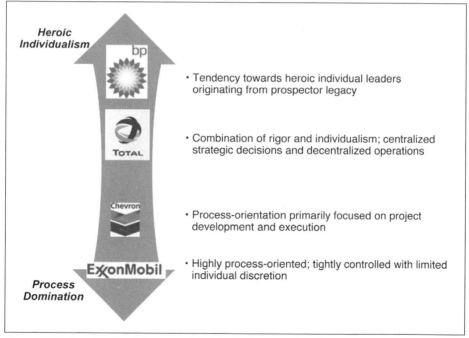

Figure 10.1: Leadership styles are related to the legacy of each company.
Source: CRA International, Inc.

Saudi Aramco is clearly building its own unique culture based on the legacy of its predecessor, Aramco, owned jointly by four U.S. major oil companies. As Abdallah S. Jum'ah told us, "Unlike any other national oil company, we continue to be proud of our connection with our American ancestors." Royal Dutch Shell seems to have moved up and down this schematic, but now leans very much towards process as it has reined in country CEOs from its former decentralized model.

Leaders tend to reflect their corporate legacy, but each one leaves a mark on the enterprise. An ExxonMobil leader is rarely charismatic but will work on further improving and rebalancing the capabilities and assets of the corporation. Lee Raymond will be remembered for his sustained attention to costs while pursing technology in difficult economic times, as much as for his successful acquisition of Mobil. Lord Browne, by contrast, will be remembered as much for his personal qualities in persuading BP to leave behind the bureaucratic, quasi-governmental aspects of its heritage and marshal the incipient individual performance aspirations of the staff as for his acquisitions of Amoco and Arco. Thierry Desmarest will be remembered for his spirit of inquiry as well as his integration of Elf and Fina into Total. Dave O'Reilly's attention to Chevron's values is at least as important as his acquisition of Texaco. Obviously, each company tends to recruit people who will fit into the corporate culture, and the desired attributes are then reinforced by teaching, training, and coaching. It is very difficult to change corporate cultures. Successful change generally comes from reinforcing desirable attributes of an existing culture and suppressing the less desirable attributes, rather than trying to impose a radical new culture.

Cultures of the independents we interviewed are equally distinctive. Marathon is one of the few survivors of the mid-size integrated oil companies for a reason. Clarence Cazalot has built on the company's historical reputation for technical excellence, making carefully-considered major asset moves to further strengthen its position. Investors have slowly come round to his view that Marathon's strong position in U.S. markets can be leveraged into global resource access. Helge Lund is adopting a similar profile at Statoil, with perhaps a higher risk tolerance. Burlington Resources also had a strong culture of excellence in developing and operating North American resource plays. Bobby Shackouls joined this operational excellence and an exceptional finance capability to assure delivery of promised results.

Several of our CEO interviewees are consciously forging new cultures. Gwyn Morgan of EnCana, Rick George of Suncor, and Bill Greehey of Valero were particularly clear on the importance of infusing their beliefs and values into their companies' cultures. It seems likely that these cultures will continue to reflect their legacies. John Wilder of TXU is taking on the Herculean task of changing a regulated utility culture into an entrepreneurial, commercial culture, based on intense respect for safety and performance.

Aubrey McClendon of Chesapeake and Frank Chapman of BG have created highly entrepreneurial cultures propelled by a strong appetite for growth. Both are proud that their companies are considered exciting places to work. Chapman commented, "There's a sort of energy and intensity when people are really engaged."

McClendon talked about the growth in his leadership team's capabilities: "They think they're good at what they do; they're proud that they used to not know how to do what they do now."

BG evolved rapidly from a legacy of a regulated utility by capitalizing on deep strengths in international natural gas distribution systems to look at the business from the market back rather than from the resource forward. TXU under John Wilder (coincidentally, both Chapman and Wilder are former Shell executives) is building on the TXU resource legacy of major coal reserves in Texas to be the low-cost power producer in Texas and surrounding states, in stark contrast to the "cost-plus" regulated business model. The process of building a new corporate genome adapted to the new environment is challenging but vital work for the leaders of the future.

All of our interviewees have been successful leaders for their time. Would they have been as successful in different times? We can never know. What is sure is that the future will bring a very different set of challenges for the next generation of leaders, and these leaders will need to reshape their organizations, building on the aspects of culture most suited to the new times and subordinating those aspects that are less well suited. First they must understand what the new times will likely bring.

Understanding the changing energy environment

The strategic assessment presented in the first four chapters identified many trends, issues, problems and opportunities. Before everything else, new leaders must recognize from this information that global energy markets are in profound transition with the convergence of two cycles: the normal energy investment cycle is moving strongly from two decades in the doldrums to at least a decade of strong performance; a new "phase change" is under way that will topple conventional oil from its position as the strategic global energy resource. In our view, there is a good chance that the next conventional business cycle downturn will be over-ridden by the phase change such that investments that would be threatened by a cyclical price downturn will, in fact, continue to earn good returns in the 2010s and beyond (fig. 10.2). The underlying reason is that China and India have become full members of the global economy, are becoming more prosperous and are growing their energy demand from what is a now a material base level. Particularly for transportation fuels, the demand structure and the energy complex to supply it that has been designed for the OECD cannot simply be scaled up to meet future demand from India and China and other non-OECD nations with high economic aspirations. The business environment for energy in the 2010s and 2020s will be quite different from the down-cycle of the 1980s and 1990s.

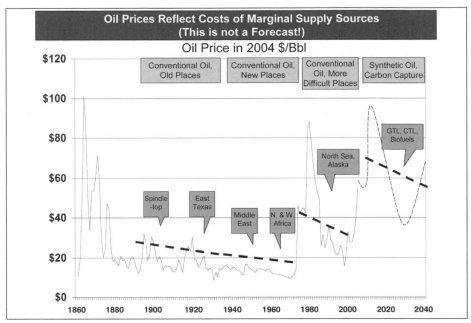

Figure 10.2: The conventional investment cycle may be overridden by the phase change. *Source: CRA International, Inc.*

Pricing economics from recent oil supply and demand trends are already bringing on a new energy supply mix, and consumption patterns are likely to change in due course. Since higher energy prices seem likely to persist for some time to come with typical volatility, price constitutes the signal that will cause the changes to happen. The next cycle will coincide with an energy phase change that will tend to sustain prices while the global energy complex is retooled.

The issue perplexing leaders of large energy companies today is determining the time over which changes will take place. The speed of change depends on seven key factors:

1. Degree of exploration success and enhanced recovery technology improvements for conventional oil and gas;
2. Progress in the development and economic competitiveness of energy sources beyond conventional oil and gas;
3. Rate of global economic growth;
4. Importance attributed to energy security;
5. Level of demand reduction through improved efficiency and conservation measures;
6. Degree of convergence between power and transportation markets through hybrid vehicles and battery technology;
7. Timing of legislative and regulatory changes targeting global warming and other social and environmental concerns.

With such a mixture of factors each containing wide-ranging probabilities associated with different outcomes, the terrain ahead is clearly *terra incognita*. Next-generation leaders will need to retain considerable flexibility and optionality as well as risk management in their thinking and monitor these trend patterns constantly.

New competitive landscape. Other patterns are coming into focus as well, some with profound impact. Demand, and therefore energy price pacesetting, is shifting from the United States to the developing countries, particularly China and India. National oil companies of producing countries (NOCs) are maturing, developing strong technical and commercial competencies and are expanding into the downstream refining and marketing sectors, challenging the IOCs' core value proposition of access to markets. More than that, some NOCs have transformed into a new breed of internationalizing companies (INOCs) in direct competition to the IOCs (chapter 2). The next-generation energy leader must not only understand this new competitive landscape but proactively develop strategies that will provide winning value propositions in the face of strong competition from companies with different drivers, different costs of capital and different business models. This change in competitive arena is at least as profound as the advent of Japanese auto manufacturers in the 1970s, and the threat to incumbent IOCs now is as serious as the threat to U.S. auto companies then.

The resource access challenge. There will likely be only a few new E&P opportunities in resource-rich countries for IOCs, and these will involve oil reserves that are very mature or that require a high level of technology to explore and develop. Further, new entrants such as INOCs and LOCs in the conventional oil business will increase competitive intensity and result in competitors bidding away the economic rent on new conventional oil projects to the benefit of resource holders. This will be less true for natural gas, as Clarence Cazalot of Marathon pointed out in chapter 7, and also less true for unconventional resources such tight gas, oil sands, coal transformation to transportation fuels, as well as biofuels. IOCs will be pushed in the direction of these new resources to provide their shareholders with growth (chapter 3). Longer-term, industry leaders will need to consider a total transformation of their value propositions to resource rich countries, ceding substantial control of the resource development and production schedule, leaving behind the charade of booking reserves owned by the host country, imaginatively addressing needs for economic development, local content, job creation and competency building, and providing access to markets. In short, successful IOCs will forge true partnerships with a range of companies, constituencies and institutions in host countries to rebuild trust by helping address the real needs of the host countries.

Changing societal expectations. The energy industry is very likely to experience a pattern of increasing external pressures relating to greenhouse gas controls and national security. This will create new competitive geo-political relationships (chapter 4). Certainly the competitive landscape of global oil is changing fast. The apparent 1990s trend toward opening of closed markets and resources appears to have been suspended, if not canceled. Historians are talking more of the "clash of civilizations" than the "End of History." The expansion of INOCs has a mercantilist flavor, which hopefully will not lead to global conflict as in the past. Politics seems again to be gaining the upper hand over economics. Technology can change that in time, but energy companies must be ready to operate in a period of geo-political turbulence.

Further, what Tom Friedman calls "the World of Disorder,"[22] will continue to confront IOCs in culturally difficult states where there are large hydrocarbon resources. There are no easy answers in these situations. It is clear that the companies alone cannot resolve situations arising from disputes among local communities, central governments, and criminal elements. As Nick Butler of BP pointed out, the role of country leader for an oil company in resource-rich countries is complex and deeply challenging, and requires diplomatic as well as technical and human competencies. The authority of this position has been diluted over time and needs to be restored.

At the same time, societal pressures continue to mount in developed countries where high expectations focus not only on what energy companies do, but also how they do it. Stakeholders demand energy supplies at affordable prices with no environmental impact. They call for companies to make bottom-line reports for social responsibility and environmental protection, as well as for financial performance. They require sensitive treatment of local communities and assurances that government oil revenues are used wisely. They will not tolerate environmental or safety lapses, and will demand greenhouse gas caps. Energy is not purchased for pleasure—it is a necessity. Consumers feel threatened by and deeply resent volatile high prices and supply uncertainty. This resentment flows into public policy, constraining energy companies' effective and efficient "license to operate." Energy leaders must step up to their responsibility to be proactive in explaining the role of the energy industry in society, and how that industry is going to address the public's concerns about prices, supply security, climate change, and a sustainable energy future. The more successful they are in repositioning themselves in the public perception, the more likely there is to be alignment around the importance of developing indigenous resources, complementing rather than competing with efforts to improve the efficiency with which energy is used.

Shareholder Value Propositions

The structure of the energy industry has always been in flux. In the last decade or so we have seen the consolidation of the majors into super-majors, the formation of a new set of North American large independents, also mainly by consolidation, a few new business models specializing in specific resource types such as LNG and oil sands, and the emergence of new INOCs as global competitors. In each case the driver has been to access a broader array of opportunities with greater efficiency, thereby providing shareholders with a distinctive proposition of growth, returns on investment and sustainability (chapter 5). We are sure that there will be a similar level of flux in the years ahead. Resources are not uniform: some are embryonic, some are in a growth stage, and others are mature or aging. The conventional oil business segment is mature, and we should expect a shake out, which is a common pattern in mature industry segments. However other segments are in the growth stage and the strategies for growth businesses should be quite different from those of mature businesses. Some companies will show spectacular success and others will be absorbed or dismantled. The quality of the company's leadership will determine whether their companies will succeed in developing a strong portfolio of businesses with a clear parenting advantage, and the quality of their investor relations programs will influence their ability to be fully valued in stock markets.

Super-majors and majors. Investors preferentially buy stocks that provide growth, returns, and sustainability. The weighting investors attach to the three factors depends on the role that they wish the stocks to play in their investment portfolios and that weighting changes with the view the investors take on the prospects for the industry. For most of the 1990s, investors saw oil and gas as a low growth sector, since most of the supply growth was provided by national oil companies in Russia and the Middle East, and they rewarded those large integrated super-majors and majors that delivered high returns. The weighting has now changed, and investors now expect growth from oil and gas super-majors and majors.

As a group, super-majors and majors will likely continue to cede market share in conventional oil as returns fall below their thresholds. It would not be surprising to see further rationalization of these assets due to falling returns in exploring for conventional oil and gas and limitations on opportunities that provide sufficient scale. We believe that this segment will be under extreme pressure from investors to rediscover the capabilities required to grow production from their current size, or to shrink to a size from which they can grow. The super-major model has to re-establish its applicability to the new business environment.

National oil companies. Flush with cash, NOCs will become more expansive and will be less inclined to invite IOCs and INOCs into their territories, except where governments see compelling value propositions that go beyond resource development. However, NOCs of resource-rich countries face their own challenges, and are subject to social and political demands that impede their effectiveness as organizations. If ever prices moderate, the battle for funds between reinvestment in the national energy complex and other government policy objectives will be reestablished. As the resource depletes, and it becomes more difficult and expensive to sustain production, the pressure to bring in foreign capital and technology will increase.

INOCs start from a much smaller base of international resources than the super-majors and will compete intensely for access to resources and may emulate Statoil in making asset acquisitions, notwithstanding CNOOC's failed effort to acquire Unocal. Some INOCs may follow the path of BP and Total to become global players, while others may founder and become largely irrelevant. Chinese INOCs will continue to expand due to their progressive business model leveraging Chinese governmental support in their value proposition to lesser-developed countries.

Independents. We debated Steve Shapiro, former executive vice president and CFO of Burlington Resources, about how the industry structure would evolve in North America. We proposed that this is a time to favor growth. He responded: "Do you have to take risk because the commodity needs you to, or because your shareholders want you to?" He went on: "So there's economics in risk, but who is best suited to take that risk?" We recalled that in the 1970s it was the majors:

> Right! They had the technology, and they were using it. Today, I think the majors are substituting geologic risk for geopolitical risk. You are seeing that transfer to their portfolios and skill sets at managing work. And then the large E&Ps are actually reducing risk because the economics of exploration works but isn't scalable. And then there are the small guys where exploration is scalable because you start so small, and so you continue to have innovation at the micro- or mini-cap level.

> And to me that drives an industry structure that has two or three very large upstream E&Ps and a bunch of small guys, with nobody in between. And an integrated model with the different levers to pull with the international geopolitical skill set to work those projects. But even that is becoming more difficult because competition from host countries is increasing. Returns are coming down. So they can't replace their assets with others that deliver the same returns they've had, which drives them really back to OECD.

Getting back into Libya doesn't do enough, return-wise. The margin per barrel isn't there like it is in the U.S.

The few large E&Ps will play more of a yield game. They will essentially become trust-like vehicles that the market will accept as an $80, $100, $120 billion dollar E&P that has slower growth, better yield. From a market point of view, it's a pure play on the commodity combined with higher yield. The value creation is capital allocation, not exploration.

Anadarko Petroleum Corp.'s acquisitions of Kerr McGee and Western Gas Resources are recent examples of moves in this direction.

Resource specialists. In addition to the three types of success models noted by Shapiro, there will also be specialists such as Suncor, EnCana, and British Gas, who, as we have seen, have created unique focused business models centered on mastery of a specific resource, as well as growth companies such as Chesapeake, who will continue to mix organic with inorganic growth until they become one of the "few large E&Ps". They will be supplemented by new entrants specializing in coal conversion and in bio-fuels.

Downstream. The consolidation strategy pursued so effectively by Valero has likely run its course. The prime opportunity in North America has shifted to leveraging refining assets to back-integrate into oil sands. This strategy was executed several years ago by majors integrating their refineries with oil sands projects in Venezuela, and has extended more recently into Canadian oil sands. There is still substantial running room in this segment. The challenge in Asia will be to add residual fuel oil conversion to existing simple refineries, which should provide opportunities for IOCs, INOCs, and NOCs to compete and collaborate in this effort. In Europe, the challenge is to reconfigure refining capacity to match the demand shift from gasoline to diesel. In each region, there are massive investment needs, providing opportunities for volume and margin growth.

We saw in chapter 1 that the role of IOCs in the petrochemical industry is in flux. With high natural gas prices in consuming countries, olefins manufacturing is migrating back to areas with abundant natural gas reserves and relatively inexpensive natural gas liquids supplies.

The biofuels segment is more problematic for IOCs with current technologies. The oil industry is not expert in running small, "cottage industry" plants. Its skills in distribution of fuel components will be brought to bear in blending and marketing the resulting transportation fuels, but the technological challenge is to bring scale to the segment so it can become a major new business line.

Power. Meanwhile, power companies are in flux too. The North American power industry is set for a shake-up following repeal of PUHCA. New leaders, such as John Wilder of TXU, want to provide real choices to consumers through new technology, with the potential to significantly lower their utility bills, and are backing up the promise with a massive investment in new low-cost, coal-fired, power generation capacity. If he is successful, then this will rejuvenate the arguments in favor of deregulation, and open up the power sector for true competition, where the winners will be those who can create an enticing value proposition to customers and back it up with low-cost supplies. The possibility of strong European competition should not be discounted. European former utilities may be developing a head start as they learn how to operate in the increasingly deregulated but highly concentrated European natural gas and power sector. Both the coal and nuclear industries are expecting revitalization and growth and there will be a battle over priorities on supply security and greenhouse gas emissions, the result of which is quite uncertain.

Pursuing new business models

The market forces impacting major energy companies are producing a variety of impacts. The forces:
- favor new value propositions for accessing conventional resources;
- stimulate technology development to monetize difficult resources;
- invite the transformation of stranded gas and solids to liquid fuels and the adaptation of refining systems to accommodate them;
- and support extending the traditional product mix to cover a full range of transportation and power products and rapid deployment of emerging technologies in renewables and greenhouse gas sequestration.

The same market forces will provide a stimulus for growth in North American and international exploration and production companies, but shortage of conventional opportunities and key skills will lead to further consolidation in the industry. It is ironic that mergers that used to be propelled by the value of euphemistic "synergy capture," or elimination of surplus workers, will now be propelled by a different form of real synergy capture designed to eliminate unnecessary work and leverage the increasingly scarce body of knowledge in the retiring experienced staff.

Beyond these observed trends, the next strategic energy driver may in fact not be any particular energy source, as in the past, but rather the "system network" which produces end-use energy fuels from a multitude of energy source options. We believe there is a strong case for new industrial complexes to transform a wide array of oil, natural gas, and even solid fuel feedstocks into transportation fuels, chemicals and power products at lower cost than existing plants. In chapter 3, we discussed the technologies and linkages that were melding into a systemic view. Essentially, technologies for fossil fuel gasification and production of Fischer Tropsch synthetic fuels could provide the important linkages. Since gasification is an exothermic reaction, the excess heat can be converted to process steam or to electricity. The hydrogen can be used in refineries or as fuel to combined-cycle power generation. The syngas can be fuel to combined-cycle power generation or feed to synthetic liquids. This suggests an asset-based convergence of the segments that convert resources into products that satisfy consumer needs for "comfort and connectivity" and mobility, while limiting emissions that could threaten existing business models.

On the demand side, hybrid vehicles are a first step in a possible convergence between mobility and power markets. This is not science fiction; all of these resource/technology combinations are economic at 2006 prices and should decrease in cost as technology advances and construction bottlenecks are removed. The new business models could threaten existing business models. The entire energy complex could be changed in the next few decades, with powerful new business models emerging that will challenge the next generation of energy leaders (fig. 10.3).

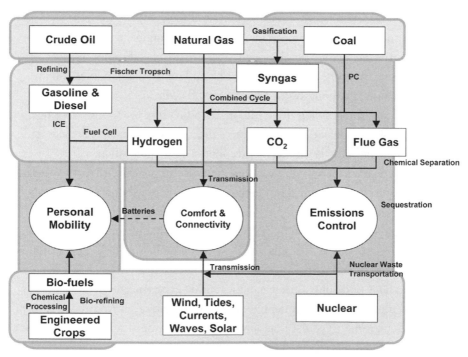

Figure 10.3: New business model possibilities. *Source: CRA International, Inc.*

Traditional business models tended to follow the verticals in figure 10.3. The shaded areas suggest existing and possible future business models. Historically, oil companies were primarily engaged in finding and developing crude oil and transforming it into fuels for clean mobility. Power companies were preoccupied with converting coal and nuclear energy into power and supplying it to captive local customers on a regulated cost-plus model.

Combining "verticals." More recently, natural gas transmission companies took advantage of partial deregulation of their industry to advocate "convergence" between gas and power, based on the expectation of low natural gas prices and the high efficiency of combined-cycle turbine technology. In the process, while several companies demonstrated that the path from regulated industry to competitive markets was fraught with danger, a number of companies have emerged with wide-ranging assets. Enormous conglomerates are emerging in Europe as that market gradually deregulates and integrates. Several of these large natural gas and power companies have already expanded internationally and are set to extend their reach further geographically and functionally into resource development. They could become important competitors to the IOCs.

Encroaching from the other side are the super-majors. They have become large international natural gas producers, and are the principal actors behind the growth in LNG. In parallel, they have become the largest natural gas marketers in North America, taking over that role from the natural gas pipeline companies, and are experimenting in power marketing and trading. European incumbent national champions are resisting market liberalization while expanding their business scope. Already we are seeing changing boundaries and new, successful business models.

Extending horizontally. We have noted that independent exploration and production companies are in a vulnerable position; they will find it difficult to sustain decent returns and growth in the current overheated market. Will any of them add coal to become pure resource plays? Or is it more likely that companies with experience in oil sands, like Suncor, will extend those mining and transformation competences into coal, and pick up some power business while they are at it. Another horizontal option is to extend the independent refiner business model into transformation of a full range of hydrocarbon resources into transportation fuels and power.

Most of the super-majors have oil sands operations and competencies in converting high-carbon feedstocks into transportation fuels. They are also in the power business as sellers of natural gas, and in some cases as power marketers. If the regulations are shaped to reward low-cost power providers, their competencies in major project management, risk management, and the chemistry of the linking technologies could make them preferred suppliers of base-load power. Most IOCs have not wanted to enter the regulated power business. In deregulated power, the issue is more about the adequacy of returns that are available in that business than any lack of competencies. There is a concern that the returns on investment are influenced by the expectations of the regulated participants and may not properly reflect the higher risk of deregulated businesses. Nevertheless Shell, BP, and Suncor are active in low-carbon power sources such as wind farms and several IOCs are also investigating opportunities in gasification of coal and coke with sequestration of the produced CO_2. It is possible that these could become important business lines in the future and position IOCs for forward integration into power markets, particularly if they see some traction in deregulation spawned by success in new business models like TXU's.

Traditional power companies are also expanding their scope. Particularly in Europe, former regulated gas and power utilities have merged into formidable enterprises, and have in many cases back-integrated into natural gas exploration and production. They are, at the same time, using their political muscle to protect their positions in their home countries, and are learning to provide new value propositions to their customers. There is also a group of companies specializing in renewable power, particularly wind. Is there a case for extending that model across the carbon neutral power sources, and even into biofuels? Companies such as Whole Foods have found a large segment of customers who value natural, non-polluting products. Green Mountain Energy has had some success in selling power to them. In the countries far up the Maslow hierarchy, this could become a big business.

Additionally, electric hybrid vehicles open up the possibility of plug-in automobiles fuelled primarily with low-cost, off-peak electricity. Power companies' mastery of coal purchasing and logistics could allow them to compete with oil companies, either through fuelling vehicles with electrons or extending from IGCC power plants into synthetic liquid-fuel manufacturing complexes.

It is our view that all the boundaries are up for redefinition, and the next generation of leaders will need to ask and find answers to some important questions:

- What are the big trends and how can we get downwind of them?
- Can we shape the way the rules of the game are written?
- What should be our direction and what goals should we set?
- What should be the boundaries of our business portfolio?
- Which areas should we focus on with greatest intensity?
- How do we leverage our existing and new competencies for competitive advantage?

Next-Generation Strategic Leadership

In every major transition, such as the phase change we see occurring in the energy industry, a new generation of leaders emerges who capture the growth segments. Wood was displaced by coal, and a new industry with new leaders, new business models and new technologies evolved during the embryonic and growth stages of coal's maturity cycle. But it was not the coal companies that captured the oil boom. A new set of competitors arose to provide their shareholders with new value propositions. The early entrepreneurs were small, and the nascent industry was disorganized and fragmented. Then Standard Oil brought scale and order.

What companies are evolving now in the 21st century energy phase change? New global leaders in biofuels, coal transformation, or perhaps in power storage technologies may be emerging today, below the radar screens of most industry observers, and they may be poised to dominate their industry segments. Even more challenging would be the emergence of strong Chinese competitors with new and better technologies for coal, biomass conversion, and hybrid vehicles, designed from necessity but perfectly adapted to the new global needs.

Understanding the changing situation

All good navigators are continuously refreshing their understanding of their current position, and the challenges of weather, currents, and shoals. The next generation leader will need to be equally aware of the changing situation. Since we cannot draw aside the mists of uncertainty that separate us from the future, we must content ourselves with asking what we believe to be key questions, the game changers. It will be up to future leaders to monitor the answers as they emerge and respond with appropriate decisions.

Society. The world is all over the board on where different countries fall on Maslow's hierarchy of needs; as societies evolve, their needs will change and leaders will be required to tailor their value propositions accordingly. We have known since Adam Smith that the prospect of wealth is a vital motivator in stimulating economic growth, which provides the opportunity to ascend the different need levels. And we have learned in the 20th century that socialism doesn't work. But how will global society deal with lifting two billion people from poverty? What will be the impact of 2.5 billion more consumers by 2050? Who will take up the late Richard Smalley's mantle and explain to people the pivotal role of energy in every area of concern for the future of the planet enabling the necessary economic growth? How can we establish dialog instead of confrontation in resolving dilemmas, to which there are no obvious solutions?

Another expression of societal needs at the high end of the Maslow hierarchy is the environment, and the critical environmental issue today is, of course, climate change. Energy companies are obliged to address the public's concern on this issue, and to partner with governments to develop cost-effective policies. When will climate change legislation be developed in the U.S.? Will we reach consensus, including China and India, that greenhouse gases should be controlled? How do we ensure that we address the causes, not just the symptoms?

Climate change, however, is only one of the issues that fall under the overall umbrella of sustainable development. Directionally, moving away from depletable fossil fuels given the world's ravenous and growing appetite for energy makes sense. But will sustainability gain more priority on legislative agendas? How will the public deal with potential tradeoffs in economics and in their lifestyles? How will geopolitical issues impact civil society's call for sustainable development?

Geopolitics. Will the "world of disorder" lead us further into a "clash of civilizations"? Is it possible to breathe new life into international agencies such that they can protect the "global commons" and intervene effectively before rather than ineffectually after genocide? Is everything relative or are there some moral absolutes governing the balance of competitive and collaborative behaviors, which are vital for the success of the human species, as some scientific studies are now suggesting?

The biggest immediate issue in this "clash" is the rise of radical extremists promoting their cultural and religious norms through terrorism. In the U.S. in particular, public fears from 9/11 and the turmoil in the Middle East have re-ignited concerns for energy security. How will the call for energy security impact the overall energy equation? Will legislative incentives turn to mandates? Is this a temporary concern or a long lasting priority on the public agenda?

Technology. The technologies that will likely matter most to the energy industries are those that affect how energy is used. These will include technologies of power storage, such as batteries, and of transforming fuels into motive power, such as fuel cells. Hydrocarbons are brilliantly designed energy carriers, allowing high energy-density packages to be transported cheaply over long and short distances. The internal combustion engine (ICE) is being constantly refined and computerized to become ever cleaner and more efficient. It will be tough to displace. But hybrids are penetrating the market; hybrids use the ICE and electricity to power vehicles. There are many combinations, from the current energy mix, to converting hydrocarbons to hydrogen for fuel cells on board or at fueling sites, to charging batteries at home. Technology advances will surely provide new options in the future. How will future changes adjust the balance between the two energy sources?

If the balance shifts towards electricity to provide motive power, how will it be generated? If gas prices stay around $7/mcf, then a wide range of potential power sources should become economic, including wind, nuclear, coal with CO_2 sequestration. Should energy companies embrace these various technologies? Gulf and Shell invested in General Atomic Corporation in the 1960s, but were unable to make a success of nuclear energy. Most oil companies acquired coal and metals mining companies in the 1980s, but later divested them without figuring out how to create value from their acquisitions. These would seem to be stern lessons of mistakes from the past. But are they? What is the same and what is different today? Do the IOCs have competencies today that they did not have in earlier times? Will there be the same loss of demand and eruption of new supplies as in the 1980s or will this be a period where demand continually advances faster than new supplies?

Technologies affecting supply of energy will be important and relatively easier to monitor. What will be the extent of further improvements in finding and developing conventional oil and gas and transforming hydrocarbons into transportation fuels? Will coal conversion to liquid fuels become commonplace? Will gas conversion to liquids reconnect the economic relationship between power and transportation markets?

On top of the blurring pace of change rushing from the information revolution, another revolution has emerged: molecular engineering in biotechnology and nanotechnology.

Scientists are venturing into the world of creation. Outside the box, genetic engineering could provide new plant forms designed perfectly for the manufacture of biofuels at a scale that could make them economic. If this happens, then the energy company competencies of large scale, complex chemical engineering will be a competitive advantage in capturing the scale economies required to make the new energy sources economic. Will this happen and in what time frame?

In this molecular revolution, change will likely become even less predictable. Things never imagined will emerge. First movers may have a dominant initial advantage. Fast followers will have the advantage of standing on the pioneers' shoulders. Slow followers will be left by the wayside. Next-generation leaders will be challenged to adapt and absorb the new technologies as they emerge.

Setting the direction

We have learned from our interviews that the successful energy companies of the future will need to reset their directions carefully, effectively execute their strategies, and demonstrate leadership that responds to the challenges ahead.

The current environment favors unconventional and technologically difficult resources and integrated value chains to bring the new resources to market. Growing resource nationalism requires IOCs to show imagination and humility in developing new value propositions tailored to the real needs of resource rich countries. Energy companies also need to work with consumer governments to help them devise sensible policies that balance their needs for national security and concerns over climate change.

There will also be interesting opportunities to reshape the boundaries of the energy sector, creating value by leveraging existing competencies into new segments, and developing new competencies to expand the scope of existing businesses. This will result in a different portfolio of businesses, and the need to prioritize so that new portfolio elements are accorded the appropriate attention and are not crowded out by the demands of more mature businesses.

Finally, in setting direction, the best companies must continue to be strongly focused on developing new business models and continuously improving their core competencies. Gwyn Morgan of EnCana asserted: "Competitive advantage is created through core competency and superior assets." This requires a deep understanding of the value chains and value drivers in the business segments that are pursued so that the competencies are targeted at the value opportunities.

This calls for a different approach to strategy development (fig. 10.4) because the future is uncertain and leadership may be far from agreement on the appropriate course.

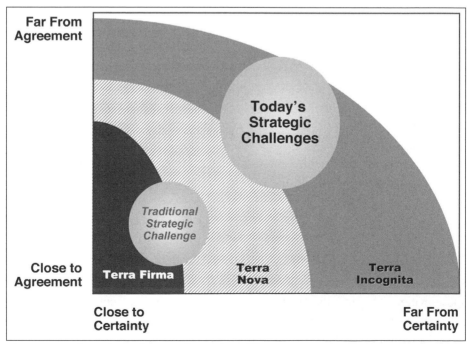

Figure 10.4: The "phase change" demands a different approach to strategy development. *Source: CRA International, Inc.*

More than ever, the future must be explored through scenarios enriched with insights from collaborative game playing to enable executives to see the future through the lenses of multiple stakeholders. Existing mental models must be challenged and different mental models explored. Leadership must be realigned toward a new direction. However, since the future is uncertain, the leadership will need to become comfortable with some ambiguity and should plan for controlled experiments to validate hypotheses. Value needs to be seen in the context of a portfolio and optionality, and will differ for players with different strategies and shareholder value propositions.

Frank Chapman of BG coined the term "asset accretion" to describe how his company leveraged assets in one part of the value chain to capture positions in other parts of the value chain. This reinforced the concept of integrated value chains and clarified that the new interpretation of integration is not linear or sequential, but much more related to real options in its design.

Executing the strategies

In executing strategies, energy company leaders will first need to secure commitment from employees and investors: employees because their commitment is required to get the job done, investors because they abhor surprises and need to be alerted to changes in advance of their execution.

Traditional IOC business models and value propositions are increasingly unpersuasive to resource owners; they must be transformed to address the real needs of multiple constituencies within the resource rich countries. These host-country needs include greater control over the pace of resource development and production, revenue predictability, demonstrable benefits to their citizens from oil and gas activity, and relationships based on trust. Current IOC offerings are weak in all of these areas of concern.

It will be easier for IOCs to build strong new platforms in unconventional resource development, where their project management, market development, and technology competencies are most valued. Mergers and acquisitions will continue to reshape the competitive arena. Our interviews persuaded us that those companies that maintain a continuous rather than episodic presence in the transaction game will be most successful.

Further, energy companies have to further improve their ability to make good choices by inverting their capital allocation process from screening projects out to screening projects in. We were impressed by the way that Total has transformed its process into a continuous conversation, and believe that many companies will be advised to rethink the whole capital project process from idea origination through approval on to construction and commissioning.

Companies must also revisit their approach to partnerships and alliances along the full value chain from opportunity origination, capture, execution, and operation, devising flexible resource architecture capable of mobilizing talent from inside and outside the corporation as needs dictate. While oil and gas companies partner extensively with each other, they are much less likely to partner with suppliers and customers than other industries. There are opportunities to partner better with host governments, national oil companies, local private oil companies, and service companies as well as with traditional service companies. With suppliers and service providers operating close to capacity, their engagement early in projects can help drive innovative solutions and high leverage of scarce skills.

Project execution will increase in importance and in complexity. Companies will need to make conscious and well thought out decisions about which competencies they should maintain internally that can truly create differential value, and which should be outsourced. Then they will need to think deeply about the nature of the relationships they wish to build with suppliers and partners. These thoughts should be guided by the principles of "coopetition"—the belief that excellence in partnering can increase the size of the overall pie, creating wealth that can be divided equitably to the benefit of all parties. The spirit of innovation that made the North Sea economic as prices fell in the 1980s needs to be exhumed and applied again to achieving productivity gains in order to accomplish more output with less input, but this time in a very different economic environment.

With a broader range of energy businesses, companies will be challenged to realign their organizations with the true value drivers of their different segments and solve the apparent conflict between functional excellence and value chain integration. Current organizational boundaries will need to change and become more permeable to increase overall productivity: some parts of the portfolio will need to be managed as integrated value chains; others can be "atomized."

Companies must work harder to go beyond merely securing a license to operate and become an integral part of the solution in satisfying the needs of the communities in which they do business. The entire energy industry must make a compelling case to the public about the importance of energy to their lives and the value of the work done by the industry. If the industry is successful in this, it will have earned the right to a productive dialog with government on risk sharing for major projects with long payouts that contribute to national security.

Leading in turbulent times

Energy company leaders in the new energy business environment must convey a sense of purpose to their organizations: creating a sense of destiny, a positive attitude, and an ability to stay the course during setbacks. This is most important and most challenging during a period of change, when continuing to do very well the same things in the same ways will not bring success.

During a period of great uncertainty, there will inevitably be disagreement on the proper course. Alignment towards a shared aspiration is vital to avoid conversations that should be focused on strategy being capsized on the shoals of operational immediacy. A critical determinant will be how the current leaders manage succession. Homegrown leaders keep the machine running on track. They can make some substantial course adjustments, and because they have the informal network, can bring the institution along with them. CEOs of the super-majors and Aramco are all homegrown. However, the process for developing new leaders must be in place.

Smaller companies find the succession process even more difficult because they have a smaller bench. Also, they cannot afford to have the disappointed internal contenders leave. External CEOs in our interview sample were BR (Bobby Shackouls), Marathon (Clarence Cazalot), Statoil (Helge Lund), Dominion Exploration and Production (Duane Radtke), and TXU (John Wilder). In each case, the company was in some trouble at the time of their accession, driving the board toward an external candidate. An external candidate will necessarily be disruptive to the status quo, and will encounter resistance. Our interviewees were all successful because they respected the competencies and past accomplishments of the organizations they inherited, but moved decisively to make the changes they determined were necessary. The other companies interviewed have or had a long-serving founder (Suncor, EnCana, Valero, BG, and Chesapeake). Of these companies, Valero and EnCana have recently arranged an orderly succession to the previous number two executive. An internal candidate seems the preferable solution if the company is on track and successful.

Energy company leaders must shape the evolution of culture and values in their organizations while completely remaking their staffs during the next decade. Dave O'Reilly, CEO of Chevron, understands this: "As CEO, it becomes very clear that that is one of the major roles, to be the keeper of the value system." This has to be done by influence and through a minimum number of critical rules. As O'Reilly says, "You can't possibly micromanage everything. One of the most important issues is communicating the value system in a manner that is fairly simple, and then checking on it periodically to make sure that the other leaders of the organization live up to it."

The new leaders must create a context in which good decisions are made naturally because the leadership team trusts each other, decision rights are understood and governance is supportive. Choosing the right leadership team and enabling the right conversations is a key leadership skill, as is effective delegation of decision rights. With massive change threatening, and a required shift from mere efficiency in cost management to effectiveness in delivering growth with a broader range of energy products, this will be a bigger challenge for the new leaders than it was for their predecessors. We suspect that a reversal of recent tendencies towards centralization in decision-making will be required. In particular, individuals charged with developing relationships in resource-rich countries must be seen to be empowered; top management is spread too thinly to own all of these important relationships.

Long-term and short-term incentive plans will need to be re-weighted to encourage a more adventurous spirit. This is not to encourage foolhardy risk taking, and certainly not to lower the bar on values. The new metrics must be strategic as well as operational and must accommodate some redundancy in organizations as new talent is brought in and developed by teaching, training and coaching programs. All natural systems require some redundancy in order to evolve successfully to changing environments, and business organizations are not exceptions. Separate budgets must be established and new approaches developed to measure the effectiveness of these programs.

Finally, an essential part of leadership today is creating an attractive place to work and to continue working beyond historical retirement dates, particularly recognizing that the industry will lose a large proportion of its experienced staff in the next decade. The increased scope of the portfolio of energy commodities and the need for stepped up investment suggests a substantial increase in total staffing dedicated to building out the new energy complex. The new leadership will become adept at attracting, motivating, and managing a more diverse workforce and orchestrating the extended enterprise of partners and suppliers that will get the work done. Indeed, it is not clear that the traditional profile of an energy CEO—a white male with a technical degree— is right for a future in which better communications with the public as well as an understanding of a diverse workforce and complex relationships with partners will be required. It will become increasingly necessary to consider diversity as an opportunity for innovation rather than as an obligation to be endured.

The leader of the future will need to develop a new set of mental models to allow him to understand the accelerating pace of change, set direction, execute strategies and provide the leadership necessary in turbulent times. We thus propose a Charter, a few big rules to guide the leaders as he or she navigates towards *terra incognita*.

A charter for the next generation

Each technology revolution has aftershocks impacting the way organizations work and the way leaders guide them. In this new molecular era, organizational structures will continue to mutate. New leaders in the new industries formed from these technologies will create new business models and new alliance schemes. Within the new industry, leadership characteristics may recycle to the early days of the industrial revolution with visionaries and strategists molding a new future. The energy industry will absorb the lessons from these new pioneers. But a molecular revolution may have much greater ramifications on societal values, as well as on business models leading to entirely new leadership challenges. This molecular revolution will be laid on top of the energy phase change creating unparalleled change in the energy industry.

Change of a scope and magnitude that are difficult to grasp will make extraordinary demands on the next generation of energy leaders. The leaders of the future will be called upon to originate, capture, and implement a series of massive growth projects. They will face great uncertainties about the demands of society, the course of geopolitics, and the progress of technology. Oil will cease to be the strategic energy form within the tenure of the next leadership generation. In prior phase changes from one strategic energy form to another, a new set of companies emerged to pioneer and develop the new form, while the previous incumbents were marginalized. This may happen again, and companies that wish to avoid this outcome will need to depart from the legacy mindsets of their predecessors by revising their mental models. We propose a Charter—a term that shares the same Latin origin as chart—with 10 areas of attention to guide leaders on their voyage of discovery to *terra incognita*:

- *Conviction:* Take a longer view with an even more extended time horizon. Be prepared to face down the doubters who think this is just another investment cycle.

- *Commodity:* Expand the company's vision of the portfolio of energy commodities and the steps in their value chains.

- *Complexity:* Understand that the business will become more complex. Energy supplies will include massive projects measured in the tens of billions of dollars, but will also include a network of small scale investments such as programs of wells in tight gas fields, local biofuels plants, and retail outlets. Some projects will be under full control of the investor, others will be operated through local proxies. Business partners will be diverse, including government-owned companies, private capital companies, and global and regional institutions, as well as local communities. Few projects will be straightforward.

- *Connectivity:* Recognize the interconnections between different energy forms and the needs met by energy supplies, as well as the needs of an increasingly diverse set of employees. There will be a requirement for new physical, information, and social networks to move molecules and electrons to where they are needed and to enable a new, diverse workforce to collaborate effectively.

- *Collaboration:* Good strategies and effective execution will require collaboration with suppliers, technology providers, and communities affected by the investments, and a commitment to collaboration will become an increasingly important value, vital for success. Value chains will continue to grow in complexity, requiring companies to assure positive interfaces at numerous nodes.

- *Consilience:* The evolutionary biologist Edward O. Wilson used this term to describe the need for common ground between the humanities and the sciences—to base strategy and policy on hard facts, but to respect aesthetics and the higher aspirations of humankind. A new mode of communication needs to emerge by which those who are inclined to reject modernity on grounds that it has failed to provide the promised benefits can be persuaded that their interests are not opposed to resource development.

- *Co-opetition:* Visualize the new competitive landscape clearly and coming up with value propositions that demonstrate competitive advantage. Competition for customers and for resources will include European continental power and gas giants, internationalizing national oil companies, national champions among the resource-rich countries, and new entrants in wind, biofuels and potentially coal resource developers. In many cases there will be opportunities to turn competition into co-opetition to create new sources of value.

- *Communication:* In times of change, communication becomes even more important. Energy companies have an abysmal record of explaining the societal benefits of energy in tranquil times; they will be required to invest more time and effort into helping a wide range of constituencies understand the changes under way and the policies that will best facilitate an orderly transition. It will be vital to explain the need to build a new energy infrastructure rather than take the old one for granted, in order to create or sustain the wealth needed for higher levels of self-actualization.

- *Competencies:* Current and future deficiencies in competence are already visible. There is inadequate capacity of skilled people to deliver the energy supply projects that have already been identified. The pipeline beyond 2010 for new projects is insufficient to meet future demand growth. There is a scarcity of good business developers willing to travel to and collaborate with resource rich countries to fill the pipeline. Companies have been unwilling to devote the time, patience and training required to build competencies in communicating with external constituencies.

Extremely important, with the graying of the workforce, will be the ability to teach, train, and coach the next generation of managers and leaders in the lessons of the past and the needs of the future.

- *Character*: Sustaining a strong ethical foundation, while conveying the sense of destiny and excitement of a new and better future, will be a test of character for the next generation. Society as a whole, their employees, and their shareholders depend on the character of the new leaders.

Energy remains an integral part of the global economy. Each new revolution builds on the former. The information revolution has transformed the way in which the energy business is carried out and continues to redesign it. The molecular revolution may morph the very fundamentals of the business. We are rapidly approaching *terra incognita*. There will be new maps and metrics to use in navigating toward and within the New World, and new leaders will come forward to sustain morale and productivity on the voyage. Several companies may perish on the journey, but the successful explorers will find new sources of value to the great benefit of their investors, employees, and society as a whole.

As Edward Gibbon observed in the 18th century: "The winds and waves are always on the side of the ablest navigators."[23]

1 Hesselbein, Goldsmith, and Beckhard, ed., 1996. *The Leader of the Future.*. New York: The Drucker Foundation, p. xi.

2 Hesselbein, p. 67.

3 Ida M. Tarbell, Aug. 1906. "John D. Rockefeller: A Character Study." *McClure Magazine.* http://www.reformation.org/john-d-rockefeller.html (Accessed Aug 31, 2006)

4 Tarbell

5 Tarbell.

6 Frederick W. Taylor, 1911. *The Principles of Scientific Management.* New York: Harper Brothers, p. 26

7 Taylor, p. 47.

8 Brad Nevin, Apr 2005. Sep 5, 2006., *The Prettiest and Arguably Most Significant Car, Ever.* Ford. com.: http://media.ford.com/newsroom/feature_display.cfm?release=20414

9 Carter McNamara, *Very Brief History of Management Theories.* Managementhelp.org. 1999. Aug 24, 2006. [http://www.managementhelp.org/mgmnt/history.htm]

10 S.K. Shahi, Aug. 2006. "The Leader of Alfred Sloan." Leadership Digest. Vol 4. Careerage.com. http://www.careerage.com/leadership/vol4/art4.shtml.

11 Bill Gates, 1999. *Business @ the Speed of Light.* New York: Warner Books, pp. 6,7.

12 http://www.pressuchicago.edu/Misc/Chicago/238040in.html

13 Richard S. Tedlow, Dec. 2005, *"The Education of Andy Grove."* Fortune, 3.

14 Jeffrey Brown, Aug. 24, 2006. Sandeep Junnarkar and Mukul Pandya, *Lasting Leadership.* Informit.com.: http://www.informit.com/articles/article.asp?p=345010&rl=1

15 Hammer Champy, 1993. Reengineering the Corporation: A Manifesto for Business Revolution. (Harper Business).

16 Hesselbein, p. 22.

17 Manz and Sims, 2001. *The New SuperLeadership*, BK, p. 5

18 Hesselbein, p. 115.

19 Hesselbein, p. 6.

20 DeGeus, 1997, *The Living Company: Habits for survival a turbulent business environment.* Boston: HBS Press, p. 4.

21 Hesselbein, p. 298.

22 Thomas L. Friedman, 1999, *The Lexus and the Oliver Tree.* New York: Farrar Straus Giroux.

23 Edward Gibbon, 1993 (1910), *History of the Decline and Fall of the Roman Empire.* New York: Knopf.

Index

H

M

S

U

About the Authors

Christopher E.H. Ross, Vice President of CRA International Inc., specializes in strategy and organization assignments in the worldwide energy industry, where he has 40 years of experience. He leads CRA's firm-wide natural gas initiative. Based successively in Europe, Africa, and the United States, he has extensive global experience helping international, national and independent oil company clients develop a shared sense of strategic direction, and execute the strategies. He has also led major transformation assignments resulting in massive reductions in costs. Ross has authored many articles providing insight into important changes in the energy business, including virtualization, social responsibility, and the 21st century shift in shareholder value creation from Return On Capital Employed to growth. Formerly, Ross was a Vice President of Arthur D. Little, and prior to that was with British Petroleum Company. He received a BS in Chemistry with Honors from London University and completed the Program for Management Development at the Harvard Business School.

Lane E. Sloan has 30 years of experience in the oil and petrochemical industries and held a number of senior leadership positions with Shell Oil, including Vice President Corporate Planning, Chief Financial Officer and President of Shell Chemical Company. He was also Regional Coordinator for all Royal Dutch Shell's operations in the Asia Pacific region and Director of the East Zone in Shell's Global Oil Products business. Today, he teaches corporate strategy and leadership to undergraduate and graduate students at the University of Houston's Bauer College of Business. He was formerly the Executive Director of the Global Energy Management Institute at the business school and is currently the Chair of the Greater Houston Energy Collaborative under the Greater Houston Partnership. Sloan has written articles for World Energy Magazine and World Energy Monthly Review. He received a BS in Business magna cum laude and an MS in Management Science from the University of Colorado. At the University of Houston, he received an MS in Accountancy and an MBA in Finance.